Synthesis and Applications of Gold Nanoparticles

Synthesis and Applications of Gold Nanoparticles

Guest Editor

Wen-Huei Chang

MDPI

Basel • Beijing • Wuhan • Barcelona • Belgrade • Novi Sad • Cluj • Manchester

Guest Editor
Wen-Huei Chang
Department of Applied
Chemistry
National Pingtung University
Pingtung
Taiwan

Editorial Office
MDPI AG
Grosspeteranlage 5
4052 Basel, Switzerland

This is a reprint of the Special Issue, published open access by the journal *Nanomaterials* (ISSN 2079-4991), freely accessible at: https://www.mdpi.com/journal/nanomaterials/special_issues/synthesis_gold_nano.

For citation purposes, cite each article independently as indicated on the article page online and as indicated below:

Lastname, A.A.; Lastname, B.B. Article Title. *Journal Name* **Year**, *Volume Number*, Page Range.

ISBN 978-3-7258-4377-0 (Hbk)
ISBN 978-3-7258-4378-7 (PDF)
https://doi.org/10.3390/books978-3-7258-4378-7

Contents

About the Editor

Wen-Huei Chang

Wen-Huei Chang earned her Bachelor's degree in Pharmacy in 1996 and obtained her Ph.D. in Biochemistry and Molecular Biology in 2002 from National Taiwan University. She is currently a professor in the Department of Applied Chemistry at National Pingtung University. She has been awarded a prestigious Fulbright Scholarship to conduct research at Yale University, reflecting her significant contributions to the field of biochemistry and her commitment to advancing scientific knowledge. With numerous publications and collaborative projects, she continues to inspire students and colleagues alike in the pursuit of excellence in research and education.

Preface

This Special Issue on the synthesis of gold nanomaterials aims to explore the latest advancements in the field, highlighting innovative methods and applications. The contributions presented herein reflect the collaborative efforts of leading researchers and practitioners dedicated to enhancing our understanding of gold nanomaterials' unique properties. We extend our gratitude to all the authors for their invaluable insights and to the reviewers for their constructive feedback, which has significantly enriched this reprint. This work is intended for scientists, engineers, and industry professionals interested in nanotechnology, materials science, and related fields. We hope it serves as a valuable resource for ongoing research and development in gold nanomaterials.

Wen-Huei Chang
Guest Editor

nanomaterials

MDPI

Article

Green Synthesis of Gold Nanoparticles Using Upland Cress and Their Biochemical Characterization and Assessment

Noah Hutchinson [1], Yuelin Wu [2,†], Yale Wang [3,†], Muskan Kanungo [4,5], Anna DeBruine [4,5], Emma Kroll [4,5], De'Jorra Gilmore [4,5], Zachary Eckrose [4,5], Stephanie Gaston [4,5], Phoebe Matel [4,5], Matey Kaltchev [4,5], Anne-Marie Nickel [4], Subha Kumpaty [6], Xiaolin Hua [2,*] and Wujie Zhang [4,5,*]

1 Department of Biomedical Engineering, Milwaukee School of Engineering, Milwaukee, WI 53202, USA; Hutchinsonn@msoe.edu
2 Department of Obstetrics, Shanghai First Maternity and Infant Hospital, Tongji University School of Medicine, Shanghai 201204, China; wuyuelin_wyl@163.com
3 Department of Mechanical Engineering, University of Milwaukee, Milwaukee, WI 53211, USA; yalewang815@gmail.com
4 Department of Physics and Chemistry, Milwaukee School of Engineering, Milwaukee, WI 53202, USA; kanungom@msoe.edu (M.K.); debruinea@msoe.edu (A.D.); krolle@msoe.edu (E.K.); gilmore@mose.edu (D.G.); zakeckrose@icloud.com (Z.E.); gastonsw@alumni.msoe.edu (S.G.); matelpr@alumni.msoe.edu (P.M.); kaltchev@msoe.edu (M.K.); nickel@msoe.edu (A.-M.N.)
5 Biomolecular Engineering Program, Milwaukee School of Engineering, Milwaukee, WI 53202, USA
6 Department of Mechanical Engineering, Milwaukee School of Engineering, Milwaukee, WI 53202, USA; kumpaty@msoe.edu
* Correspondence: xiaolin_hua@tongji.edu.cn (X.H.); zhang@msoe.edu (W.Z.)
† Contributed equally.

Abstract: This research focuses on the plant-mediated green synthesis process to produce gold nanoparticles (Au NPs) using upland cress (Barbarea verna), as various biomolecules within the upland cress act as both reducing and capping agents. The synthesized gold nanoparticles were thoroughly characterized using UV-vis spectroscopy, surface charge (zeta potential) analysis, scanning electron microscopy-energy-dispersive X-ray spectroscopy (SEM-EDX), atomic force microscopy (AFM), attenuated total reflection Fourier transform infrared spectroscopy (ATR-FTIR), and X-ray diffraction (XRD). The results indicated the synthesized Au NPs are spherical and well-dispersed with an average diameter ~11 nm and a characteristic absorbance peak at ~529 nm. EDX results showed an 11.13% gold content. Colloidal Au NP stability was confirmed with a zeta potential (ζ) value of −36.8 mV. X-ray diffraction analysis verified the production of crystalline face-centered cubic gold. Moreover, the antimicrobial activity of the Au NPs was evaluated using Gram-negative *Escherichia coli* and Gram-positive *Bacillus megaterium*. Results demonstrated concentration-dependent antimicrobial properties. Lastly, applications of the Au NPs in catalysis and biomedicine were evaluated. The catalytic activity of Au NPs was demonstrated through the conversion of 4-nitrophenol to 4-aminophenol which followed first-order kinetics. Cellular uptake and cytotoxicity were evaluated using both BMSCs (stem) and HeLa (cancer) cells and the results were cell type dependent. The synthesized Au NPs show great potential for various applications such as catalysis, pharmaceutics, and biomedicine.

Keywords: green synthesis; nanoparticles; catalysis; cytotoxicity; upland cress

1. Introduction

Gold nanoparticles have potential applications in numerous fields including chemical/biochemical industries, textiles, energy, bioremediation, bioimaging, biosensorics, controlled delivery of therapeutic agents, and agriculture due to their unique properties [1,2]. In particular, the characteristic chemical, biological, optical, and electrical properties of gold nanoparticles have been utilized for nano-catalysis and nanotheranostics, to provide novel routes for chemical reactions, and to offer a personalized approach for medical diagnosis and treatment [3–7]. Specifically, biocompatible gold nanoparticles exhibit unique surface interactions and can be functionalized through the attachment of numerous components (drugs, targeting moieties, radioisotopes, DNA, fluorescent dyes, linkers, polyethylene glycol, and others) to offer multimodal approaches to modern biomedical processes [8,9].

Several traditional methods for the production of gold nanoparticles, including chemical and physiochemical methods, have been utilized for these purposes. However, many of these pathways draw concern due to their adverse environmental impact, high cost and energy consumption, as well as the potentially limited applications of the produced nanoparticles. Alternatively, biological and green synthesis methods utilize microorganisms as "biomachinery" or naturally occurring biomolecules from plant extract that serve as reducing and capping agents in a bottom-up synthesis approach [10]. This is advantageous over traditional methods as it provides versatile, biocompatible nanomaterials through an environmentally-conscious and cost-effective approach. To date, various organisms including plants, bacteria, algae, and fungi have been found to contain the phytochemicals necessary for the green synthesis and stabilization of nanoparticles. Extracts used for green synthesis of Au NPs include *Coffea arabica* (coffee), *Solanum nigrum* (black nightshade), *Nasturtium officinale* (watercress), *Brazilian red propolis* (honeybee product), *Litsea cubeba* (May Chang), *Chlorella vulgaris* (algae), *Mimosa tenuiflora* (Jurema), and *Ziziphus zizyphus* (Jujube) [6,11–15].

Upland cress (*Barbarea verna*) is a widely available biennial leafy green in the *Brassicaceae* family which has been found to contain a high content of synthetically viable phytochemicals including ascorbic acid, carotenoids, and tocopherols that are potentially useful for the reduction and stabilization of gold nanoparticles [16]. Phenolics such as flavonols mainly contribute to the reducing process [17]. Phenolics and other plant metabolites (such as sugars and enzymes) are responsible for stabilization and capping [18]. Recently, upland cress has been used to successfully synthesize silver nanoparticles [19]. In this work, a novel green synthesis process to produce gold nanoparticles (Au NPs) using upland cress, for different applications from silver nanoparticles, was developed and optimized. It was hypothesized that the nanoparticles formed using this method would show favorable stability due to the presence of phytochemical based capping agents present in the upland cress extract. Furthermore, the synthesized Au NPs would show antimicrobial properties and biocompatibility, as well as serve as a catalyst. This novel green synthesis method offers a cost-effective and environmentally friendly counterpart to traditional methods for biocompatible nanoparticle synthesis.

2. Materials and Methods

2.1. Materials

Upland cress (B&W Quality Growers, Fellsmere, FL, USA) was purchased from Whole Foods, a local grocery store, and stored at 4 °C. Gold (III) chloride trihydrate ($HAuCl_4$; 520918), 4-nitrophenol ($C_6H_5NO_3$; 241326), Folin-Ciocalteu reagent (F9252), gallic acid (G7384), and sodium borohydride ($NaBH_4$; 80637300) were obtained from Millipore Sigma (St. Louis, MO, USA) and were used as received.

2.2. Analysis of the Upland Cress Extract

The total phenolic content of the sample was determined by the Folin-Ciocalteu method [20]. A 20 μL aliquot of upland cress extract was added to 1.58 mL water and 100 μL of the Folin-Ciocalteu reagent. After 5 min, 300 μL of 20% sodium carbonate solution was added to the mixture and agitated for 10 min. After sitting in the dark for 2 h at 22 °C, the absorbance was measured at 765 nm using UV spectrophotometer (Genesys™ 150; Thermo Fisher Scientific, Waltham, MA, USA). Different concentrations of gallic acid were used to prepare the calibration curve (R^2 = 0.9915). The total phenolic content was expressed as μg gallic acid equivalent (GAE) per mL extract.

The concentration of ascorbic acid (vitamin C) in the upland cress was analyzed through iodometric titration [21]. Briefly, 4 mL of upland cress extract, 0.2 mL 0.5% starch, 1 mL 0.6 M potassium iodide, 1 mL of 1 M HCl, and 30 mL of DI water were mixed in a flask. The mixture was titrated using 0.002 M potassium iodate with the first permanent trace of blue-black color as an indicator of the endpoint.

2.3. Green Synthesis of Au NPs

The green synthesis and purification protocols(Scheme 1) were based on previous work [19] with modifications, in the extraction process (blending instead of boiling was used). Briefly, 10 g of upland cress were blended in 100 mL deionized (DI) water for 15 s. The mixture was then vacuum filtered twice using Whatman filter paper and centrifuged at 4000 rpm for 5 min to produce the extract (supernatant). Extract (5 mL), 10 mM gold (III) chloride trihydrate (0.4 mL), and DI water (35 mL) were mixed and then placed in a shaker (215 rpm) at 37 °C.

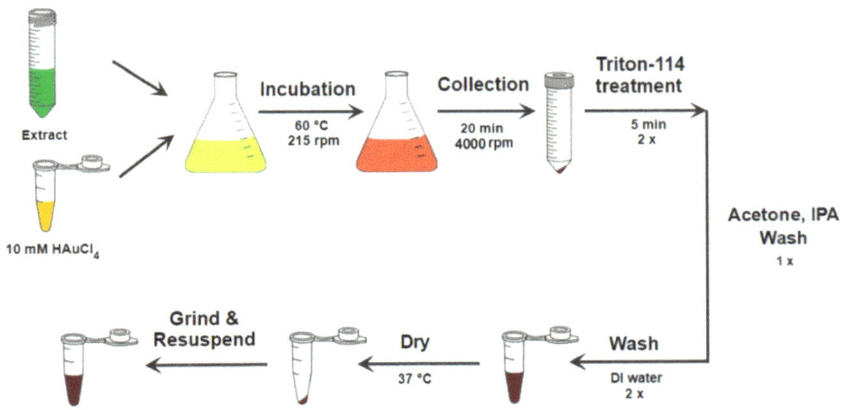

Scheme 1. Process for the green synthesis of Au NPs using upland cress.

For purification, sonication and two more solvents were utilized to achieve a more thorough process in comparison to previous work [19]. The nanoparticles were collected through centrifugation (4000 rpm, 20 min). The particles were suspended in DI water (10 mL) and sonicated for 5 min. The suspension was then centrifuged (4000 rpm, 20 min). Subsequently, the collected nanoparticles were subjected to a series of wash/vortex and centrifugation cycles using Triton X-114 (0.75 μL/mL DI water), acetone, isopropyl alcohol, and DI water respectively. Finally, the nanoparticles were oven dried at 45 °C, ground using a mortar and pestle, and stored at −20 °C for future use.

2.4. Characterization of Au NPs

UV-Vis spectroscopy (Genesys^{TM} 150; Thermo Fisher Scientific, Waltham, MA, USA) was used to study the effects of incubation time on the green synthesis process. The peak absorbance wavelength was determined using the scanning mode (450–650 nm). Zeta (ζ) potential of the Au NPs (0.4 mM) was measured using a Zetasizer Nano ZS (Malvern, Westborough, PA, USA). Scanning electron microscopy (FE-SEM, Hitachi S-4800 ultra-high resolution cold cathode field emission scanning electron microscope, Kefeld, Germany) was used to image Au NPs (40 mM) that were dried and mounted to aluminum stubs. At the same time, energy dispersive X-ray spectroscopy (EDX, Noran (Si(Li))detector, Thermo Fisher Scientific, Waltham, MA, USA) was used to verify the elemental composition of the Au NPs. Atomic force microscopy (AFM) images, using 1.25 mM Au NP solution, were taken using contact mode on a Bruker MultiMode atomic force microscope (Billerica, MA, USA) with a Veeco Nanoscope IIIa controller (Santa Barbara, CA, USA). Au NPs (oven-dried at 45 °C) were also analyzed using attenuated total reflection Fourier transform infrared (ATR-FTIR; MIRacle 10, IR-Tracer 100; Shimadzu, Kyoto, Japan) spectroscopy. Lastly, powder X-ray diffraction analysis was performed using a Bruker D8 Discover X-ray diffractometer (Billerica, MA, USA) to confirm the crystalline structure of the Au NPs.

2.5. Catalysis of the Reduction of 4-Nitrophenol

Sodium borohydride (200 mM; over ice) and 4-nitrophenol (2.0 mM) solutions were freshly prepared. Then, 50 µL 4-nitrophenol, 5 µL Au NPs (80 mM), and 2 mL DI water were gently mixed in a cuvette. After this, 25 µL $NaBH_4$ was added into the cuvette immediately before starting measurements. Scans were performed every minute for 30 min using a UV-Vis spectrophotometer (250–550 nm) [22].

2.6. Antibacterial Activity Testing

The antibacterial effects of Au NPs on both Gram-negative *Escherichia coli* (Item #: 470176-528, Ward's Science, Rochester, NY, USA) and Gram-positive *Bacillus megaterium* (Item #: 15-4900, Carolina Biological Supply Company, Burlington, NC, USA) bacteria were evaluated using the agar disc diffusion method [23,24]. *E. coli* and *B. megaterium* were cultured in nutrient broth media (37 °C, 24 h) and inoculated onto agar plates (Mueller-Hinton growth medium). Diffusion discs were dipped into varying concentrations of Au NP solution (0.50, 0.25, 0.10, 0.05 mM) and placed on the inoculated plates. Ampicillin discs (10 mcg, AMP10-1815; Carolina Biological Supply Company, Burlington, NC, USA) and blank discs were added to the plates and served as positive and negative controls, respectively. The plates were incubated (37 °C, 24 h) and the antibacterial inhibition zones were analyzed.

2.7. Cytotoxicity and Cellular Uptake Studies

Bone marrow mesenchymal stem cells (BMSCs; MUBMX-01001, Cyagen, Santa Clara, CA, USA) and HeLa cells (Shanghai Key Laboratory of Maternal Fetal Medicine, Shanghai, China) were cultured in Dulbecco's Modified Eagle's Medium (DMEM)/F-12 (Gibco, Grand Island, NY, USA) and DMEM, separately. The medium was supplemented with 10% fetal bovine serum, 100 U ml^{-1} penicillin, and 100 mg/L streptomycin (Gibco, Grand Island, NY, USA). The cells were cultured in a humidified CO_2 incubator (5%) at 37 °C. For the cytotoxicity study, cells were seeded in 96-well plates at a density of 10,000 cells/well. After 24 h incubation, culture medium was replaced using fresh medium containing various concentrations of Au NPs (0.10, 0.25, 0.50, 1.00, 1.50, 2.00, 2.50 mM). The cells were then rinsed twice with PBS after 24 h and medium containing CCK8 (10 µL/100 µL medium; Beyotime Institute of Biotechnology, Shanghai, China) was added. After 2 h, 100 µL medium of each well was transferred to a new 96-well plate and the absorbance was determined at 450 nm using a micro plate reader (Bio-Rad 680, Bio-Rad; Hercules, CA, USA). Cell viability was determined using the absorbance ratio of an experiment well to the average of the control wells (i.e., cell culture medium only).

To assess cellular uptake, at 70–80% confluency, cells were cultured in fresh medium containing 1 mM Au NPs for 24 h. Then, cells were detached using 0.25% trypsin-ethylenediaminetetraacetic acid (EDTA) (Gibco, Grand Island, NY, USA) digestion, rinsed twice in cold PBS, and collected through centrifugation. Cells were then fixed using 2.5% glutaraldehyde for 2 h. Subsequently, the cells were washed and fixed with cacodylate buffer and osmium tetroxide (2%), respectively, dehydrated with 70–100% acetone and embedded and cut in a film (70 nm) using an ultra-microtome. After a uranyl acetate-lead citrate double staining, the samples were observed under TEM (H-600, Hitachi; Tokyo, Japan).

3. Results and Discussion

3.1. Total Phenolic and Ascorbic Acid Content

When plants contain higher total phenolic content, they possess stronger antioxidant activity [25]. Ascorbic acid also contributes to high levels of antioxidant capacity of upland cress [19]. Hence, it is critical to determine the total phenolic and ascorbic acid content. The total phenolic and ascorbic acid concentrations were determined to be 163.3 ± 1.5 μg GAE/mL extract and 24.5 ± 1.1 μg/mL extract, respectively. The high phenolic and ascorbic acid content potentially contributed to reduction and capping during nanoparticle synthesis.

3.2. Effects of Incubation Time on the Green Synthesis Process

To improve the extraction efficiency and consistency of the synthesis process, the previous procedure [19] was modified. Specifically, blending was used to prepare the extract rather than boiling, as boiling showed inconsistencies. The unique optical properties of Au NPs are primary indicators for confirming the successful synthesis of nanoparticles. The apparent color transition from pale green to wine red color (Figure 1) is indicative of the formation of gold nanoparticles due to the surface plasmon resonance phenomenon [26]. As the nanoparticles grow, the absorption wavelengths become longer and redder. The color and intensity changes reflect the formation and growth of Au NPs, respectively [26]. There was no significant color change observed after 4 h. To further investigate the effects of incubation time on the green synthesis process, 2-, 4-, and 6-h-long periods were selected based on a previous study [19]. As shown in Figure 2A, a characteristic absorption peak is visible for each nanoparticle sample prepared with different incubation times. A right shift and intensity increase of the characteristic absorbance peak between the 2- and 4-h samples indicate the growth of Au NPs and potential formation of agglomerates. After 4 h, there was no significant peak shift. Comparing with the control (0 h incubation, insert of Figure 2A), the absorbance remains in the low (450–650 nm) wavelength range. No absorbance peak is visible ~530 nm. After 4 h, there was no significant peak shift. Based on the results, a 6-h incubation time was chosen for the nanoparticle synthesis. UV-Vis spectroscopy results (Figure 2B) indicated a characteristic absorption peak of about 529 nm, which is within the characteristic range for gold nanoparticles (~500–550 nm). The concentration-dependence of particle formation was also observed when various concentrations of Au NPs (0.5, 1.0, 2.5 mM) were used (Figure 2B). The observed time-dependence and concentration-dependence of Au NP formation is consistent with literature [27,28].

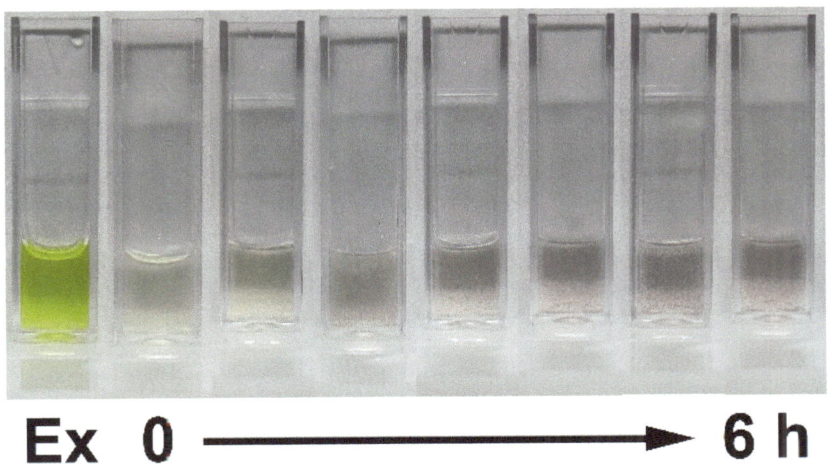

Figure 1. Solution color change of the upland cress extract and gold (III) chloride trihydrate mixture during the incubation period. Time interval: 1 h.

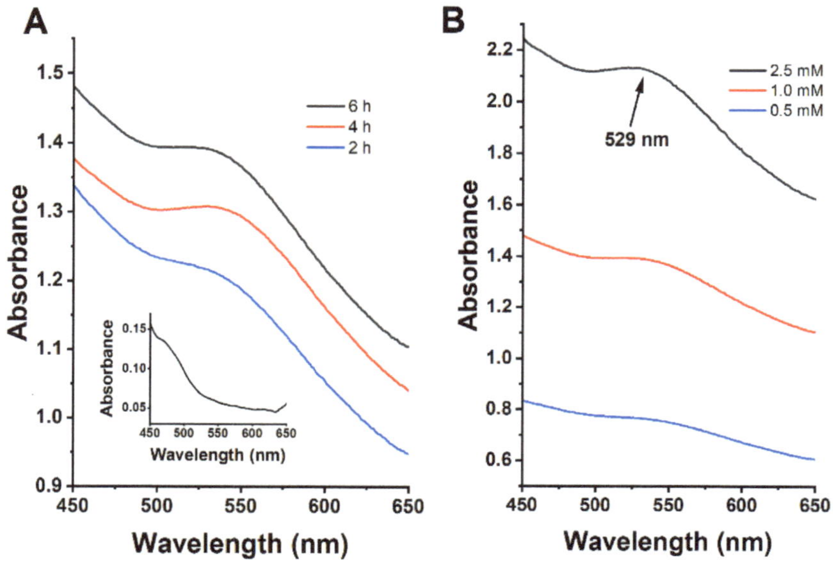

Figure 2. UV-Vis spectra of the synthesized Au NPs. (**A**) different incubation times during the green synthesis process (1.0 mM); insert: 0 h incubation (i.e., mixture of extract and $HAuCl_4$ in DI water). (**B**) different concentrations of synthesized Au NPs (6-h incubation).

3.3. Morphology, Chemical Composition, and Surface Charge

Both SEM and AFM were used to examine the morphology and size/size distribution of the Au NPs. SEM and AFM offer straightforward visualization of metallic nanoparticles due to their high resolution. Au NPs are indicated by the AFM and SEM as the bright spots in either image. SEM imaging (Figure 3A) revealed the production of spherical Au NPs with uniform size (10.7 $\pm$ 2.2 nm) and without aggregation. AFM imaging (Figure 3B) further confirmed the production of well dispersed spherical nanoparticles with a narrow size distribution. EDX results (Figure 3C) indicate an 11.13% gold composition by mass. Carbon and oxygen peaks within the spectrum indicate the presence of phytoconstituents, organic capping agents associated with the upland cress extract. Other inorganic elemental species such as calcium, potassium, chlorine, and magnesium were observed and their presence can be attributed to their high content in upland cress [12,29]. The presence of copper in the EDX spectrum was observed due to the conductive adhesive used for SEM imaging. Zeta potential measurements were performed to assess the surface charge and stability of the synthesized Au NPs. Zetasizer readings provided an average zeta (ζ) potential of -36.8 mV, implying favorable colloidal stability [30]. The apparent stability is most likely due to the phytochemical capping of the Au NPs.

The ATR-FTIR spectrum of Au NPs is shown in Figure 4. Functional groups were assigned to the corresponding spectral bands based on literature [15,31]. The bands include 3286 cm^{-1} (–OH stretching of phenolics and other phytochemicals); 2928 cm^{-1} and 2875 cm^{-1} (–CH stretching of alkanes); 1636 cm^{-1} (including –NH bending and –C=O); 1512 cm^{-1} (–CH of alkanes and –NO of nitro-compounds); 1454 cm^{-1} (including –OH bending –C=O of phenolics and other phytochemicals); and 1387 cm^{-1} (–CN stretching of aromatic amine group). Based on the results, the presence of stabilizing/capping agents (phenolics and other phytochemicals of upland cress extract) on the Au NPs was confirmed [15].

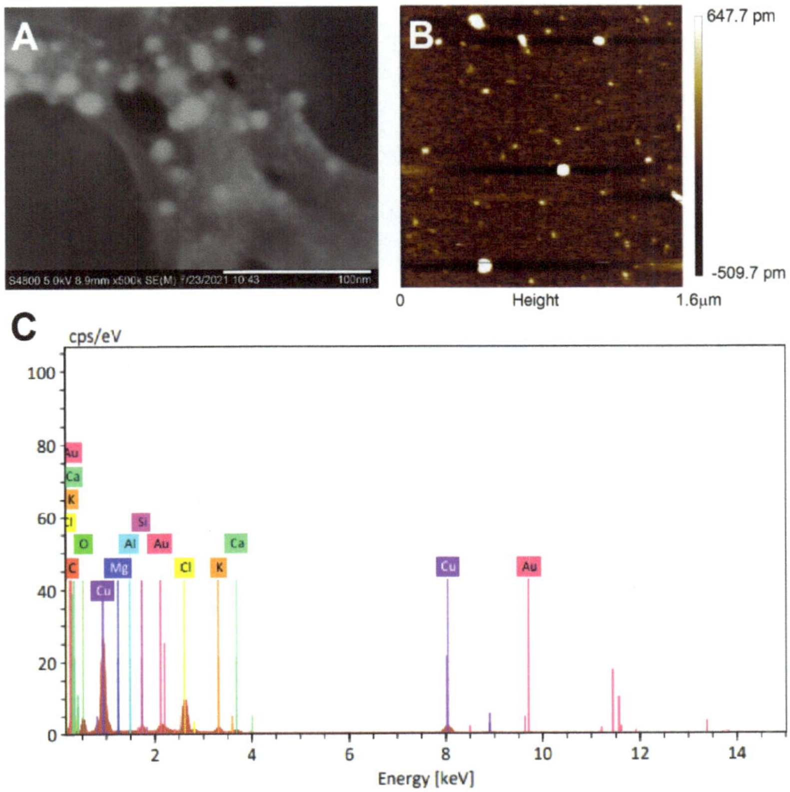

Figure 3. SEM (**A**) and AFM (**B**) images and EDX spectrum (**C**) of the synthesized Au NPs. SEM image scale bar: 100 nm.

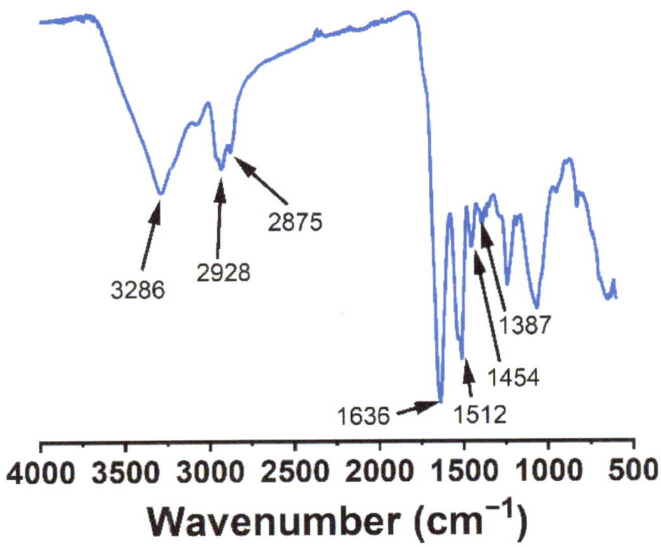

Figure 4. ATR-FTIR spectrum of the synthesized Au NPs.

3.4. Crystal Structure of Au NPs

Powder XRD analysis (Figure 5) provided diffraction peaks at 2θ angles 38.20°, 43.73°, 64.77°, 77.72°, and 82.09° corresponding to the crystalline gold atomic planes (111), (200), (220), (311), and (222) confirming the expected face-centered cubic structure (JCPDS Card No. 96-901-1613). Unassigned diffraction peaks are presumed to be related to the production of bio-organic crystallite phases on the surface of the Au NPs [32,33]. Peak broadening observable in the XRD pattern can be attributed to the scale of the measured crystallites as explained by the Scherrer equation (Equation (1)) [19,34].

$$D = \frac{k\lambda}{\beta\ \cos(\theta)} \tag{1}$$

where $k = 1$, $\lambda = 0.1542$, β is the full width at half maximum, and θ is the diffraction angle. Using 2θ values of 38.20° and 64.77°, the nanoparticle size was determined to be approximately 13 nm, which is similar to the SEM and AFM results.

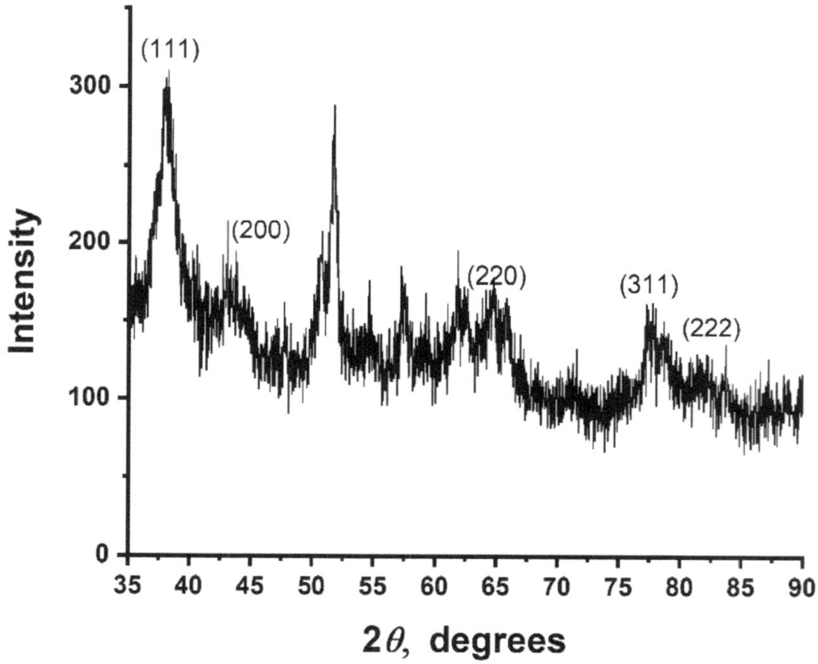

Figure 5. XRD spectrum of the synthesized Au NPs.

3.5. Catalysis of the Reduction of 4-Nitrophenol

The catalysis of the reduction of 4-nitrophenol is a commonly used reaction when testing the catalytic ability of nanoparticles [12,35–37]. Verification of nanoparticle catalysis is centered on the analysis of the extinction of the absorbance peak of 4-nitrophenol (400 nm) which indicates the catalyzed reduction of 4-nitrophenol to 4-aminophenol. Results show that over a 30-min measurement interval, the characteristic peak of 4-nitrophenol decreased substantially, verifying successful catalytic ability (Figure 6A). The catalysis was also verified through the reaction system color change from bright yellow to pale pink (color of Au NPs). The kinetics of the catalyzed reaction were analyzed according to the Langmuir-Hinshelwood mechanism for bimolecular surface reactions [6,38]. According to this general model, the reduction reaction occurs on the surface of the Au NPs. Here, borohydride ions (BH_4^-) adsorb to the surface and hydrogen species are formed via electron transfer. At the same time, 4-nitrophenol adsorbs to the surface. The 4-nitrophenol is then reduced to 4-aminophenol on the surface before detachment from the catalyst site [12]. The pseudo-first order equation used to analyze this process is shown as Equation (2).

$$ln^{(\frac{A_t}{A_0})} = -k_{app}t \quad (2)$$

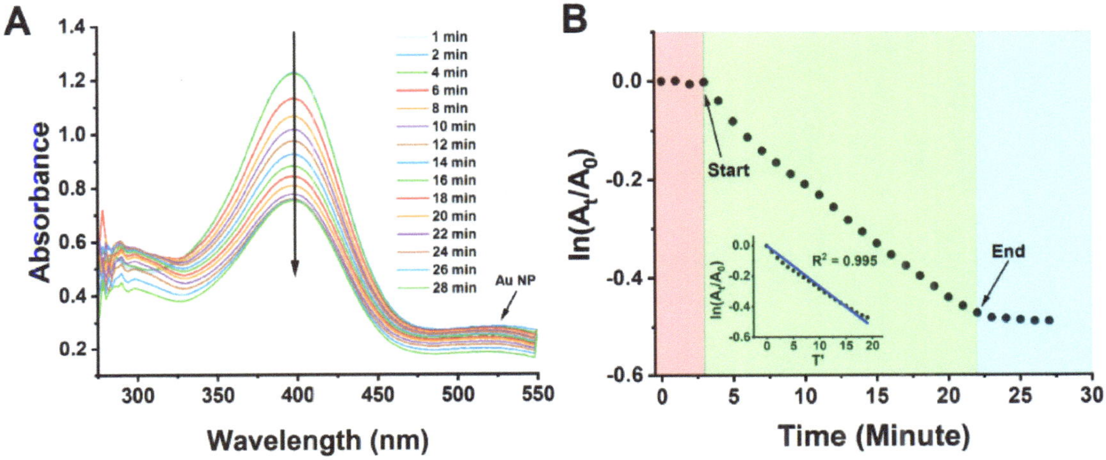

Figure 6. UV-vis spectra (**A**) at different time points and reaction kinetics. (**B**) The Au NP catalyzed reduction of 4-nitrophenol. T′ = T − 4.

Application of this relationship reveals an initial period of no reaction (4 min), followed by first-order reaction kinetics (linear region). The rate constant was derived from the linear region and was found to be 0.0267 min^{-1} (Figure 6B) with an R^2 of 0.995. The initial period of no reaction is most likely attributed to the blockage of potential catalysis sites by capping molecules. This size of Au NPs is preferred for catalysis according to reference [39,40].

3.6. Antibacterial Activity

The analysis of inhibitory zones shows a dose-dependent and species dependent antibacterial effect on Gram-negative *E. coli* and Gram-positive *B. megaterium* (Figure 7). Measurements of bacterial growth inhibition indicate the largest zones of inhibition occurring at Au NP doses of 0.25 mM (9 mm) and 0.5 mM (7.25 mm) for *E. coli* and *B. megaterium*, respectively. Analysis of the average inhibition zones across all trials indicate a higher antibacterial activity in the *B. megaterium* with an average of 7.3 mm versus an average of 6.6 mm seen with *E. coli*. The Au NPs appear to have lower antibacterial activity than Ag NPs using the same synthesis method [19]. It is reported that nanoparticles have been demonstrated to show a size and surface ligand dependent cytotoxic effect [24,41]. The small size of the synthesized Au NPs and the presence of bioorganic surface ligands related to the upland cress extract may contribute to the observed antibacterial activity. This antibacterial activity of Au NPs may be attributed to one or multiple mechanisms of Au NPs including the direct disruption of major internal cell function (ATP production, DNA replication, enzyme inhibition), the formation of toxic reactive oxygen species (ROS), as well as direct damage to the cellular membrane [41,42]. The antibacterial activity of Au NPs show promise for their biomedical applications with added infection prevention benefits.

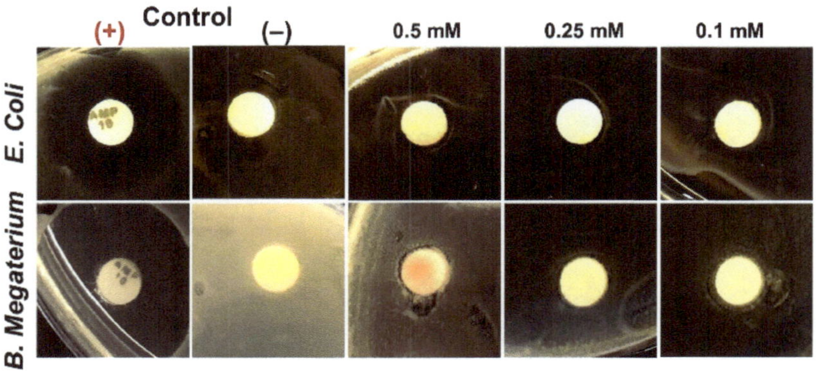

Figure 7. Antimicrobial testing results: top: *E. coli* and bottom: *B. megaterium*. The ampicillin and blank discs were used as positive and negative control, separately.

3.7. Cytotoxicity and Cellular Uptake

Considering the potential biomedical applications, both stem (BMSCs) and cancer (HeLa) cells were used to study the cellular uptake and cytotoxicity of Au NPs (Figure 8). BMSCs were studied for potential stem-cell based medicines, while HeLa cells were examined for prospective cancer treatment applications. A broad range (0.1–2.5 mM) of nanoparticle concentrations were tested and LC_{50} (50% lethal concentration—the concentration that kills 50% of the cells) were determined. The results showed that BMSCs (LC_{50} = 2 mM) are more sensitive to the Au NPs than HeLa cells (LC_{50} > 2.5 mM). The cell viability of the BMSCs dropped lower than 70% when the Au NP concentration reaches 0.1 mM while the HeLa viability remains higher than 70% until the concentration surpasses 1.0 mM. A recent study showed about 80% cell viability of PC3 cancer cells using Au NPs of similar size, which is close to our results (84%) [43]. A cell viability of ~79% for BMSCs was reported with spherical Au NPs (18 nm; 0.09 mM) [44] which is comparable to our results as well (77%; 0.1 mM). A recent study indicated that 15 nm Au NPs could affect the characteristic MSC marker expression (e.g., CD 105) and cell differentiation, especially when the concentration is higher than 9 μg/mL [45]. Cell damage and apoptosis may be explained by the generation of reactive oxidative stresses (ROS) [46]. Based on the cytotoxicity results a concentration of 1.0 mM was used for cellular uptake studies. TEM images of the cellular studies (Figure 9) indicate that nanoparticles are visible within the cells. The nanoparticles are mostly within membrane-bound vesicles such as endosomes (formed due to endocytosis). These results are consistent with previous publications which state that endocytosis (most likely receptor-mediated) is the primary route for cellular uptake [43,47]. For potential applications, the size of the Au NPs can be fine-tuned to adjust toxicity and cellular uptake, as it has been well studied that the size greatly affects the nanoparticle-cell interactions (e.g., cellular uptake efficiency and mechanism) (size-dependent cellular uptake and localization profiles of silver nanoparticles; size- and cell type-dependent cellular uptake, cytotoxicity and in vivo distribution of gold nanoparticles).

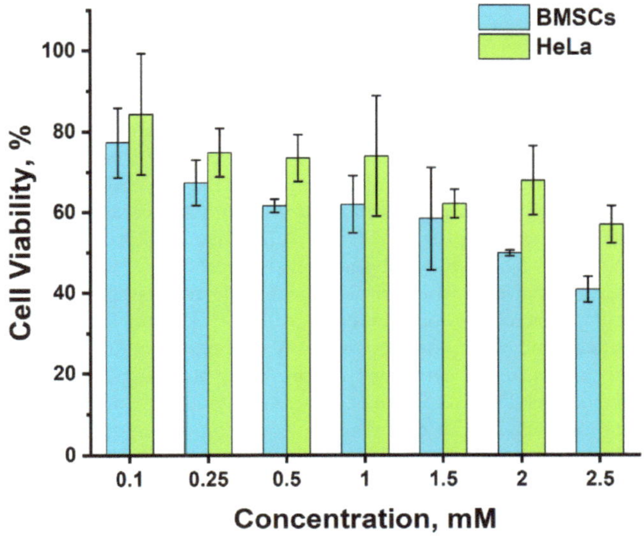

Figure 8. Cytotoxicity of Au NPs using both BMSCs and HeLa cells.

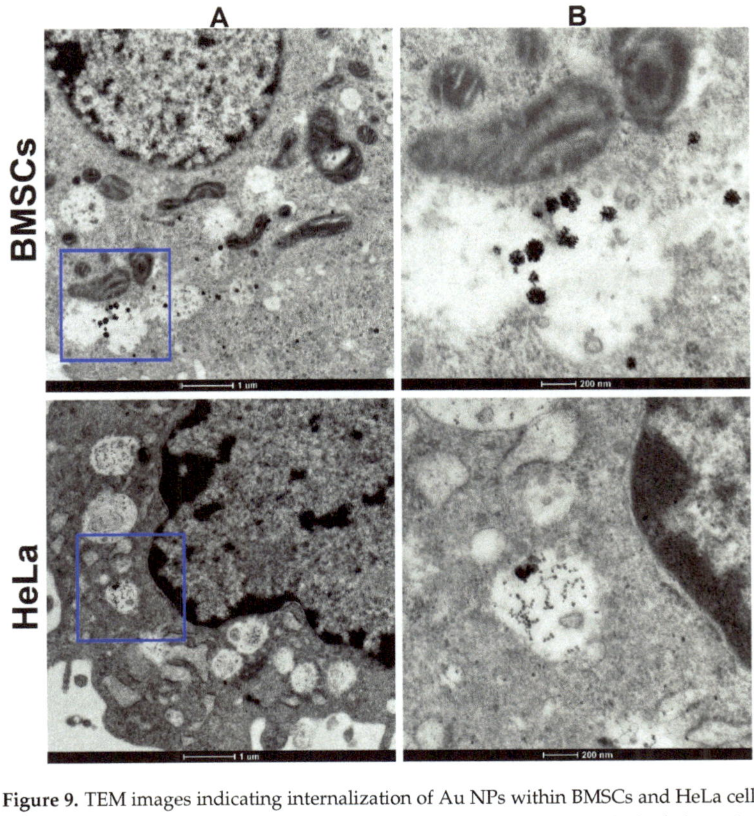

Figure 9. TEM images indicating internalization of Au NPs within BMSCs and HeLa cells. Enlarged images of the highlighted regions in column A are shown in column B. Scale bars: 1 μm (**A**) and 200 nm (**B**).

4. Conclusions

The study focused on the novel use of upland cress as a green synthesis agent for the production of gold nanoparticles. A blending method was used to extract the necessary phytochemicals (e.g., phenolics and ascorbic acid) within the upland cress to serve as a reducing agent for the reduction of gold (III) chloride trihydrate and a capping agent. The resultant purified particles were characterized using UV-Visible spectroscopy, SEM-EDX, AFM, Zetasizer, ATR-FTIR, and XRD. The results indicated the successful synthesis of Au NPs which were found to be spherical, well dispersed with an average diameter ~11 nm, and a characteristic absorbance peak at ~529 nm. A negative ζ-value of −36.8 mV indicated stability of the Au NPs while XRD analysis verified the production of crystalline face-centered cubic gold. Furthermore, the antimicrobial testing results demonstrated concentration-dependent antimicrobial properties of the Au NPs. The catalytic ability of Au NPs was demonstrated through the conversion of 4-nitrophenol to 4-aminophenol (first-order kinetics). Cellular uptake and cytotoxicity studies using both BMSCs and HeLa cells indicated the uptake of Au NPs into cells and cell type dependent cytotoxicity. This green synthesis method provides a simple, cost effective, green solution to produce gold nanoparticles and provides a promising source of functionalized nanoparticles for use in various applications.

Author Contributions: N.H., Y.W. (Yuelin Wu), Y.W. (Yale Wang), M.K. (Muskan Kanungo), A.D., E.K., D.G., Z.E., S.G., P.M., M.K. (Matey Kaltchev) performed the experiments, while N.H., Y.W. (Yuelin Wu), Y.W. (Yale Wang) and W.Z. performed data analysis. N.H. and W.Z. took the lead in writing. W.Z., X.H., A.-M.N. and S.K. supervised the project and/or designed experiments. All authors have read and agreed to the published version of the manuscript.

Funding: This work is financially supported by the National Natural Science Foundation of China (81873816 and 82071629).

Data Availability Statement: The data that support the findings of this study are available from the corresponding authors upon reasonable request.

Acknowledgments: The author would like to acknowledge Michael Navin for his technical support.

Conflicts of Interest: The authors declare no conflict of interest.

References

1. Dykman, L.A.; Khlebtsov, N.G. Gold nanoparticles in biology and medicine: Recent advances and prospects. *Acta Nat.* **2011**, *3* 34–55. [CrossRef]
2. Keijok, W.J.; Pereira, R.H.A.; Alvarez, L.A.C.; Prado, A.R.; da Silva, A.R.; Ribeiro, J.; de Oliveira, J.P.; Guimarães, M.C.C. Controlled biosynthesis of gold nanoparticles with Coffea arabica using factorial design. *Sci. Rep.* **2019**, *9*, 16019. [CrossRef] [PubMed]
3. Abdel-Raouf, N.; Al-Enazi, N.M.; Ibraheem, I.B.M. Green biosynthesis of gold nanoparticles using Galaxaura elongata and characterization of their antibacterial activity. *Arabian J. Chem.* **2017**, *10*, S3029–S3039. [CrossRef]
4. Khan, I.; Saeed, K.; Khan, I. Nanoparticles: Properties, applications and toxicities. *Arabian J. Chem.* **2019**, *12*, 908–931. [CrossRef]
5. Küünal, S.; Rauwel, P.; Rauwel, E. Plant extract mediated synthesis of nanoparticles. In *Emerging Applications of Nanoparticles and Architecture Nanostructures*; Barhoum, A., Makhlouf, A.S.H., Eds.; Elsevier: Amsterdam, The Netherlands, 2018; Chapter 14; pp. 411–446. [CrossRef]
6. Rodríguez-León, E.; Rodríguez-Vázquez, B.E.; Martínez-Higuera, A.; Rodríguez-Beas, C.; Larios-Rodríguez, E.; Navarro, R.E. López-Esparza, R.; Iñiguez-Palomares, R.A. Synthesis of gold nanoparticles using mimosa tenuiflora extract, assessments of cytotoxicity, cellular uptake, and catalysis. *Nanoscale Res. Lett.* **2019**, *14*, 334. [CrossRef] [PubMed]
7. Wang, L.; Hu, C.; Shao, L. The antimicrobial activity of nanoparticles: Present situation and prospects for the future. *Int. J Nanomed.* **2017**, *12*, 1227–1249. [CrossRef]
8. Navyatha, B.; Nara, S. Gold nanotheranostics: Future emblem of cancer nanomedicine. *Nanobiomedicine* **2021**, *8*, 18495435211053945 [CrossRef]
9. Tiwari, P.M.; Vig, K.; Dennis, V.A.; Singh, S.R. Functionalized gold nanoparticles and their biomedical applications. *Nanomaterials* **2011**, *1*, 31–63. [CrossRef] [PubMed]
10. Ahmed, S.; Annu; Ikram, S.; Yudha, S.S. Biosynthesis of gold nanoparticles: A green approach. *J. Photochem. Photobiol. B Biol.* **2016**, *161*, 141–153. [CrossRef] [PubMed]

11. Botteon, C.E.A.; Silva, L.B.; Ccana-Ccapatinta, G.V.; Silva, T.S.; Ambrosio, S.R.; Veneziani, R.C.S.; Bastos, J.K.; Marcato, P.D. Biosynthesis and characterization of gold nanoparticles using brazilian red propolis and evaluation of its antimicrobial and anticancer activities. *Sci. Rep.* **2021**, *11*, 1974. [CrossRef] [PubMed]
12. Doan, V.-D.; Thieu, A.T.; Nguyen, T.-D.; Nguyen, V.-C.; Cao, X.-T.; Nguyen, T.L.-H.; Le, V.T. Biosynthesis of gold nanoparticles using litsea cubeba fruit extract for catalytic reduction of 4-nitrophenol. *J. Nanomater.* **2020**, *2020*, 4548790. [CrossRef]
13. Annamalai, J.; Nallamuthu, T. Characterization of biosynthesized gold nanoparticles from aqueous extract of Chlorella vulgaris and their anti-pathogenic properties. *Appl. Nanosci.* **2015**, *5*, 603–607. [CrossRef]
14. Aljabali, A.A.A.; Akkam, Y.; Al Zoubi, M.S.; Al-Batayneh, K.M.; Al-Trad, B.; Abo Alrob, O.; Alkilany, A.M.; Benamara, M.; Evans, D.J. Synthesis of gold nanoparticles using leaf extract of ziziphus zizyphus and their antimicrobial activity. *Nanomaterials* **2018**, *8*, 174. [CrossRef] [PubMed]
15. Pourhassan-Moghaddam, M.; Zarghami, N.; Mohsenifar, A.; Rahmati-Yamchi, M.; Gholizadeh, D.; Akbarzadeh, A.; de la Guardia, M.; Nejati-Koshki, K. Watercress-based gold nanoparticles: Biosynthesis, mechanism of formation and study of their biocompatibility in vitro. *Micro Nano Lett.* **2014**, *9*, 345–350. [CrossRef]
16. Xiao, Z.; Rausch, S.R.; Luo, Y.; Sun, J.; Yu, L.; Wang, Q.; Chen, P.; Yu, L.; Stommel, J.R. Microgreens of brassicaceae: Genetic diversity of phytochemical concentrations and antioxidant capacity. *LWT* **2019**, *101*, 731–737. [CrossRef]
17. Dzimitrowicz, A.; Jamroz, P.; diCenzo, G.C.; Gil, W.; Bojszczak, W.; Motyka, A.; Pogoda, D.; Pohl, P. Fermented juices as reducing and capping agents for the biosynthesis of size-defined spherical gold nanoparticles. *J. Saudi Chem. Soc.* **2018**, *22*, 767–776. [CrossRef]
18. Nadeem, M.; Abbasi, B.H.; Younas, M.; Ahmad, W.; Khan, T. A review of the green syntheses and anti-microbial applications of gold nanoparticles. *Green Chem. Lett. Rev.* **2017**, *10*, 216–227. [CrossRef]
19. Johnson, D.L.; Wang, Y.; Stealey, S.T.; Alexander, A.K.; Kaltchev, M.G.; Chen, J.; Zhang, W. Biosynthesis of silver nanoparticles using upland cress: Purification, characterisation, and antimicrobial activity. *Micro Nano Lett.* **2020**, *15*, 110–113. [CrossRef]
20. Sir Elkhatim, K.A.; Elagib, R.A.A.; Hassan, A.B. Content of phenolic compounds and vitamin c and antioxidant activity in wasted parts of sudanese citrus fruits. *Food Sci. Nutr.* **2018**, *6*, 1214–1219. [CrossRef]
21. Dioha, I.J.; Olugbemi, O.; Onuegbu, T.; Shahru, Z. Determination of ascorbic acid content of some tropical fruits by iodometric titration. *Int. J. Biol. Chem. Sci.* **2012**, *5*, 2180–2184. [CrossRef]
22. Serrà, A.; Artal, R.; Pozo, M.; Garcia-Amorós, J.; Gómez, E. Simple environmentally-friendly reduction of 4-nitrophenol. *Catalysts* **2020**, *10*, 458. [CrossRef]
23. Mohamed, M.M.; Fouad, S.A.; Elshoky, H.A.; Mohammed, G.M.; Salaheldin, T.A. Antibacterial effect of gold nanoparticles against Corynebacterium pseudotuberculosis. *Int. J. Vet. Sci. Med.* **2017**, *5*, 23–29. [CrossRef] [PubMed]
24. Shamaila, S.; Zafar, N.; Riaz, S.; Sharif, R.; Nazir, J.; Naseem, S. Gold nanoparticles: An efficient antimicrobial agent against enteric bacterial human pathogen. *Nanomaterials* **2016**, *6*, 71. [CrossRef] [PubMed]
25. Zhang, Y.-J.; Gan, R.-Y.; Li, S.; Zhou, Y.; Li, A.-N.; Xu, D.-P.; Li, H.-B. Antioxidant phytochemicals for the prevention and treatment of chronic diseases. *Molecules* **2015**, *20*, 21138–21156. [CrossRef] [PubMed]
26. Amendola, V.; Pilot, R.; Frasconi, M.; Maragò, O.M.; Iatì, M.A. Surface plasmon resonance in gold nanoparticles: A review. *J. Phys. Condens. Matter Inst. Phys. J.* **2017**, *29*, 203002. [CrossRef] [PubMed]
27. Zuber, A.; Purdey, M.; Schartner, E.; Forbes, C.; van der Hoek, B.; Giles, D.; Abell, A.; Monro, T.; Ebendorff-Heidepriem, H. Detection of gold nanoparticles with different sizes using absorption and fluorescence based method. *Sens. Actuators B Chem.* **2016**, *227*, 117–127. [CrossRef]
28. Haiss, W.; Thanh, N.T.K.; Aveyard, J.; Fernig, D.G. Determination of size and concentration of gold nanoparticles from uv−vis spectra. *Anal. Chem.* **2007**, *79*, 4215–4221. [CrossRef]
29. Guo, M.; Li, W.; Yang, F.; Liu, H. Controllable biosynthesis of gold nanoparticles from a Eucommia ulmoides bark aqueous extract. *Spectrochim. Acta Part A Mol. Biomol. Spectrosc.* **2015**, *142*, 73–79. [CrossRef]
30. Pochapski, D.J.; Carvalho dos Santos, C.; Leite, G.W.; Pulcinelli, S.H.; Santilli, C.V. Zeta potential and colloidal stability predictions for inorganic nanoparticle dispersions: Effects of experimental conditions and electrokinetic models on the interpretation of results. *Langmuir* **2021**, *37*, 13379–13389. [CrossRef]
31. Ningaraju, S.; Munawer, U.; Raghavendra, V.B.; Balaji, K.S.; Melappa, G.; Brindhadevi, K.; Pugazhendhi, A. Chaetomium globosum extract mediated gold nanoparticle synthesis and potent anti-inflammatory activity. *Anal. Biochem.* **2021**, *612*, 113970. [CrossRef] [PubMed]
32. Amargeetha, A.; Velvan, S. X-ray diffraction (XRD) and Energy Dispersive Spectroscopy (eds) analysis of silver nanoparticles synthesized from Erythrina indica flowers. *Nanosci. Technol. Open Access* **2018**, *5*, 1–5. [CrossRef]
33. Muthuvel, A.; Adavallan, K.; Balamurugan, K.; Krishnakumar, N. Biosynthesis of gold nanoparticles using Solanum nigrum leaf extract and screening their free radical scavenging and antibacterial properties. *Biomed. Prev. Nutr.* **2014**, *4*, 325–332. [CrossRef]
34. Holder, C.F.; Schaak, R.E. Tutorial on powder x-ray diffraction for characterizing nanoscale materials. *ACS Nano* **2019**, *13*, 7359–7365. [CrossRef]
35. Iben Ayad, A.; Luart, D.; Ould Dris, A.; Guénin, E. Kinetic analysis of 4-nitrophenol reduction by "water-soluble" palladium nanoparticles. *Nanomaterials* **2020**, *10*, 1169. [CrossRef] [PubMed]
36. Majumdar, R.; Bag, B.G.; Ghosh, P. Mimusops elengi bark extract mediated green synthesis of gold nanoparticles and study of its catalytic activity. *Appl. Nanosci.* **2016**, *6*, 521–528. [CrossRef]

37. O'Neill, M.; Raghuwanshi, V.S.; Wendt, R.; Wollgarten, M.; Hoell, A.; Rademann, K. Gold nanoparticles in novel green deep eutectic solvents: Self-limited growth, self-assembly & catalytic implications. *Z. Phys. Chem.* **2015**, *229*, 221–234. [CrossRef]
38. Thawarkar, S.R.; Thombare, B.; Munde, B.S.; Khupse, N.D. Kinetic investigation for the catalytic reduction of nitrophenol using ionic liquid stabilized gold nanoparticles. *RSC Adv.* **2018**, *8*, 38384–38390. [CrossRef]
39. Suchomel, P.; Kvitek, L.; Prucek, R.; Panacek, A.; Halder, A.; Vajda, S.; Zboril, R. Simple size-controlled synthesis of Au nanoparticles and their size-dependent catalytic activity. *Sci. Rep.* **2018**, *8*, 4589. [CrossRef] [PubMed]
40. Fenger, R.; Fertitta, E.; Kirmse, H.; Thünemann, A.F.; Rademann, K. Size dependent catalysis with CTAB-stabilized gold nanoparticles. *Phys. Chem. Chem. Phys.* **2012**, *14*, 9343–9349. [CrossRef]
41. Sukhanova, A.; Bozrova, S.; Sokolov, P.; Berestovoy, M.; Karaulov, A.; Nabiev, I. Dependence of nanoparticle toxicity on their physical and chemical properties. *Nanoscale Res. Lett.* **2018**, *13*, 44. [CrossRef]
42. Sani, A.; Cao, C.; Cui, D. Toxicity of gold nanoparticles (AuNPs): A review. *Biochem. Biophys. Rep.* **2021**, *26*, 100991. [CrossRef]
43. Carnovale, C.; Bryant, G.; Shukla, R.; Bansal, V. Identifying Trends in gold nanoparticle toxicity and uptake: Size, shape, capping ligand, and biological corona. *ACS Omega* **2019**, *4*, 242–256. [CrossRef]
44. Fan, J.-H.; Li, W.-T.; Hung, W.-I.; Chen, C.-P.; Yeh, J.-M. Cytotoxicity and differentiation effects of gold nanoparticles to human bone marrow mesenchymal stem cells. *Biomed. Eng. Appl. Basis Commun.* **2011**, *23*, 141–152. [CrossRef]
45. Volkova, N.; Pavlovich, O.; Fesenko, O.; Budnyk, O.; Kovalchuk, S.; Goltsev, A. Studies of the influence of gold nanoparticles on characteristics of mesenchymal stem cells. *J. Nanomater.* **2017**, *2017*, 6934757. [CrossRef]
46. Surapaneni, S.K.; Bashir, S.; Tikoo, K. Gold nanoparticles-induced cytotoxicity in triple negative breast cancer involves different epigenetic alterations depending upon the surface charge. *Sci. Rep.* **2018**, *8*, 12295. [CrossRef] [PubMed]
47. Xia, Q.; Huang, J.; Feng, Q.; Chen, X.; Liu, X.; Li, X.; Zhang, T.; Xiao, S.; Li, H.; Zhong, Z.; et al. Size and cell type-dependent cellular uptake, cytotoxicity and in vivo distribution of gold nanoparticles. *Int. J. Nanomed.* **2019**, *14*, 6957–6970. [CrossRef]

nanomaterials

MDPI

Article

Tetrahedral DNA Framework-Programmed Electrochemical Biosenors with Gold Nanoparticles for Ultrasensitive Cell-Free DNA Detection

Chenguang Wang [1,2,†], Wei Wang [3,†], Yi Xu [1], Xiaoshuang Zhao [4], Shuainan Li [1,2], Qiuling Qian [1,2] and Xianqiang Mi [1,2,4,5,6,*]

1 Shanghai Advanced Research Institute, Chinese Academy of Sciences, Shanghai 201210, China; wangcg@sari.ac.cn (C.W.); xuyi@sari.ac.cn (Y.X.); lishuainan2019@sari.ac.cn (S.L.); qianqiuling2018@sari.ac.cn (Q.Q.)
2 University of Chinese Academy of Sciences, Beijing 100049, China
3 Shanghai Pudong New District Zhoupu Hospital, Shanghai University of Medicine & Health Sciences Affiliated Zhoupu Hospital, Shanghai 201318, China; doctorww1978@163.com
4 State Key Laboratory of Functional Material for Informatics, Shanghai Institute of Microsystem and Information Technology, Chinese Academy of Sciences, Shanghai 200050, China; zxsh1217@mail.sim.ac.cn
5 CAS Center for Excellence in Superconducting Electronics (CENSE), Shanghai 200050, China
6 Key Laboratory of Systems Health Science of Zhejiang Province, Hangzhou Institute for Advanced Study, University of Chinese Academy of Sciences, Chinese Academy of Sciences, Hangzhou 310024, China
* Correspondence: mixq@mail.sim.ac.cn
† These authors contributed equally to this work.

Abstract: Tumor-associated cell-free DNA (cfDNA) is a dynamic biomarker for genetic analysis, early diagnosis and clinical treatment of cancers. However, its detection has limitations because of its low abundance in blood or other complex bodily fluids. Herein, we developed an ultrasensitive cfDNA electrochemical biosensor (E-cfDNA sensor) based on tetrahedral DNA framework (TDF)-modified gold nanoparticles (Au NPs) with an interface for cfDNA detection. By accurately controlling the numbers of base pairs on each DNA framework, three types of TDFs were programmed: 26 base pairs of TDF; 17 base pairs of TDF; and 7 base pairs of TDF (TDF-26, TDF-16 and TDF-7, respectively). We also combined the TDF with hybridization chain reaction (HCR) to achieve signal amplification. Under optimal conditions, we detected the breast cancer susceptibility gene 1 (BRCA-1), a representative cfDNA closely related to breast cancer. An ultra-low detection limit of 1 aM with a linear range from 1 aM to 1 pM by TDF-26 was obtained, which was superior to the existing methods. Each type of TDF has excellent discrimination ability, which can distinguish single mismatch. More significantly, we also detected BRCA-1 in mimic serum samples, demonstrating that the E-cfDNA sensor has potential use in clinical research.

Keywords: tetrahedral DNA framework; gold nanoparticles; hybridization chain reaction; cell-free DNA; electrochemical biosenors

1. Introduction

Tumor-associated cell-free DNA (cfDNA) is a dynamic biomarker derived from different release mechanisms, such as necrosis, apoptosis and active release from carcinoma cells [1,2]. cfDNA can be used for early cancer diagnosis, gene mutation diagnosis, assisted targeted therapy and prognosis, etc. [3,4]. However, due to its extremely low content in blood or other bodily fluids, it is necessary to develop an ultrasensitive detection system for the analysis of cfDNA.

In recent decades, a certain number of technologies have been reported for the analysis of cfDNA, including polymerase chain reaction, DNA sequencing [5,6], microarray [7], nanomaterial-based biosensors [8,9], etc. Among these technologies, electrochemical-based biosensors are regarded as a promising direction for cfDNA detection because of their advantages of high sensitivity, low cost, and miniaturization [10–12]. In order to improve the detection performance of electrochemical analysis, quite a few methods have been studied. For example, as an important branch of electrochemical biosensors, screen-printed carbon electrodes (SPCE) are not only reproducible and inexpensive, but can also be prepared into multi-channel electrodes to achieve high-throughput detection and improve detection efficiency [13]. However, the screen-printed carbon surface is highly rough and prone to nonspecific adsorption [14]. Inorganic nanomaterials such as gold nanoparticles (AuNPs) can be modified on the electrode surface, providing excellent electrical conductivity, high surface area to volume ratio, superior catalytic capability and stability, and the ability to control the electrode microenvironment [15–17].

AuNPs are used to boost the fixation of the capture probe through the Au-S bond [18,19]. The traditional capture probes are usually thiolated single-stranded DNA [20]. Nevertheless, it is hard to modulate the density and orientation of the capture probes at the interface, which affect their recognition and binding of target molecules [21]. In recent years, tetrahedral DNA framework (TDF), a kind of three-dimensional programmable soft lithography nanomaterial, has been used as a capture probe to be fixed on the surface of the gold electrode [22]. Due to its rigid structure and controllable size, the TDF could precisely control the direction of the capture probes and the distance between the probes, avoiding molecular entanglements, providing a solution-phase-like environment, and improving the sensitivity of electrochemical sensors [23–25]. Lin et al. used millimeter-sized gold electrodes modified with different sizes of TDFs for DNA detection, which achieved attomolar sensitivity [26]. Our group has successfully combined TDFs with poly-adenine-based AuNPs for the analysis of BRCA1 with a detection limit of 0.1 fM [27].

In order to further improve the detection performance of electrochemical cfDNA biosensors, various signal amplification strategies are introduced into the biosensors, such as enzyme amplification strategies [28,29], and nanoparticle-based amplification strategies [30,31]. The hybridization chain reaction (HCR) is a toehold-mediated amplification reaction that could be applied to solid-state interfaces, such as electrodes, nanoparticles, glass slides or microfluidic chips, etc. [32]. The HCR products combine with other output moleculars to achieve the ultrasensitive detection of nucleic acid. Yang et al. reported an HCR-based electrochemical genosensor, in which the capture probe was immobilized on the electrode substrate through Au-S bonds, realizing BRCA1 detection with ultrahigh sensitivity [33]. Ge et al. developed an electrochemical biosensor based on the synergistic effect of the TDF and the HCR. TDF could control the density and direction of the capture probes and targets, while the targets triggered the HCR that combined with HRP to achieve sensitive detection of miRNAs [34].

Herein, given the synergistic superiority of TDF and HCR, we exploited home-made Au NPs modified with a multichannel electrochemical biosensor for ultrasensitive detection of cell-free DNA (E-cfDNA sensor). Using this platform, we detected the breast cancer susceptibility gene 1 (BRCA1, a representative cfDNA closely related to breast cancer) that achieved a detection limit at the attomolar level. In addition, this E-cfDNA sensor also realized the discrimination of a single-base mismatch. Moreover, this platform provides a universal tool for other cfDNA, and has great potential in clinical liquid biopsy.

2. Materials and Methods

2.1. Materials and Instruments

All nucleotide sequences used (Tables 1 and S1, ordered from Sangon Biotech, Shanghai, China): tris (2-carboxyethyl) phosphine hydrochloride, named TCEP solution (Sigma-Aldrich, Shanghai, China); poly-HRP40 (streptavidin modified), named SA-polyHRP (Fitzgerald Industries International Inc., New Castle, DE, USA); TMB Substrate (Neogen, KY, USA); $HAuCl_4$ (99.8% Au, Strem Chemicals Inc., Bischheim, France); fetal bovine serum, named FBS (Invitrogen, Carlsbad, CA, USA); all chemical buffer (Sangon Biotech, Shanghai, China). We also used: 16-mutichannel screen-printed carbon electrode, named 16-SPCE (CH Instruments, Inc., Shanghai, China); CHI-660C electrochemical workstation

(CH Instruments, Inc., Shanghai, China); NanoDrop One (Thermo Fisher Scientific Inc., Tumwater, WA, USA); Nova nanoSEM 450 instrument (FEI Company, Rockville, MD, USA); and AFM Multimode 8 instrument (Bruker, Billerica, CA, USA).

Table 1. Nucleotide sequences for selective ability of the E-cfDNA sensor.

Name	Sequence (5′-3′)
Mismatch-1	TGGTAACAGTGTGAGGTTTAACGGAACAAATGGAAGAAAATC
Mismatch-3	TGGTAACAGTGTGAGGTTTAACGGAACAAGATGATGAAAATC
Random	TGGTAACAGTGTGAGGTTTAACGTCGATGCCTGATCTTGGTA
Target-BRCA-1	TGGTAACAGTGTGAGGTTTAACGGAACAAAAGGAAGAAAATC

Human serum samples were derived from healthy human blood, which was donated by volunteers. First, the whole blood was centrifuged at 3500 rpm for 10 min, then the supernatant was collected and kept at 4 °C for later use.

2.2. Synthesis and Characterization of TDFs

TDFs of different sizes were synthesized according to the improved process [35]. Briefly, single-stranded A (1 μL), three-thiolated single-stranded B, C, D (1 μL, respectively) and TCEP (10 μL) were added to 86 μL T-buffer (20 mM Tris, 50 mM $MgCl_2$, pH 8.0). The mixture was assembled in T100^{TM} PCR Thermal Cycler (heated to 95 °C for 10 min then cooled to 4 °C for 20 min) for later use. The self-assembled TDFs were characterized by 8% native polyacrylamide gel electrophoresis (PAGE), and running condition was 100 V for 120 min in R-buffer (40 mM Tris, 1 mM Ethylene Diamine Tetraacetic Acid (EDTA), pH 8.0), which was visualized under UV light.

2.3. Synthesis and Characterization of HCR Structures

Biotin-H1 (10 μM) and biotin-H2 (10 μM) formed hairpin structures, respectively (heated to 95 °C for 10 min, and cooled to 4 °C for 20 min). Then, a mixture containing H1 (1 μM), H2 (1 μM) and different concentrations of initiator (target BRCA-1) were prepared in T-buffer. After that, the mixture was incubated for 2 h at room temperature to form HCR products. The products were characterized by 2% agarose gels (running condition was 150 V for 60 min in R-buffer) and visualized under UV light.

2.4. Development of E-cfDNA Sensor

Firstly, the SPCE was pretreated with the electrochemical workstation to generate AuNPs on the working electrode. Before deposition, the SPCE was cleaned by P-buffer (10 mM phosphate buffer, 0.14 M NaCl, 2.7 mM KCl, pH 7.4) and dried by N_2. Then, 60 μL $HAuCl_4$ solution was dribbled onto the electrode surface to form SPGE, the electrodeposition condition was as follows: scan rate, 100 mV/s; deposition time, 300 s; deposition potential, −200 mV. After that, excess $HAuCl_4$ solution was removed by P-buffer for later use. Secondly, 10 μL fresh TDF solution (1 μM) was dribbled onto the electrode and incubated at 30 °C over 8 h, and then rinsed by P-buffer. The TDF-modified SPGE was prepared for the following experiments.

2.5. cfDNA Detection by E-cfDNA Sensor

Under optimal conditions, the target BRCA-1 was first hybridized with the capture probe (extended chain of TDF) at the SPGE surface in T-buffer for 2 h. At the same time, biotin-H1 (1 μM) and biotin-H2 (1 μM) were heated and annealed as the before condition. After the target BRCA-1 hybridization was completed, H1/H2 mixture (100 nM) was dribbled onto the modified electrode and incubated for 2 h (room temperature). The extra solution was rinsed with P-buffer, and 10 μg/mL SA-polyHRP (3 μL) incubated the electrode surface for 15 min at room temperature. Finally, the electrodes were measured by the CHI-660C, the procedure for testing samples in FBS (50%) and human serum (50%) was the same as above.

3. Results and Discussion

3.1. Principle of the E-cfDNA Sensor

The design principle of the E-cfDNA sensor for cfDNA detection was based on the redox reaction that convert chemicals signals into electrical signals. As illustrated in Figure 1, firstly, Au NPs were deposited on the surface of the 16-SPCE to form homemade screen-printed gold electrodes (SPGE). Secondly, the TDF contained three thiol group-modified vertices that could be immobilized on the surface of the SPGE through Au-S bonds, and another vertex of the TDF carried a pendant DNA probe that could bind to the target DNA. Thirdly, to further address the limitation of electrochemical biosensors in terms of specificity and sensitivity, we introduced HCR to achieve signal amplification. We used target DNA as the initiator and two biotin-labeled hairpin structures (biotin-H1 and biotin-H2) as the fuel chains. When the target DNA was present, the promoter sequence on target DNA hybridized to H1, forming a cascade reaction to produce HCR products. Finally, SA-polyHRP was attached to the biotin-tagged HCR products and the reduction of H_2O_2 was catalyzed in the presence of TMB, resulting in quantitative electrochemical signals.

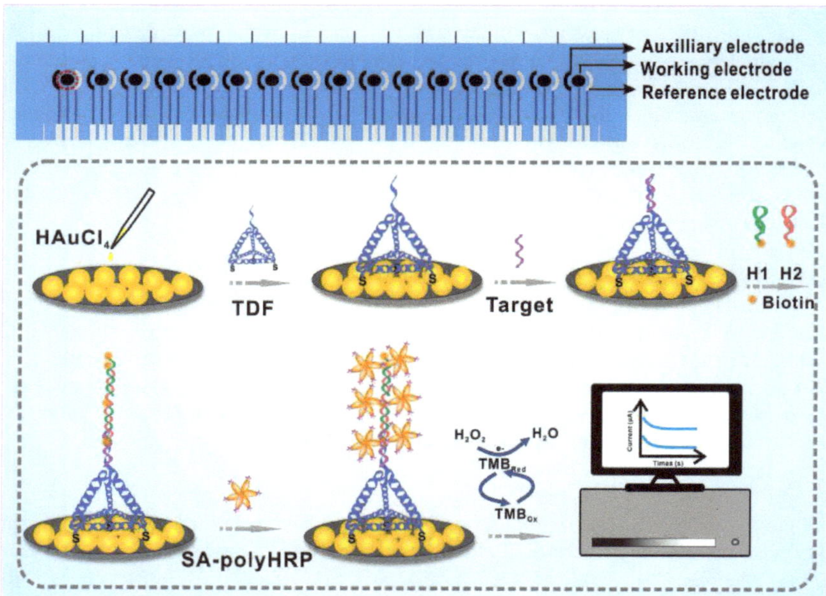

Figure 1. Schematic interpretation of the E-cfDNA sensor. Amperometric current (IT) and cyclic voltammetry (CV) were employed to investigate the performance of this platform.

3.2. Characterization and Optimization of Treated Electrode

16-SPCE is easily modified by various nanomaterials (gold, silver, graphene, etc.) to optimize the detection performance [36,37]. Here, we characterized the size and the morphology of the SPCE before and after electrodeposition by SEM. The pristine morphology of the SPCE surface was characterized in Figure 2a. After the bare electrode was deposited, Au NPs were deposited on the electrode surface in an aggregated state (Figure 2b,c). The number and the size of Au NPs on the electrode surface increased with the prolongation of the electrode deposition time, which increased the specific surface area (insert images shown in Figure 2a–c). Clearer SEM images of SPCE and SPGE with different deposition time were provided in Figure S1. The diameter of AuNPs was approximately 123.62 nm, ranging from 56 to 194 nm when the deposition time was 300 s. A larger specific surface area facilitated the subsequent immobilization of the TDFs. To improve the detection performance of SPGE, we optimized the electrodeposition time and the concentration of $HAuCl_4$.

As shown in Figure S2, the peak current in the CV curve increased with the deposition time and reached the highest value at 300 s, which was taken as the optimal deposition time. Next, the concentration of $HAuCl_4$ was optimized using the electrodeposition method. As shown in Figure S3, when the concentration of $HAuCl_4$ was 50 µg/mL, the current no longer increased, and we took this concentration as the optimal concentration for preparing electrodes.

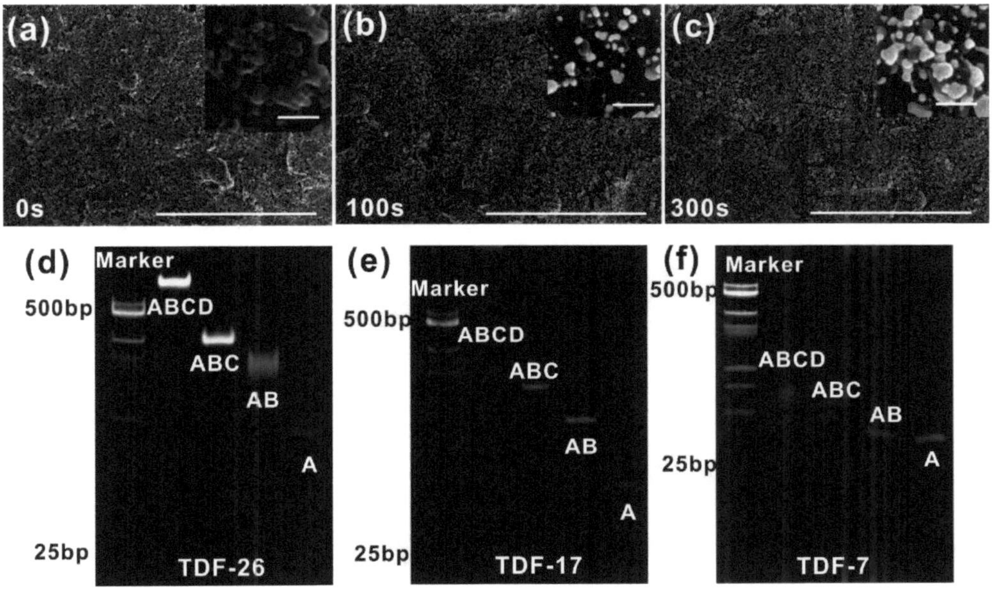

Figure 2. The SEM images of the bare SPCE (**a**) and SPGE with different deposition time: 100 s (**b**), 300 s (**c**). The scale bar was 20 µm. Inserts: Clearer SEM images of SPCE and SPGE. The scale bar was 200 nm. PAGE analysis of the formation of the TDF: (**d**) TDF-26, (**e**) TDF-17 and (**f**) TDF-7.

In order to subsequently explore the performance of TDFs on the SPGE surface, we first synthesized three different sizes of TDFs: TDF-26, TDF-17 and TDF-7. As shown in Figure 2d–f, native polyacrylamide gel electrophoresis (PAGE) analysis proved that TDFs were successfully assembled independently. Taking TDF-17 as an example, we selected TDF-17, triple-stranded, double-stranded and single-stranded DNA (ABCD, ABC, AB, A) to assemble DNA nanostructures. ABCD shifted slower than ABC, AB and A combinations, proving that the additional sequences and thiol groups did not interfere with the assembly. We also characterized the morphology of TDFs (TDF-17) on the mica surface by using atomic force microscopy (AFM), which indicated that the programmed structure of TDFs with pyramidal configuration was uniform and no aggregates appeared on the surface (Figure S4). The average edge length of the TDF-17 measured was about 6.194 nm, which was close to the theoretical value (Figure S4).

Furthermore, we also investigated the HCR reaction in solution. As shown in Figure S5, in the absence of an initiator (target BRCA-1), biotin-H1 and biotin-H2 maintained hairpin structures when introducing the target BRCA-1 sequence, a partial sequence of this target hybridized to the biotin-H1 strand, and the HCR reaction was triggered by alternately adding biotin-H1 and biotin-H2 to form long HCR products. It has been previously reported that the length of HCR products are inversely proportional to the concentration of the initiator. When the concentration of the target BRCA-1 was 1 µM, the length of the product was about 500 bp. When the concentration of the target BRCA-1 was 0.1 µM, the length of the product was about 1000 bp. Later, these long HCR products could bind more SA-polyHRP, ultimately achieving efficient signal amplification.

3.3. Comparison of Capture Performance among Different Probes

To evaluate the capture performance of the TDF-HCR-based E-cfDNA sensor, we first employed TDF-17 as the capture probe. The single-strand group and the single-TDF group were designed as controls. As shown in Figure 3a, the amperometry was used to directly characterize the electrochemical process of different probes. At an initial potential of 100 mV, we immediately obtained an attenuation curve of current (I) versus time (t), which leveled out within 100 s. As shown in Figure 3b, the blank current of the E-cfDNA sensor was as low as 0.33 μA, demonstrating little non-specific adsorption of nucleic acid or enzymes. The current was 1.65 μA for the single-strand group and 4.18 μA for single-TDF group, while the current was 8.09 μA for TDF-HCR group. The same trend appeared in the cyclic voltammetry (CV) tests (Figure S6). Compared with the single-stranded DNA capture probe, three-dimensional TDF as a rigid scaffold could be anchored on the surface of gold electrodes, and the extended strand at the vertex of the TDF maintained an ordered and upright orientation. In addition, TDF has a spatial structure that can enlarge the distance between probes, avoiding intermolecular entanglement. Furthermore, compared with the single-TDF group, the TDF-HCR strategy showed more than two times higher than the current signals. This result was attributed to the target DNA initiating the cascade HCR to form a long product, and the biotin-labeled product provided numerous binding sites for binding multiple avidin-labeled polyHRP, resulting in a significant increase in signal.

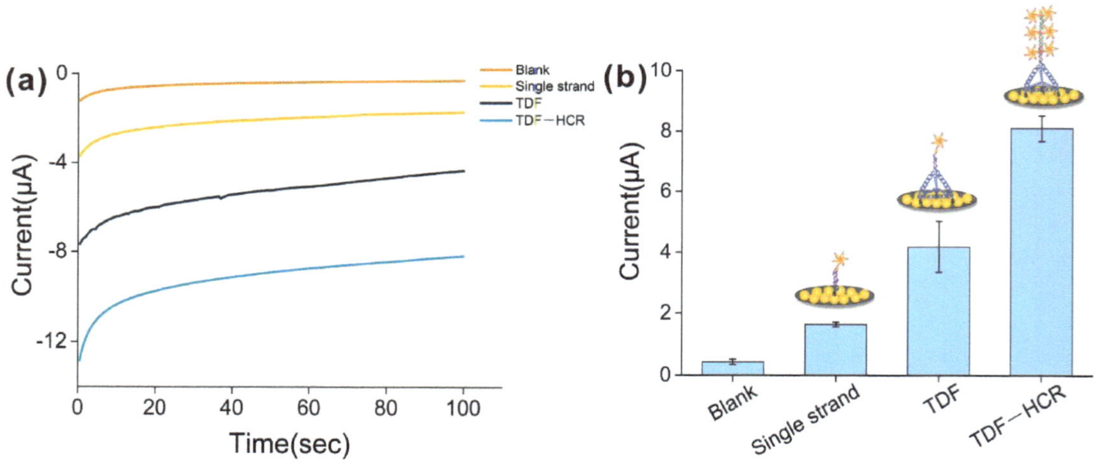

Figure 3. (**a**) Typical I-T curves for three kinds of probes (single strand group, TDF group and TDF-HCR group) modified on SPGE at target concentration of 1 nM. The potential was held at 100 mV and the reduction current was recorded at 100 s. (**b**) The corresponding current of three kinds of probes when the scan time was 100 s. Error bars represent the SD of at least 3 independent experiments.

3.4. Performance Verification of the E-cfDNA Sensor Mediated by TDF Regulation

Modulating the orientation and density of capture probes can improve hybridization efficiency between the target DNA and probe [38,39]. Here, we designed and programmed differently sized TDFs to accurately modulate the density of the capture probes on the SPGE surface (Figure 4a,d,g). Three sizes of TDFs were used: TDF-26, TDF-17 and TDF-7, each of which contained 26, 17, and 7 base pairs on each edge, and the corresponding theoretically calculated edge lengths were 8.8 nm, 5.8 nm, 2.4 nm, respectively. As shown in Figure 4, the response current signal increased significantly as the concentration of the target BRCA-1 increased from 0 nM to 1 nM, which suggested that the electrochemical signal closely depended on the concentration of the target (Figure 4b,e,h). However, each group had different detection limits and linear ranges for the target. For the DTF-26 group

(Figure 4c), the linear detection ranged from 1 aM to 1 pM with a 1 aM limit of detection (LOD), and the regression equation was Y = 0.4154 Log(X) + 1.7414 (R^2 = 0.9826). For the DTSP-17 group (Figure 4f), the linear detection ranged from 10 aM to 1 pM with a 10 aM LOD, the regression equation was Y = 0.4427Log(X) + 1.4654 (R^2 = 0.9445). For the DTSP-17 group (Figure 4i), the linear detection ranged from 1 fM to 1 pM with a 1 fM LOD, and the regression equation was Y = 0.1613Log(X) + 0.5294 (R^2 = 0.9371). Based on the above results, we demonstrated that as the size of the TDFs increased, the concentration of the lowest detectable target molecule decreased, and the sensitivity increased. In previously reported studies, Lin et al. found that the assembly density of DNA tetrahedral probes was inversely proportional to their size, and the hybridization efficiency of probes on the interface also heavily depended on the distance between probes [26]. That is, within a certain range, the larger the size of the DNA tetrahedron, the longer distance between the probes, resulting in a higher hybridization efficiency of the probes. In this work, due to the jointly optimized Au deposition substrate and HCR amplification system, the lowest detection limit can be as low as 1 aM, which is far superior to previous reports (Table 2).

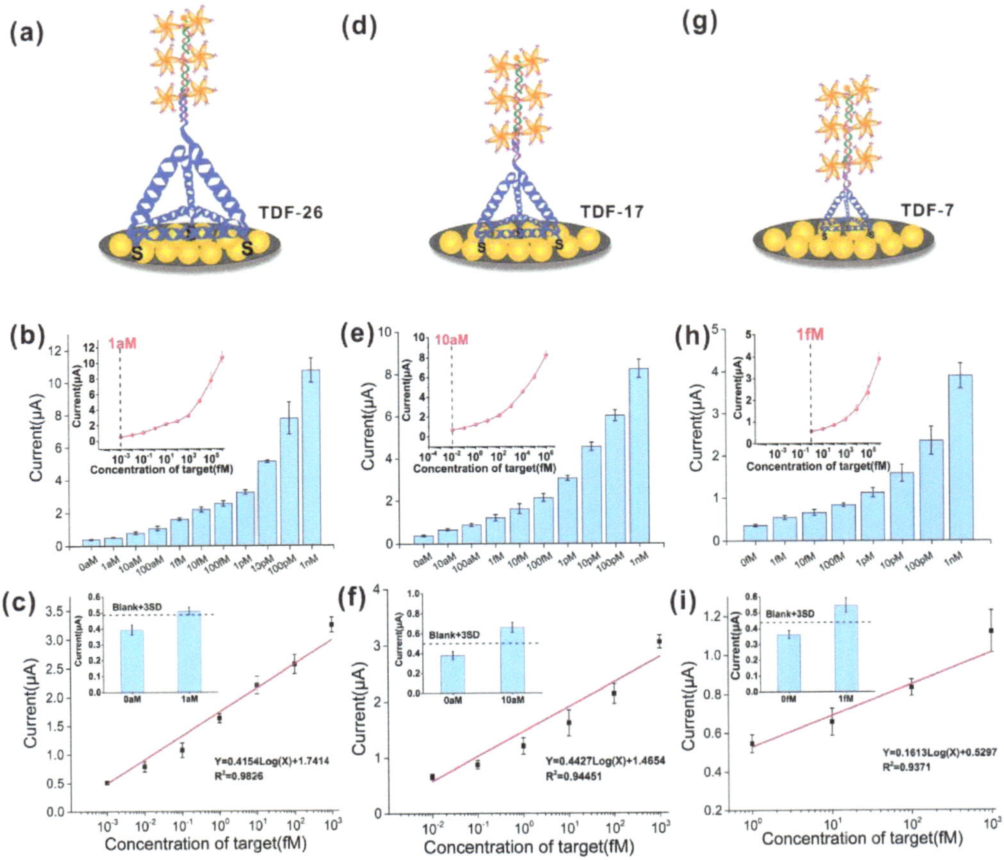

Figure 4. Sensitivity of the E-cfDNA sensor mediated by differently sized TDF (TDF-26, TDF-17, TDF-7). (**a**,**d**,**g**) Scheme illustration. (**b**,**e**,**h**) Amperometric current amplification with corresponding increased concentration (from 0 nM to 1 nM) of target DNA. Insert: a dose–response curve between DNA concentration and current. (**c**,**f**,**i**) Linear calibration curves. Insert: Limits of detection. Error bars represent the SD of at least 3 independent experiments.

Table 2. Comparison of the E-cfDNA sensor with other techniques for DNA detection.

Techniques	Name	Linear Range	LOD	Refs.
Fluorescence Biosensors	CRISPR-Cas12a-based cfDNA biosensing system	1 fM to 100 pM	0.34 fM	[8]
	DNA tetrahedral-based fluorescent microarray platform	100 aM to 1 pM	10 aM	[39]
Electrochemical Biosensors	Label-free electrochemical biosensor	0.01 fM to1 pM	2.4 aM	[28]
	HCR and DNA nanostructure-based electrochemical biosensor	1 fM to 100 pM	100 aM	[34]
Electrochemiluminescence Biosensors	DNA walk-based electrochemiluminescence biosensing	1 fM to 100 pM	0.18 fM	[40]
	Cas12a-based electrochemiluminescence biosensor	1 pM to 10 nM	0.48 pM	[41]
	E-cfDNA sensor	1 aM to 1 pM	1 aM	This work

3.5. Selective Ability of the E-cfDNA Sensor

The selectivity and specificity of the proposed E-cfDNA sensor was tested by using a perfectly matched sequence, a single-base mismatch DNA sequence (Mismatch-1), a three-base mismatched DNA sequence (Mismatch-3), and a random DNA sequence (Random) as detection targets (Table 1). As shown in Figure 5, the three types of TDF/HCR-based sensors could easily distinguish the target sequence from the mismatched sequence. Taking TDF-26 as an example, the amperometric current corresponding to the perfectly matched target (1 nM) was 10.335 μA, whereas the amperometric current corresponding to Mismatch-1, Mismatch-3 and Random were 1.749 μA, 0.751 μA, 0.459 μA, respectively. The current signals corresponding to target DNA were significantly higher than that of mismatched DNA, up to 22-fold, demonstrating the high specificity of the sensor. The results were mainly due to the uniqueness of the design, as shown in Figure S1. HCR cannot form a cascade reaction without the target sequence.

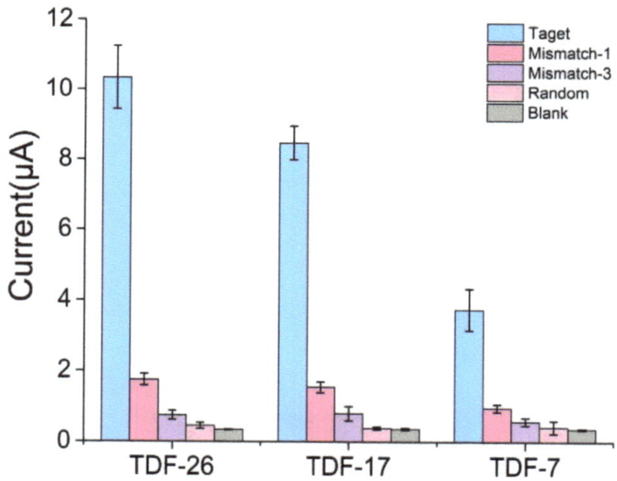

Figure 5. Specificity investigation of differently sized TDFs for target DNA (1 nM) and other mismatch DNA (1 nM). Error bars represent the SD of at least 3 independent experiments.

3.6. Application to the Clinical Utility

To further demonstrate the superiority of our electrochemical biosensor, the sensor was used to detect cell-free DNA in serum to demonstrate capture performance in complex components. We spiked the 1 nM of target BRCA-1 into fetal bovine serum (50%) and human serum (50%). As shown in Figure 6, the functionalized electrodes exhibited negligible signal changes (less than 7%) in either fetal bovine serum or human serum compared with PBS buffer, which indicated that the possibility of the E-cfDNA sensor can be used in clinical samples.

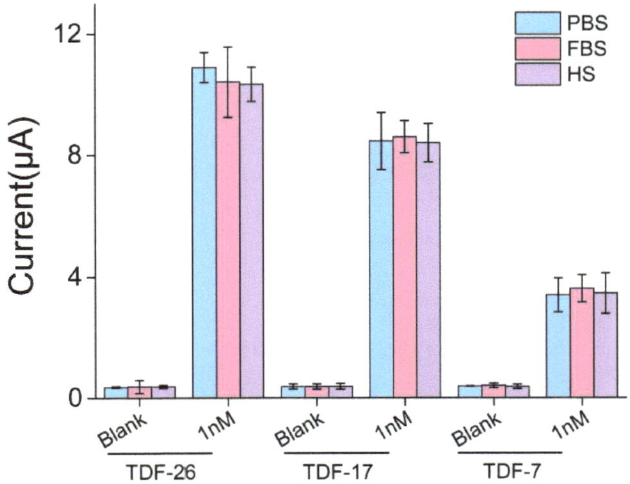

Figure 6. Performance verification of E-cfDNA sensor in PBS buffer, 50% fetal bovine serum (FBS) and 50% human serum (HS). Error bars represent the SD of at least 3 independent experiments.

4. Conclusions

In conclusion, we fabricated a home-made AuNP-deposited multi-channel electrode, and then exploited the ultrasensitive E-cfDNA sensor for specific cell-free DNA detection based on the programmed TDF and HCR. The TDF nanostructures immobilized on the electrode interface served as rigid scaffolds, and hybridized with the target sequence to trigger the HCR reaction, which achieved signal amplification. Compared to the traditional single-stranded capture probe, our TDF-HCR strategy showed over eight-fold increased amperometric current signals for detection of the target BRCA-1 gene. To further improve the sensitivity of the E-cfDNA sensor, we programmed TDFs of different sizes to precisely control the orientation of the capture probes and the distance probe-to-probe. Each of the TDF group exhibited a linear response to its target DNA, especially TDF-26, which showed the highest amperometric current signals with an ultra-low detection limit of 1 aM. In addition, our E-cfDNA sensor also maintained ultrahigh sensitivity in complex matrices, revealing promise for clinical early tumor detection, mutation screening and prognosis.

Supplementary Materials: The following are available online at https://www.mdpi.com/article/10.3390/nano12040666/s1: Table S1: DNA Sequences used in the work; Figure S1: The SEM images of AuNPs deposited on SPGE; Figure S2: Optimizing the deposition time of $HAuCl_4$ on SPCE; Figure S3: Optimizing the deposition concentration of $HAuCl_4$ on SPCE; Figure S4: The AFM image of TDF-17; Figure S5: Characterization of the HCR products; Figure S6: Cyclic voltammetry (CV) of different capture probes.

Author Contributions: Conceptualization, C.W., W.W., X.M.; methodology, C.W., W.W., Y.X.; validation, C.W., W.W.; formal analysis, C.W., X.Z.; investigation, C.W., S.L., Q.Q.; data curation, C.W.; writing—original draft preparation, C.W.; writing—review and editing, W.W., Y.X., X.M.; visualization, W.W.; supervision, X.M.; project administration, X.M.; funding acquisition, W.W., Y.X., X.M. All authors have read and agreed to the published version of the manuscript.

Funding: This work was funded by the Strategic Priority Research Program of the Chinese Academy of Sciences (In vitro diagnostic technology and equipment); Program of Shanghai Academic/Technology Research Leader, grant number 20XD1404600; Shanghai Municipal Science and Technology Commission, grant numbers 20511107600, 19511107100 and 19511107102; Chinese Academy of Science, grant number KFJ-STS-QYZD-2021-08-002; National Key Research and Development Program of China, grant number 2016YFC0100600; and Hundred-Talent Program (Shanghai University of Medicine & Health Sciences), grant number ZPBRK-18-04.

Institutional Review Board Statement: Not applicable.

Informed Consent Statement: Not applicable.

Data Availability Statement: The data is available on reasonable request from the corresponding author.

Conflicts of Interest: The authors declare no conflict of interest.

References

1. Polski, A.; Xu, L.Y.; Prabakar, R.K.; Kim, J.W.; Shah, R.; Jubran, R.; Kuhn, P.; Cobrinik, D.; Hicks, J.; Berry, J.L. Cell-Free DNA Tumor Fraction in the Aqueous Humor Is Associated With Therapeutic Response in Retinoblastoma Patients. *Transl. Vis. Sci. Technol.* **2020**, *9*, 30. [CrossRef] [PubMed]
2. Zhang, Q.; Gerratana, L.; Zhang, Y.; Flaum, L.; Shah, A.; Davis, A.; Behdad, A.; Gradishar, W.; Platanias, L.; Cristofanilli, M. ESR1 mutation in cell free DNA (cfDNA) is associated with significantly increased circulating tumor cell (CTC)-clusters and progress in stage III/IV breast cancer after systemic treatments. *Cancer Res.* **2019**, *79* (Suppl. S4), P4-01-04. [CrossRef]
3. Sato, K.A.; Nishizuka, S.S.; Iwaya, T.; Kume, K.; Otsuka, K.; Wakabayashi, G. Tumor-unique mutation detection in cell-free DNA to monitor colorectal tumor burden using a cancer-associated gene sequencing panel. *Cancer Res.* **2015**, *75* (Suppl. S15), 5234. [CrossRef]
4. Kustanovich, A.; Schwartz, R.; Peretz, T.; Grinshpun, A. Life and death of circulating cell-free DNA. *Cancer Biol. Ther.* **2019**, *20*, 1057–1067. [CrossRef]
5. Miller, D.E.; Hawley, S.; Miller, D.E.; Hawley, S. Isolation, Sequencing and Analysis of Human Cell-Free DNA (Cfdna). *J. Investig. Med.* **2014**, *62*, 732–733.
6. Assi, R.E.; Albitar, M.; Ma, W.L.; Patel, K.; Takahashi, K.; Yilmaz, M.; Short, N.J.; Jabbour, E.; Garcia-Manero, G.; Kantarjian, H.M. et al. Comparison of somatic mutations profiles from next-generation sequencing (NGS) of cell-free DNA (cfDNA) versus bone marrow (BM) in acute myeloid leukemia (AML). *J. Clin. Oncol.* **2018**, *36* (Suppl. S15), 7051. [CrossRef]
7. Tang, W.W.; Fu, K.; Sun, H.D.; Rong, D.W.; Wang, H.J.; Cao, H.Y. CircRNA microarray profiling identifies a novel circulating biomarker for detection of gastric cancer. *Mol. Cancer* **2018**, *17*, 137. [CrossRef]
8. Choi, J.H.; Lim, J.; Shin, M.; Paek, S.H.; Choi, J.W. CRISPR-Cas12a-Based Nucleic Acid Amplification-Free DNA Biosensor via Au Nanoparticle-Assisted Metal-Enhanced Fluorescence and Colorimetric Analysis. *Nano Lett.* **2021**, *21*, 693–699. [CrossRef]
9. Ma, S.H.; Zhang, Y.P.; Ren, Q.Q.; Wang, X.F.; Zhu, J.H.; Yin, F.; Li, Z.G.; Zhang, M. Tetrahedral DNA nanostructure based biosensor for high-performance detection of circulating tumor DNA using all-carbon nanotube transistor. *Biosens. Bioelectron.* **2022**, *197*, 113785. [CrossRef]
10. Abi, A.; Mohammadpour, Z.; Zuo, X.L.; Safavi, A. Nucleic acid-based electrochemical nanobiosensors. *Biosens. Bioelectron.* **2018**, *102*, 479–489. [CrossRef]
11. Bo, X.J.; Zhou, M.; Guo, L.P. Electrochemical sensors and biosensors based on less aggregated graphene. *Biosens. Bioelectron.* **2017**, *89*, 167–186. [CrossRef]
12. Sharifi, M.; Avadi, M.R.; Attar, F.; Dashtestani, F.; Ghorchian, H.; Rezayat, S.M.; Saboury, A.A.; Falahati, M. Cancer diagnosis using nanomaterials based electrochemical nanobiosensors. *Biosens. Bioelectron.* **2019**, *126*, 773–784. [CrossRef] [PubMed]
13. Torre, R.; Costa-Rama, E.; Nouws, H.P.A.; Delerue-Matos, C. Screen-Printed Electrode-Based Sensors for Food Spoilage Control Bacteria and Biogenic Amines Detection. *Biosensors* **2020**, *10*, 139. [CrossRef] [PubMed]
14. Su, J.; Liu, W.H.; Chen, S.X.; Deng, W.P.; Dou, Y.Z.; Zhao, Z.H.; Li, J.Y.; Li, Z.H.; Yin, H.; Ding, X.T.; et al. A Carbon-Based DNA Framework Nano-Bio Interface for Biosensing with High Sensitivity and a High Signal-to-Noise Ratio. *ACS Sens.* **2020**, *5*, 3979–3987. [CrossRef] [PubMed]
15. Kanyong, P.; Rawlinson, S.; Davis, J. Simultaneous electrochemical determination of dopamine and 5-hydroxyindoleacetic acid in urine using a screen-printed graphite electrode modified with gold nanoparticles. *Anal. Bioanal. Chem.* **2016**. [CrossRef]
16. Gutierrez-Sanchez, C.; Pita, M.; Vaz-Dominguez, C.; Shleev, S.; De Lacey, A.L. Gold Nanoparticles as Electronic Bridges for Laccase-Based Biocathodes. *J. Am. Chem. Soc.* **2012**, *134*, 17212–17220. [CrossRef]
17. Wang, J.X.; Drelich, A.J.; Hopkins, C.M.; Mecozzi, S.; Li, L.J.; Kwon, G.; Hong, S. Gold nanoparticles in virus detection: Recent advances and potential considerations for SARS-CoV-2 testing development. *Wiley Interdiscip. Rev. Nanomed. Nanobiotechnol* **2022**, *14*, e1754. [CrossRef]
18. Yang, F.; Li, Q.; Wang, L.H.; Zhang, G.J.; Fan, C.H. Framework-Nucleic-Acid-Enabled Biosensor Development. *ACS Sens.* **2018**, *3*, 903–919. [CrossRef]

19. Chand, R.; Neethirajan, S. Microfluidic platform integrated with graphene-gold nano-composite aptasensor for one-step detection of norovirus. *Biosens. Bioelectron.* **2017**, *98*, 47–53. [CrossRef]
20. Senel, M.; Dervisevic, M.; Kokkokoglu, F. Electrochemical DNA biosensors for label-free breast cancer gene marker detection. *Anal. Bioanal. Chem.* **2019**, *411*, 2925–2935. [CrossRef]
21. Zeng, D.D.; Zhang, H.; Zhu, D.; Li, J.; San, L.L.; Wang, Z.H.; Wang, C.G.; Wang, Y.S.; Wang, L.H.; Zuo, X.L.; et al. A novel ultrasensitive electrochemical DNA sensor based on double tetrahedral nanostructures. *Biosens. Bioelectron.* **2015**, *71*, 434–438. [CrossRef] [PubMed]
22. Pei, H.; Lu, N.; Wen, Y.L.; Song, S.P.; Liu, Y.; Yan, H.; Fan, C.H. A DNA Nanostructure-based Biomolecular Probe Carrier Platform for Electrochemical Biosensing. *Adv. Mater.* **2010**, *22*, 4754–4758. [CrossRef] [PubMed]
23. Squires, T.M.; Messinger, R.J.; Manalis, S.R. Making it stick: Convection, reaction and diffusion in surface-based biosensors. *Nat. Biotechnol.* **2008**, *26*, 417–426. [CrossRef] [PubMed]
24. Chao, J.; Zhu, D.; Zhang, Y.N.; Wang, L.H.; Fan, C.H. DNA nanotechnology-enabled biosensors. *Biosens. Bioelectron.* **2016**, *76*, 68–79. [CrossRef]
25. Mitchell, N.; Schlapak, R.; Kastner, M.; Armitage, D.; Chrzanowski, W.; Riener, J.; Hinterdorfer, P.; Ebner, A.; Howorka, S. A DNA Nanostructure for the Functional Assembly of Chemical Groups with Tunable Stoichiometry and Defined Nanoscale Geometry. *Angew. Chem. Int. Ed.* **2009**, *48*, 525–527. [CrossRef] [PubMed]
26. Lin, M.H.; Wang, J.J.; Zhou, G.B.; Wang, J.B.; Wu, N.; Lu, J.X.; Gao, J.M.; Chen, X.Q.; Shi, J.Y.; Zuo, X.L.; et al. Programmable Engineering of a Biosensing Interface with Tetrahedral DNA Nanostructures for Ultrasensitive DNA Detection. *Angew. Chem. Int. Ed.* **2015**, *54*, 2151–2155. [CrossRef]
27. Feng, D.Z.; Su, J.; He, G.F.; Xu, Y.; Wang, C.G.; Zheng, M.M.; Qian, Q.L.; Mi, X.Q. Electrochemical DNA Sensor for Sensitive BRCA1 Detection Based on DNA Tetrahedral-Structured Probe and Poly-Adenine Mediated Gold Nanoparticles. *Biosensors* **2020**, *10*, 78. [CrossRef]
28. Wang, H.F.; Ma, R.N.; Sun, F.; Jia, L.P.; Zhang, W.; Shang, L.; Xue, Q.W.; Jia, W.L.; Wang, H.S. A versatile label-free electrochemical biosensor for circulating tumor DNA based on dual enzyme assisted multiple amplification strategy. *Biosens. Bioelectron.* **2018**, *122*, 224–230. [CrossRef]
29. Luo, Z.W.; Xu, Y.; Huang, Z.J.; Chen, J.M.; Wang, X.Q.; Li, D.; Li, Y.X.; Duan, Y.X. A rapid, adaptative DNA biosensor based on molecular beacon-concatenated dual signal amplification strategies for ultrasensitive detection of p53 gene and cancer cells. *Talanta* **2020**, *210*, 120638. [CrossRef]
30. Chen, S.C.; Chen, K.T.; Jou, A.F.J. Polydopamine-gold composite-based electrochemical biosensor using dual-amplification strategy for detecting pancreatic cancer-associated microRNA. *Biosens. Bioelectron.* **2021**, *173*, 112815. [CrossRef]
31. Chen, Y.; Xiang, J.Y.; Liu, B.; Chen, Z.B.; Zuo, X. Gold nanoparticle-engineered electrochemical aptamer biosensor for ultrasensitive detection of thrombin. *Anal. Methods* **2020**, *12*, 3729–3733. [CrossRef] [PubMed]
32. Bi, S.; Yue, S.Z.; Zhang, S.S. Hybridization chain reaction: A versatile molecular tool for biosensing, bioimaging, and biomedicine. *Chem. Soc. Rev.* **2017**, *46*, 4281–4298. [CrossRef] [PubMed]
33. Yang, H.; Gao, Y.; Wang, S.Q.; Qin, Y.; Xu, L.; Jin, D.; Yang, F.; Zhang, G.J. In situ hybridization chain reaction mediated ultrasensitive enzyme-free and conjugation-free electrochemcial genosensor for BRCA-1 gene in complex matrices. *Biosens. Bioelectron.* **2016**, *80*, 450–455. [CrossRef] [PubMed]
34. Ge, Z.L.; Lin, M.H.; Wang, P.; Pei, H.; Yan, J.; Sho, J.Y.; Huang, Q.; He, D.N.; Fan, C.H.; Zuo, X.L. Hybridization Chain Reaction Amplification of MicroRNA Detection with a Tetrahedral DNA Nanostructure-Based Electrochemical Biosensor. *Anal. Chem.* **2014**, *86*, 2124–2130. [CrossRef]
35. Wen, Y.L.; Liu, G.; Pei, H.; Li, L.Y.; Xu, Q.; Liang, W.; Li, Y.; Xu, L.; Ren, S.Z.; Fan, C.H. DNA nanostructure-based ultrasensitive electrochemical microRNA biosensor. *Methods* **2013**, *64*, 276–282. [CrossRef] [PubMed]
36. Sharma, S.; Singh, N.; Tomar, V.; Chandra, R. A review on electrochemical detection of serotonin based on surface modified electrodes. *Biosens. Bioelectron.* **2018**, *107*, 76–93. [CrossRef]
37. Govindasamy, M.; Mani, V.; Chen, S.M.; Chen, T.W.; Sundramoorthy, A.K. Methyl parathion detection in vegetables and fruits using silver@graphene nanoribbons nanocomposite modified screen printed electrode. *Sci. Rep.* **2017**, *7*, 46471. [CrossRef]
38. Li, F.Q.; Mao, X.H.; Li, F.; Li, M.; Shen, J.L.; Ge, Z.L.; Fan, C.H.; Zuo, X.L. Ultrafast DNA Sensors with DNA Framework-Bridged Hybridization Reactions. *J. Am. Chem. Soc.* **2020**, *142*, 9975–9981. [CrossRef]
39. Zhang, H.; Liu, X.X.; Zhang, C.X.; Xu, Y.; Su, J.; Lu, X.L.; Shi, J.Y.; Wang, L.H.; Landry, M.P.; Zhu, Y.; et al. A DNA tetrahedral structure-mediated ultrasensitive fluorescent microarray platform for nucleic acid test. *Sens. Actuators B-Chem.* **2020**, *321*, 128538. [CrossRef]
40. Lv, H.Y.; Chen, A.Y.; Cheng, W.Q.; Kong, L.S.; Zhao, M.; Ding, S.J.; Ju, H.X.; Cheng, W. Efficient DNA Walker Guided with Well-Regulated Interfacial Tracks for Ultrasensitive Electrochemiluminescence Biosensing. *Anal. Chem.* **2020**, *92*, 15624–15631. [CrossRef]
41. Liu, P.F.; Zhao, K.R.; Liu, Z.J.; Wang, L.; Ye, S.Y.; Liang, G.X. Cas12a-based electrochemiluminescence biosensor for target amplification-free DNA detection. *Biosens. Bioelectron.* **2021**, *176*, 112954. [CrossRef] [PubMed]

nanomaterials

MDPI

Article

Efficacy of Gold Nanoparticles against Drug-Resistant Nosocomial Fungal Pathogens and Their Extracellular Enzymes: Resistance Profiling towards Established Antifungal Agents

Abobakr Almansob, Ali H. Bahkali and Fuad Ameen *

Department of Botany & Microbiology, College of Science, King Saud University, Riyadh 11451, Saudi Arabia; aalmansob@ksu.edu.sa (A.A.); abahkali@ksu.edu.sa (A.H.B.)
* Correspondence: fuadameen@ksu.edu.sa

Abstract: Drug resistance of filamentous fungi to the commonly used antifungal agents is a major concern in medicine. Therefore, an effective approach to treat several opportunistic fungal infections is the need of the hour. *Mentha piperita* is used in home remedies to treat different disorders. Isolates of fungi were taken from hospitals in Riyadh, Saudi Arabia, and identified using molecular tools. Amphotericin B, Voriconazole, and Micafungin were applied to screen the resistance of these isolates using both disc and broth microdilution techniques. An aqueous extract of *Mentha piperita* was utilized to synthesize AuNPs and the nanoparticles were characterized using UV-Vis, FTIR, TEM, EDAX, and XRD. The AuNPs were tested for antifungal activity against the nosocomial fungal pathogens and the activity of extracellular enzymes of such pathogens were analyzed after treatment with AuNPs. We conclude that AuNPs synthesized using *Mentha piperita* do not possess especially effective antifungal properties against multi-drug resistant Aspergillus species. Five out of eighteen isolates were inhibited by AuNPs. When inhibition was observed, significant alterations in the activity profile of extracellular enzymes of the nosocomial fungi were observed.

Keywords: *Mentha piperita*; nosocomial fungi; AuNPs; antifungal; extracellular enzymes

1. Introduction

Aspergillus is an omnipresent, filamentous, mycotoxigenic fungus classified into a group of pathogens termed environmental opportunistic pathogens (EOPs) [1]. It is notably recognized for causing nosocomial invasive aspergillosis, a rigorous health hazard among immune-compromised patients [2,3]. Although limited reports are available on nosocomial invasive aspergillosis, it is observed to be concomitant with the maintenance activities of buildings in and around hospitals with limited chances of incidence among common inhabitants [4].

Lower respiratory infections result in several deaths and pose a serious concern in countries with developing economies [5]. Invasive aspergillosis is a lower respiratory infection primarily of nosocomial origin [6,7]. Members of *Aspergillus* spp., such as *Aspergillus niger*, *Aspergillus flavus*, *Aspergillus fumigatus*, and *Aspergillus terreus* are established pathogens isolated in patients with invasive pulmonary aspergillosis [8]. Recent studies propose using metallic nanoparticles for the management of other lower respiratory infections like COVID-19 [9].

Nanotechnology has become an established field of research over the recent decade and the use of nanomaterials is pivotal for this upsurge in medicine and diagnostics [10–13]. Due to the multiple advantages they possess, biosynthesis using a bottom-up approach is recently preferred over chemical or physical methods [14,15]. In particular, biosynthesis using plant extracts is gaining importance compared to microbe-assisted synthesis, due to the added advantages of the large volume of capping agents, laborious expenditures associated with microbial cultures, and separation techniques along with the reduced extent of toxicity and time. In addition, the phytochemicals of plant extracts can stabilize the nanomaterials, leading to relatively higher stability [16–18]. In some cases, plant enzymes like bromelain and vegetable wastes or their extracts have been used to synthesize AuNPs [19–21]. Adding to these benefits, metallic nanoparticles, notably AuNPs, synthesized for applications in medicine are known to possess antifungal properties, particularly against pathogens capable of causing respiratory infections [22–27].

Although research continues to combat resistance to antifungal agents due to their extensive usage, fungi continue to improve the resistance mechanisms. Based on their location of action, antifungal agents are categorized into the following classes: polyenes, azoles, allylamines, flucytosine, and echinocandins [28,29]. Most antifungal agents target the formation and functioning of ergosterol, the significant constituent of the fungal cell wall, and inhibit the synthesis of macromolecules [30]. As antifungal resistance has turned out to be a critical issue, new agents to stand against infections caused by fungi have become the need of the hour.

Owing to this background, AuNPs were synthesized and used in this study as antifungal agents against clinical pathogens responsible for nosocomial infections. In addition, this is the first-ever international report with regard to a few aspects: 1. On testing the efficacy of AuNPs on filamentous fungi, especially against multi-drug resistant *Aspergillus* species. 2. In evaluating the antifungal activity of AuNPs by both disc diffusion and broth microdilution methods for applications in nanomedicine against filamentous fungi. 3. In comparing the antifungal effects of AuNPs with existing standard drugs.

2. Materials and Method

2.1. Collection and Identification of Fungal Pathogens

2.1.1. Sample Collection and Ethical Clearance

Blood samples were obtained from patients with aspergillosis in the microbiology departments of both King Khalid University Hospital and Regional Laboratory and Blood Bank of King Saud Medical City hospital over a period of 1 month. The fungal isolates were collected from the departments according to the guidance of the Ministry of Health-Kingdom of Saudi Arabia (MOH-KSA) and after the acceptance of the Institutional Review Board of King Khalid University Hospital. The isolates were collected, preserved in sterile bags, and transported to the laboratory under the most ideal conditions possible.

2.1.2. Retrieval and Identification of Fungi

Eighteen isolates obtained from the source departments were revived, sub-cultured on Sabouraud's agar (Oxoid Ltd., Basingstoke, UK), and incubated for 5 days at 28 °C. The pure cultures were stored in glycerol at −80 °C for further analysis. Mat morphology and microscopy were used for primary identification of the isolates. However, precise identification of the isolates was done on the basis of 18S rRNA sequencing and a comparison of homology with the sequences of the existing genotypes available in databanks.

2.1.3. Screening for Antifungal Resistance

The fungal isolates were screened to establish the resistance to key antifungal agents: Amphotericin B (Cayman, Ann Arbor, MI, USA), Voriconazole (TCI, Tochigi, Japan), and Micafungin (Sigma-Aldrich, St. Louis, MO, USA). The Clinical and Laboratory Standards Institute (CLSI) M38-A standards were used as a reference to compare the minimum inhibitory concentration (MIC) obtained by Kirby–Bauer disc diffusion using modified the Espinel-Ingroff et al. technique [31–33]. Discs impregnated with DMSO were used as a control. The reference for each standard drug was: 1. Amphotericin B (10 µg); reporting resistance (R) when ZI $\leq$ 12 mm, Intermediate (I): ZI = 13–14 mm, and Susceptible (S): ZI $\geq$ 15 mm. 2. Voriconazole (10 µg); resistance (R) when: ZI $\leq$ 25 mm, Intermediate (I): ZI = 26–27 mm, and Susceptible (S): ZI $\geq$ 28 mm. 3. Reporting resistance to Micafungin was done resembling caspofungin (5 µg) applied in the Espinel-Ingroff et al. technique, resistance (R) when ZI $\leq$ 13 mm, Intermediate (I): ZI = 14–16 mm, and Susceptible (S): ZI $\geq$ 17 mm.

2.2. Synthesis and Characterization of AuNPs

2.2.1. Preparation of the Plant Extract

M. piperita leaves were collected and identified by Government Botanical Survey and the voucher specimen was stored. The plant leaves were dried in an incubator (Memmert GmbH, Burladingen, Germany) for 7 days at 25 °C. The dried leaves were ground (AL-SAIF ELEC grinder, 90,582) and washed thoroughly with double distilled water in order to remove any surface contaminants. The decoction extraction method was applied for the preparation of the aqueous extract. The extract was prepared in a specific 1:16 ratio (plant to solvent). A total of 10 gms of the powdered material was dissolved in 160 mL of double-distilled water. Overall, the concentration of extract was 0.06 g/mL (62.5 mg/mL or 62,500 ppm). The extracts were transferred to an incubator shaker (Gallenkamp, Gillingham, UK) and kept overnight at 25 °C. Later, filtration of the extract was achieved in two stages; with sterile large 124 cm filter papers (Whatman type No. 1, Gillingham, UK) and Acrodisc Syringe Filters (0.45 µm pore size/ Pall, New York, NY, USA) to eliminate any possible contamination.

2.2.2. Synthesis of AuNPs

The aqueous plant extract was blended with 1 mM $HAuCl_4 \cdot 3H_2O$ (abcr GmbH, Burladingen, Germany) prepared under aqueous conditions in ratios of 1:1, 1:5, 1:10, and 1:15 and incubated at room temperature until a visible color change is observed. The control sample was prepared using the same test procedures, ratio, and conditions. Yet, sterilized distilled water was used in place of $HAuCl_4 \cdot 3H_2O$. The purple solution obtained after incubation was centrifuged at higher centrifugal speeds (MiniSpin®, Eppendorf AG, Burladingen, Germany). The pellet was suspended in Milli-Q water and dried in an incubator for 24 hrs at 48–50 °C. Stocks of AuNPs were prepared at a concentration of 1000 ppm and serial dilutions of this stock were performed to obtain working solutions with concentrations of 200 ppm, 100 ppm, 50 ppm, and 25 ppm.

2.2.3. Characterization of AuNPs

The biosynthesized AuNPs were initially characterized using a UV-visible spectrophotometer (UV-Vis, JASCO, New York, NY, USA). Possible encapsulates were identified using Fourier-transform infrared spectroscopy (FTIR, PerkinElmer Spectrum 100, New York, NY, USA). Elemental analysis was performed using energy-dispersive X-ray spectroscopy (EDX, JEOL, Tokyo, Japan). Morphology was observed using a transmission electron microscope (TEM, JEM 1011, Akishima, Tokyo, Japan). The crystalline nature was studied using X-ray diffraction (XRD, Rigaku Ultima IV, New York, NY, USA).

2.3. Antifungal Activity of AuNPs

Varying concentrations of AuNPs (25, 50, 100, 200, and 1000 ppm) were prepared and 10 µL of the solutions were loaded onto sterile discs. The antifungal activity was measured as a zone of inhibition in mm (mean $\pm$ SD). The broth microdilution technique was adopted following the CLSI M38-A protocol as a reference method to screen the accuracy and sensitivity of the disc diffusion method.

2.4. Effect of AuNPs on Extracellular Fungal Enzymes

The *A. flavus* (AM11) strain was used to test the influence of AuNPs on fungal metabolism represented by the enzymatic activity of the fungi. The profiling of nineteen extracellular enzymes was performed through API-Zym (Biomeruex, Craponne, France) strips test. The protocol was based on the standard method adopted by Pietrzak et al. [34], with slight modifications. The experiment was done by assigning two groups; test and control groups. In the test group, fungal inoculant (spore's suspension: equivalent to

0.5 McFarland) in media (30 mL malt extract broth; MEB) was treated with AuNPs (1 mL of AuNPs for determining the minimum effective concentration—MEC) whereas, the control group was the same as the test group but without the AuNPs.

After treatment with AuNPs, the media was incubated for 7 days at 25 °C in 100 mL flasks. Subsequently, the fungal biomass was filtered through filter paper (Whatman type 1, Gillingham, UK). The filtrates were quantified for enzyme activity according to the standard API-Zym protocol depending on an increase in the intensity of color among the test sample in comparison to the control under identical conditions. A 0–5 score system was used, in which 0 determines no activity, while 5 determines maximum liberation of the hydrolyzed substrate. The concentration of enzymes was directly proportional to the intensity of the color. According to the scores 0, 1, 2, 3, 4, and 5, the enzyme activity will be 0, 5, 10, 20, 30, and ≥40 nmol, respectively.

2.5. Statistical Analysis

All experiments were performed in triplicate and represented as mean ± SD. Two-way ANOVA was applied and $p < 0.05$ was considered to be significant.

3. Results and Discussion

3.1. Identification of the Fungal Pathogens

Eighteen members of *Aspergillus* spp., such as *Aspergillus niger*, *Aspergillus flavus*, *Aspergillus fumigatus*, and *Aspergillus terreus* were identified in total. Seventeen isolates were identified by 18S rRNA sequencing. Each identified isolate was given an accession number and stored in NCBI. The other isolate termed *Aspergillus terreus* 8 was identified by mat morphology and microscopic observations. However, the identification of the isolates by use of molecular tools and phylogeny was assisted by the maximum likelihood tree method and bootstrapping by Molecular Evolutionary Genetic Analysis (Mega X) software [35] (Figure 1 and Table 1).

Table 1. List of fungi isolated and their NCBI accession numbers.

NO.	Strain Code	Fungi (Similarity: 100%)	NCBI Accession No.
1	(AM1)	*Aspergillus flavus*	OK396684
2	(AM2)	*Aspergillus flavus*	OK396685
3	(AM3)	*Aspergillus flavus*	OK396686
4	(AM4)	*Aspergillus flavus*	OK396687
5	(AM5)	*Aspergillus flavus*	OK396688
6	(AM6)	*Aspergillus fumigatus*	OK396689
7	(AM7)	*Aspergillus fumigatus*	OK396690
8	(AM8)	*Aspergillus niger*	OK396691
9	(AM9)	*Aspergillus niger*	OK396692
10	(AM10)	*Aspergillus terreus*	OK396693
11	(AM11)	*Aspergillus flavus*	OK396694
12	(AM12)	*Aspergillus flavus*	OK396695
13	(AM13)	*Aspergillus flavus*	OK396696
14	8	*Aspergillus terreus* 8	-
15	(AM14)	*Aspergillus flavus*	OK396697
16	(AM15)	*Aspergillus flavus*	OK396698
17	(AM16)	*Aspergillus flavus*	OK396699
18	(AM17)	*Aspergillus niger*	OK396700

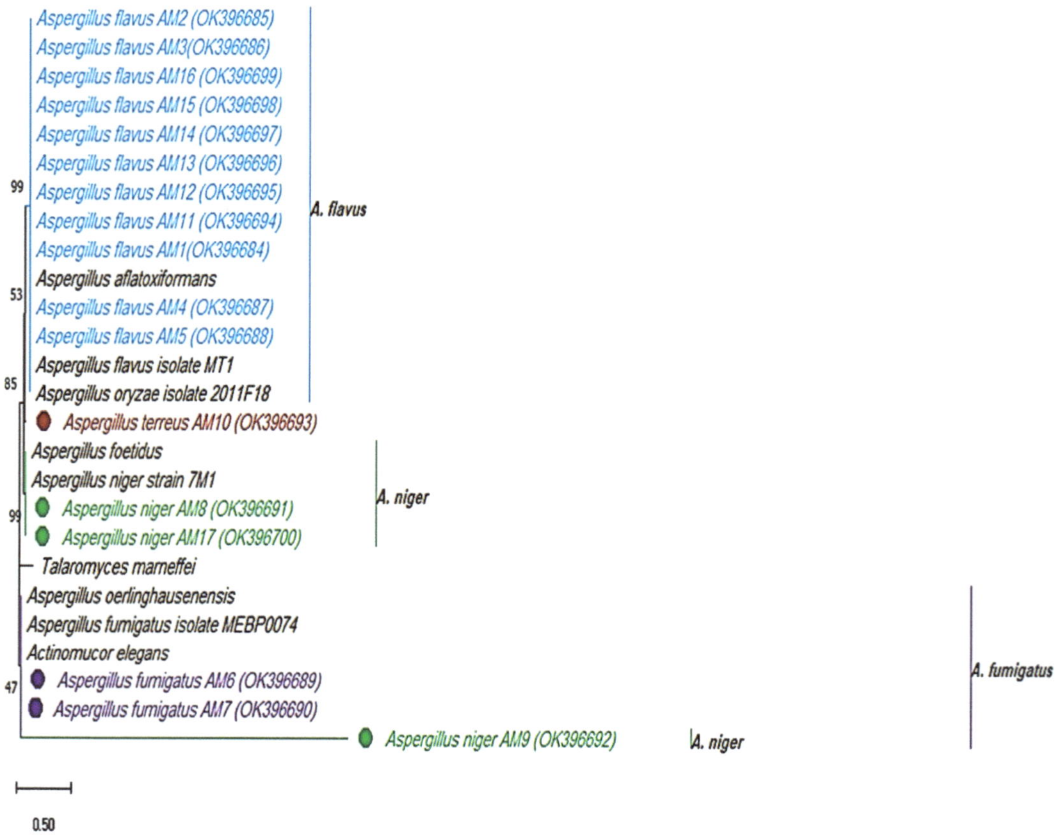

Figure 1. Identification of the clinical fungal pathogens by 18S rRNA sequencing. Maximum likelihood tree method and Tamura 3-parameter mode, 1000 bootstrapping by Mega X software.

3.2. Screening for Fungal Resistance to Standard Antifungals

Considering the 300 pathogenic fungi of *Aspergillus* spp., *A. fumigatus*, *A. flavus*, *A. niger*, and *A. terreus* are considered significant in causing infections amongst immunocompromised patients [36,37]. These strains were isolated in patients with invasive aspergillosis of nosocomial origin, as per this study. Although infections caused by these pathogens are dependent on topographical and climatic conditions of countries, *A. flavus* is the most prevalent (61.1%) eukaryotic microbe in countries with hot and arid climatic conditions such as Saudi Arabia [38,39].

Among the eighteen tested strains in the disc method, seventeen strains (94.4%) were resistant (except *A. flavus* (AM15), which was sensitive (5.6%)) to Amphotericin B. *A. fumigatus* (AM6) was the most sensitive strain to this strong antifungal agent (Table 2). All tested strains were Voriconazole resistant (100%). In particular, *A. terreus* 8 was the most resistant (Table 3). Similar observations were made for Micafungin (Table 4). According to the standard broth microdilution technique, the susceptibility of fungal strains to Amphotericin B, Voriconazole, and Micafungin are shown in Tables 5–7. Owing to the CLSI method, all strains (18) were resistant to Amphotericin B, especially *A. terreus* (AM10), *A. flavus* (AM11), *A. flavus* (AM3), and *A. terreus* 8 with MIC >16 μg/mL. With respect to Voriconazole, most isolates were resistant, (72.2% (13)) especially *A. flavus* (AM4), *A. flavus* (AM14,) and *A. flavus* (AM15), whereas intermediate and high sensitivity profiles to Voriconazole were 16.7% (3) and 11.1% (2), respectively. Micafungin was not effective against most fungi tested 94.4% (17).

Table 2. Susceptibility of fungal species to different concentrations of Amphotericin B (disc method).

No.	Isolate	Inhibition Zone (mm)			
		1 μg/mL	2 μg/mL	4 μg/mL	10 μg/mL
1	*A. flavus (AM1)*	2 ± 3.5	6.7 ± 0.6	7.7 ± 0.6	9.39 ± 0.6 (R)
		9	9	9	9
2	*A. flavus (AM2)*	2± 3.5	7.7 ± 0.6	8.7 ± 0.6	9.7 ± 0.6 (R)
		9	9	9	9
3	*A. flavus (AM3)*	2.3 ± 1.3	7.7 ± 0.5	8.7 ± 0.5	10.7 ± 0.5 (R)
		10.0	10.0	10.0	10.0
4	*A. flavus (AM4)*	2 ± 1.2	6.3 ± 0.5	9 ± 0	9.79 ± 0.5 (R)
		8	8	8	8
5	*A. flavus (AM5)*	9.3 ± 0.6	8.7 ± 0.6	9 ± 0	9 ± 0 (R)
		10	10	10	10
6	*A. fumigatus (AM6)*	7.7 ± 0.6	10.7 ± 0.6	11.7 ± 0.6	19.0 ± 1 (R)
		10	10	10	10
7	*A. fumigatus (AM7)*	7.7 ± 0.6	10.7 ± 0.6	12.3 ± 0.6	18.7 ± 0.6 (R)
		10	10	10	10
8	*A. niger (AM8)*	0 ± 0.6	0 ± 0.6	6.7 ± 0.6	12.7 ± 0.6 (R)
		0	0	0	0
9	*A. niger (AM9)*	2 ± 3.5	6.7 ± 0.6	9 ± 2	18.7 ± 0.6 (R)
		7	7	7	7
10	*A. terreus (AM10)*	6.7 ± 0.6	7.7 ± 0.6	8.7 ± 0.6	14.7 ± 0.6 (R)
		9	9	9	9
11	*A. flavus (AM11)*	5.7 ± 0.6	7.7 ± 0.6	9.7 ± 0.6	10.7 ± 0.6 (R)
		9	9	9	9
12	*A. flavus (AM12)*	2 ± 0.6	6.7 ± 0.6	7.3 ± 0.6	9.7± 0.6 (R)
		8	8	8	8
13	*A. flavus (AM13)*	0 ± 0	0 ± 0	6.7 ± 0.6	9.3 ± 0.6 (R)
		0	0	0	0
14	*A. terreus* 8	7 ± 0	6.7 ± 0.6	7.7 ± 0.6	11.7 ± 0.6 (R)
		10	10	10	10
15	*A. flavus (AM14)*	7.7 ± 0.6	8 ± 0	9 ± 0	12.3 ± 0.6 (R)
		9	9	9	9
16	*A. flavus (AM15)*	0 ± 0	0 ± 0	0 ± 0	0 ± 0 (R)
		0	0	0	0
17	*A. flavus (AM16)*	6.7 ± 0.6	7.3 ± 1.2	8.7 ± 0.6	9.7 ± 0.6 (R)
		7	7	7	7
18	*A. niger (AM17)*	0 ± 0	6 ± 0	12.7 ± 0.6	16.7 ± 0.6 (S)
		0	0	0	0

R: Resistant, S: Sensitive.

Table 3. Susceptibility of fungal species to different concentrations of Voriconazole (disc method).

N	Isolate	Inhibition Zone (mm)			
		1 µg/mL	2 µg/mL	4 µg/mL	10 µg/mL
1	*A. flavus (AM1)*	0 ± 0	6.3 ± 0.6	6.7 ± 0.6	7 ± 0 (R)
		7	7	7	7
2	*A. flavus (AM2)*	0 ± 0	6 ± 0	6.3 ± 0.6	6.3 ± 0.6 (R)
		6	6	6	6
3	*A. flavus (AM3)*	0 ± 0	0 ± 0	0 ± 0	6.0 ± 0 (R)
		0	0	0	0
4	*A. flavus (AM4)*	0 ± 0	0 ± 0	0 ± 0	6.7 ± 0.6 (R)
		0	0	0	0
5	*A. flavus (AM5)*	6 ± 0	6.7 ± 0.6	6.7 ± 0.6	8.7 ± 0.6 (R)
		8	8	8	8
6	*A. fumigatus (AM6)*	0 ± 0	0 ± 0	7.7 ± 0.6	10.3 ± 0.6 (R)
		0	0	0	0
7	*A. fumigatus (AM7)*	0 ± 0	0 ± 0	0 ± 0	9 ± 0 (R)
		0	0	0	0
8	*A. niger (AM8)*	0 ± 0	0 ± 0	0 ± 0	8 ± 0 (R)
		0	0	0	0
9	*A. niger (AM9)*	0 ± 0	6 ± 0	7.7 ± 0.6	13.3 ± 0.6 (R)
		0	0	0	0
10	*A. terreus (AM10)*	6 ± 0	6 ± 0	6.7 ± 0.6	7 ± 0 (R)
		7	7	7	7
11	*A. flavus (AM11)*	6 ± 0	6 ± 0	6.3 ± 0.6	7 ± 0 (R)
		7	7	7	7
12	*A. flavus (AM12)*	0 ± 0	0 ± 0	0 ± 0	7.7 ± 0.6 (R)
		7	7	7	7
13	*A. flavus (AM13)*	0 ± 0	0 ± 0	6.3 ± 0.6	7 ± 0.6 (R)
		7	7	7	7
14	*A. terreus 8*	0 ± 0	0 ± 0	0 ± 0	0 ± 0 (R)
		20	20	20	20
15	*A. flavus (AM14)*	0.0 ± 0	0.0 ± 0	0.0 ± 0	14.7 ± 0.6 (R)
		0	0	0	0
16	*A. flavus (AM15)*	6.7 ± 0.6	7 ± 0	8.7 ± 0.6	9.7 ± 0.6 (R)
		9	9	9	9
17	*A. flavus (AM16)*	6.7 ± 0.6	7 ± 0	8 ± 0	8.7 ± 0.6 (R)
		8	8	8	8
18	*A. niger (AM17)*	0 ± 0	0 ± 0	6.7 ± 0.6	16.7 ± 0.6 (R)
		0	0	0	0

R: Resistant.

Table 4. Susceptibility of fungal species to different concentrations of Micafungin.

N	Isolate	Inhibition Zone (mm)			
		1 µg/mL	2 µg/mL	4 µg/mL	10 µg/mL
1	*A. flavus (AM1)*	0 ± 0	0 ± 0	0 ± 0	6.3 ± 0.6 (R)
		0	0	0	0
2	*A. flavus (AM2)*	0 ± 0	0 ± 0	0 ± 0	0 ± 0 (R)
		0	0	0	0
3	*A. flavus (AM3)*	0 ± 0	0 ± 0	0 ± 0	0 ± 0 (R)
		2	2	2	2
4	*A. flavus (AM4)*	0 ± 0	0 ± 0	0 ± 0	7 ± 0 (R)
		0	0	0	0
5	*A. flavus (AM5)*	0 ± 0	0 ± 0	0 ± 0	6.7 ± 0.6 (R)
		0	0	0	6.6
6	*A. fumigatus (AM6)*	2 ± 3.5	4 ± 3.5	6.7 ± 0.6	8 ± 1.7 (R)
		0	0	0	0
7	*A. fumigatus (AM7)*	6 ± 0	6.3 ± 0.6	6.7 ± 0.6	7.7 ± 1.5 (R)
		0	0	0	0
8	*A. niger (AM8)*	0 ± 0	0 ± 0	0 ± 0	3.3 ± 5.8 (R)
		0	0	0	0
9	*A. niger (AM9)*	0 ± 0	6.7 ± 0.6	6.7 ± 0.6	9.7 ± 0.6 (R)
		11	11	11	11
10	*A. terreus (AM10)*	0 ± 0	0 ± 0	0 ± 0	0 ± 0 (R)
		9	9	9	9
11	*A. flavus (AM11)*	0 ± 0	0 ± 0	0 ± 0	0 ± 0 (R)
		0	0	0	0
12	*A. flavus (AM12)*	0 ± 0	0 ± 0	6.3 ± 0.6	5 ± 4.4 (R)
		0	0	0	0
13	*A. flavus (AM13)*	0 ± 0	0 ± 0	0 ± 0.6	7.3 ± 1.5 (R)
		0	0	0	0
14	*A. terreus 8*	0 ± 0	0 ± 0	0 ± 0	9.7 ± 0.6 (R)
		0	0	0	0
15	*A. flavus (AM14)*	0 ± 0	0 ± 0	0 ± 0	6.3 ± 0.6 (R)
		0	0	0	0
16	*A. flavus (AM15)*	0 ± 0	0 ± 0	0 ± 0	0 ± 0 (R)
		4.3	4.3	4.3	4.3
17	*A. flavus (AM16)*	0 ± 0	0 ± 0	0 ± 0	6.3 ± 0.6 (R)
		0	0	0	0
18	*A. niger (AM17)*	0 ± 0	0 ± 0	0 ± 0	0 ± 0 (R)
		4.3	4.3	4.3	4.3

R: Resistant.

Table 5. Susceptibility of fungal species to different concentrations of Amphotericin B (broth microdilution method).

Isolate	Growth Rate/Score				
	1 μg/mL	2 μg/mL	4 μg/mL	8 μg/mL	16 μg/mL
A. flavus (AM1)	3 ± 0.6	2 ± 0	1 ± 0	0± 0 (R)	0 ± 0
	4	4	4	4	4
A. flavus (AM2)	3 ± 0	1 ± 0	0 ± 0 (R)	0 ± 0	0 ± 0
	4	4	4	4	4
A. flavus (AM3)	4 ± 0	3 ± 0.6	2 ± 1.2	1 ± 1.2	1 ± 0.6 (R)
	4	4	4	4	4
A. flavus (AM4)	4 ± 0	4 ± 0	4 ± 0.6	2 ± 1.7	0 ± 0 (R)
	4	4	4	4	4
A. flavus (AM5)	4 ± 0	4 ± 0	3 ± 1.2	1 ± 0	0 ± 0 (R)
	4	4	4	4	4
A. fumigatus (AM6)	4 ± 0	4 ± 0.6	2 ± 1.2	1 ± 0	0 ± 0.6 (R)
	4	4	4	4	4
A. fumigatus (AM7)	4 ± 0	3 ± 0.6	3 ± 2.3	0 ± 0 (R)	0 ± 0
	4	4	4	4	4
A. niger (AM8)	4 ± 0	3 ± 2.3	1 ± 1.2	1 ± 0.6	0 ± 0 (R)
	4	4	4	4	4
A. niger (AM9)	4 ± 0	4 ± 0	0 ± 0 (R)	0 ± 0	0 ± 0
	4	4	4	4	4
A. terreus (AM10)	4 ± 0	4 ± 0	4 ± 0	4 ± 0	3 ± 1.2 (R)
	4	4	4	4	4
A. flavus (AM11)	4 ± 0	4 ± 0	4 ± 0	3 ± 0	2 ± 0 (R)
	4	4	4	4	4
A. flavus (AM12)	4 ± 1	3 ± 1.2	3 ± 1.7	1 ± 1.5	0 ± 0 (R)
	4	4	4	4	4
A. flavus (AM13)	4 ± 0	4 ± 0	3 ± 0.6	0 ± 0.6 (R)	0 ± 0
	4	4	4	4	4
A. terreus 8	4 ± 0	4 ± 0	4 ± 0	4 ± 0.6	2 ± 0.6(R)
	4	4	4	4	4
A. flavus (AM14)	4 ± 0	3 ± 0	2 ± 0	0 ± 0 (R)	0 ± 0
	4	4	4	4	4
A. flavus (AM15)	4 ± 0	3 ± 1.2	3 ± 2.3	3 ± 2.3	0 ± 0 (R)
	4	4	4	4	4
A. flavus (AM16)	4 ± 0	2 ± 0	0 ± 0 (R)	0 ± 0	0 ± 0
	4	4	4	4	4
A. niger (AM17)	2 ± 0	2 ± 0	1 ± 0	0 ± 0 (R)	0 ± 0
	4	4	4	4	4

R: Resistant.

Table 6. Susceptibility of fungal species to different concentrations of Voriconazole (broth microdilution method).

Isolate	Growth Rate/Score				
	1 μg/mL	2 μg/mL	4 μg/mL	8 μg/mL	16 μg/mL
A. flavus (AM1)	4 ± 0	3 ± 0	3 ± 0	2 ± 0	2 ± 0 (R)
	4	4	4	4	4
A. flavus (AM2)	4 ± 0	4 ± 0	4 ± 0.6	3 ± 0	1 ± 0 (R)
	4	4	4	4	4
A. flavus (AM3)	4 ± 0	3 ± 0	2 ± 0	2 ± 0	1 ± 0 (R)
	4	4	4	4	4
A. flavus (AM4)	4 ± 0	3 ± 0	3 ± 0	3 ± 0	3 ± 0 (R)
	4	4	4	4	4
A. flavus (AM5)	3 ± 0	3 ± 0	2 ± 0	1 ± 0	0 ± 0 (R)
	4	4	4	4	4
A. fumigatus (AM6)	2 ± 0	0 ± 0 (I)	0 ± 0	0 ± 0	0 ± 0
	4	4	4	4	4
A. fumigatus (AM7)	3 ± 0	0 ± 0 (I)	0 ± 0	0 ± 0	0 ± 0
	4	4	4	4	4
A. niger (AM8)	2 ± 0	2 ± 0	2 ± 0	1 ± 0	1 ± 0 (R)
	4	4	4	4	4
A. niger (AM9)	0 ± 0 (S)	0 ± 0	0 ± 0	0 ± 0	0 ± 0
	4	4	4	4	4
A. terreus (AM10)	3 ± 0	3 ± 0	2 ± 0	1 ± 0	1 ± 0 (R)
	4	4	4	4	4
A. flavus (AM11)	2 ± 0	2 ± 0	2 ± 0	2 ± 0	0 ± 0 (R)
	4	4	4	4	4
A. flavus (AM12)	3 ± 0	3 ± 0	2 ± 0	1 ± 0	0 ± 0 (R)
	4	4	4	4	4
A. flavus (AM13)	3 ± 0	3 ± 0	2 ± 0	2 ± 0	1 ± 0 (R)
	4	4	4	4	4
A. terreus 8	3 ± 0	3 ± 0	3 ± 0	3 ± 0	1 ± 0.6 (R)
	4	4	4	4	4
A. flavus (AM14)	3 ± 0	3 ± 0	3 ± 0	3 ± 0	2 ± 0 (R)
	4	4	4	4	4
A. flavus (AM15)	4 ± 0	4 ± 0	4 ± 0	3 ± 0	2 ± 0 (R)
	4	4	4	4	4
A. flavus (AM16)	0 ± 0 (S)	0 ± 0	0 ± 0	0 ± 0	0 ± 0
	4	4	4	4	4
A. niger (AM17)	2 ± 0	0 ± 0 (I)	0 ± 0	0 ± 0	0 ± 0
	4	4	4	4	4

R: Resistant, I: Intermediate, and S: Sensitive.

Table 7. Susceptibility of fungal species to different concentrations of Micafungin (broth microdilution method).

Isolate	Growth Rate/Score				
	1 μg/mL	2 μg/mL	4 μg/mL	8 μg/mL	16 μg/mL
A. flavus (AM1)	4 ± 0	4 ± 0	4 ± 0	3 ± 0	1 ± 0 (R)
	4	4	4	4	4
A. flavus (AM2)	4 ± 0	4 ± 0	4 ± 0	4 ± 0	0 ± 0 (R)
	4	4	4	4	4
A. flavus (AM3)	4 ± 0	4 ± 0	4 ± 0	4 ± 0	3 ± 0 (R)
	4	4	4	4	4
A. flavus (AM4)	4 ± 0	4 ± 0	4 ± 0	4 ± 0	2 ± 0 (R)
	4	4	4	4	4
A. flavus (AM5)	4 ± 0	4 ± 0	4 ± 0	4 ± 0	1 ± 0 (R)
	4	4	4	4	4
A. fumigatus (AM6)	4 ± 0	4 ± 0	4 ± 0	4 ± 0	3 ± 0 (R)
	4	4	4	4	4
A. fumigatus (AM7)	4 ± 0	4 ± 0	4 ± 0	4 ± 0	3 ± 0 (R)
	4	4	4	4	4
A. niger (AM8)	4 ± 0	3 ± 0	2 ± 0	1 ± 0	1 ± 0 (R)
	4	4	4	4	4
A. niger (AM9)	4 ± 0	3 ± 0	2 ± 0	1 ± 0	1 ± 0 (R)
	4	4	4	4	4
A. terreus (AM10)	4 ± 0	4 ± 0	4 ± 0	4 ± 0	2 ± 0 (R)
	4	4	4	4	4
A. flavus (AM11)	4 ± 0	4 ± 0	4 ± 0	3 ± 0	1 ± 0 (R)
	4	4	4	4	4
A. flavus (AM12)	4 ± 0	4 ± 0	4 ± 0	3 ± 0	1 ± 0 (R)
	4	4	4	4	4
A. flavus (AM13)	4 ± 0	4 ± 0	3 ± 0	2 ± 0	1 ± 0 (R)
	4	4	4	4	4
A. terreus 8	4 ± 0	4 ± 0	4 ± 0	4 ± 0	3 ± 0 (R)
	4	4	4	4	4
A. flavus (AM14)	4 ± 0	4 ± 0	4 ± 0	4 ± 0	1 ± 0 (R)
	4	4	4	4	4
A. flavus (AM15)	4 ± 0	4 ± 0	4 ± 0	4 ± 0	3 ± 0 (R)
	4	4	4	4	4
A. flavus (AM16)	3 ± 0	3 ± 0	2 ± 0.6	1 ± 0	0 ± 0 (R)
	4	4	4	4	4
A. niger (AM17)	0 ± 0 (S)	0 ± 0	0 ± 0	0 ± 0	0 ± 0
	4	4	4	4	4

R: Resistant, S: Sensitive.

Overall, a broad resistance was observed and 55.6% of these resistant fungi opposed the antifungal activity as witnessed through the MIC ≥ 16 μg/mL of Amphotericin B. Furthermore, the resistance recorded with the other two antifungal agents was MIC ≥ 16 μg/mL in all resistant fungi. This grave matter of concern was documented to be higher than the expected and noticed values of other similar studies conducted either with polyene or azole antifungals [40–42]. Adding to this, the significant resistance recorded among *Aspergillus* spp. can either be intrinsic or of an acquired type. The latter type can occur as an outcome of the long-term habit of using antifungals for chronic aspergillosis, especially against *A. fumigatus*, *A. flavus*, or *A. terreus* [43]. Although Micafungin and Voriconazole have emerged recently as alternatives for the management of invasive aspergillosis, the resistance observed in this study was in contrast to the reports that support the susceptibility of fungi to these two agents [44–46].

3.3. Characterization of AuNPs

The visible color change to purple after incubating the precursor $HAuCl_4{\cdot}3H_2O$ with the plant extract is an initial confirmation for the reducing abilities of *M. piperita* (Figure 2) [47]. Surface plasmon resonance (SPR) is a band that occurs on the surface of metallic nanomaterials. Noticeable peaks in the range of 530 to 540 nm correspond to the SPR of AuNPs [48–50]. UV-Vis analysis indicates the formation of AuNPs by the plant extract as per the characteristic wavelength observed around 530 nm (Figure 3).

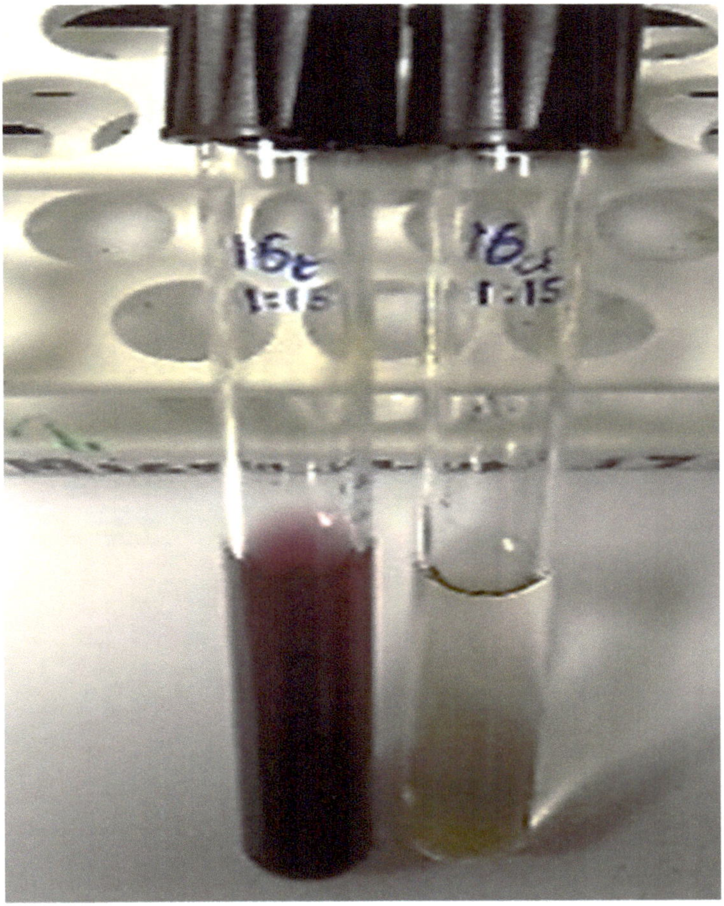

Figure 2. Initial confirmation of synthesis of AuNPs by a visible color change.

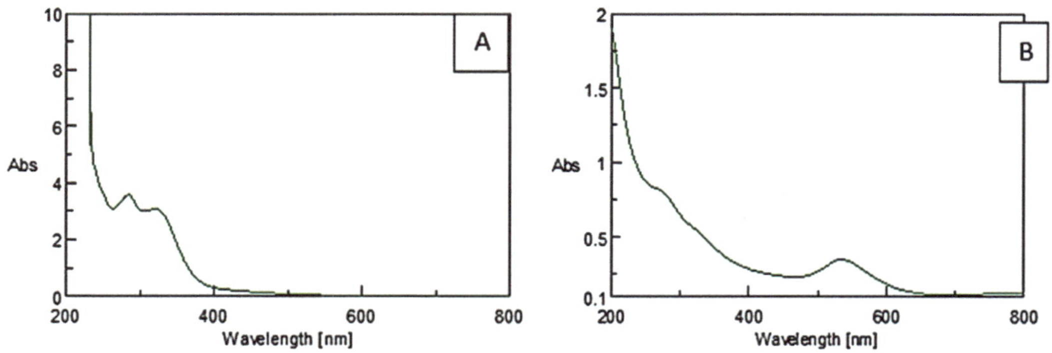

Figure 3. UV-Vis spectra of (**A**) plant extract and (**B**) AuNPs synthesized using *M. piperita*.

To further confirm the synthesis of AuNPs by *M. piperita* extract, FTIR was adopted (Figure 4). The FTIR peaks obtained in this study are supported well by previous reports which indicate the possible reducing and encapsulating agents on the AuNPs [51,52]. Extracts from plants of the genus *Mentha* rich in phenolic acids and essential oils are known to possess antifungal properties and are used as food preservatives [53,54]. According to a widely accepted hypothesis, a series of antioxidants, enzymes, and phenolics present in a plant extract can reduce the cations of gold to zerovalent gold. Consequently, the assemblage of gold atoms leads to the formation of AuNPs [55]. Phenylpropenes such as Apiol and Isoeugenol, Terpenoids such as Spathulenol, Ledene, α-Guaiene, and Pinene, and cyclohexanones such as Menthone are the active compounds of *M. piperita* extract. These compounds might act as reducing and stabilizing agents for the AuNPs [56].

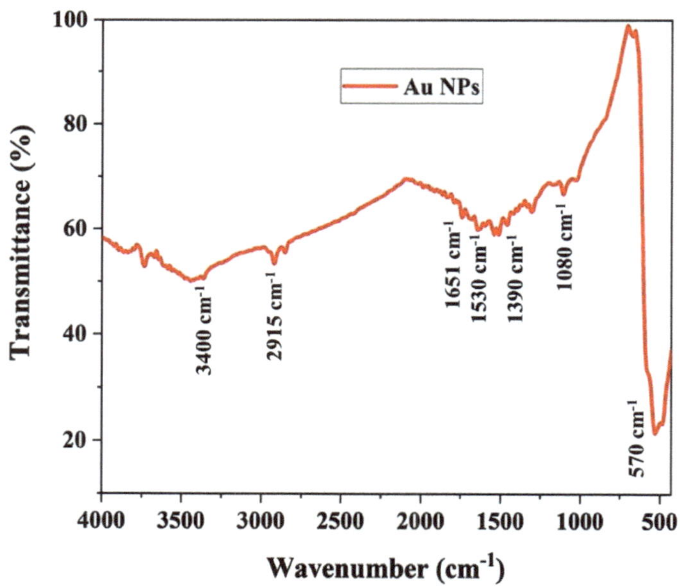

Figure 4. FTIR analysis of the hydrosol.

The AuNPs synthesized by the *M. piperita* extract formed strong prominent bands at 510, 1080, 1390, 1530, 1651, 2915, and 3400 cm^{-1}. The bands at 3400 cm^{-1} and 2915 cm^{-1} suggest the presence of stretching vibrations (O–H), and aldehydic C–H stretching, respectively. Again, their corresponding N–H bending vibration was seen at 1651 cm^{-1} and

1530 cm^{-1}, respectively. In addition, the weak bands at 1390 cm^{-1} and 1080 cm^{-1} were assigned to C–N stretching vibrations of aromatic and aliphatic amines. The C–N stretching vibrations of aromatic and aliphatic amines were obtained at 1390 cm^{-1} and 1080 cm^{-1}. The band observed at 570 cm^{-1} belongs to C–Br stretching vibrations which are consistent with the previous study [57].

After the possible encapsulates were identified, the morphology of AuNPs was studied using TEM. TEM is a widely applied technique to study the morphology of nanomaterials at various magnifications [58]. According to this technique, the synthesized AuNPs were predominantly spherical in shape (Figure 5). ImageJ software (1.8.0) predicted the size of the nanomaterials to be in the range of 38.5 ± 10.6 nm. After the morphology was studied, elemental mapping was performed by EDAX to determine the purity of the AuNPs. The composition of C (15.65%), O (0.29%), and Au (84.06%) indicates that the synthesized AuNPs were predominantly metallic gold forms. A strong and distinctive peak for AuNPs at 2.1 KeV was observed (Figure 6) [59]. After the morphology and elemental analyses were performed, the crystalline nature of the AuNPs was determined using XRD (Figure 7). The XRD peaks i.e., (111), (200), (220), (311), and (222) observed at their respective 2θ values, confirm the formation of face-centered cubic (fcc) structure of metallic gold which matched with the JCPDS No. 04-0784 [60,61]. The peak corresponding to the (111) plane is more intense than the other planes confirming that the plane (111) is the predominant orientation. The lattice constant (a) of AuNPs is calculated using the following formula (Equation (1)):

$$\frac{1}{d^2} = \frac{h^2 + k^2 + l^2}{a^2} \quad (1)$$

where interplanar spacing (d), can be calculated by using Bragg's law ($2d\ sin\theta = \lambda$) and (hkl) are the Miller indices of the diffraction planes. The highest intensity diffraction peak belonging to the (111) plane is selected for calculation of the lattice parameter i.e., found to be 4.077 Å. The average crystallite size (D) was calculated from the XRD analysis using the Debye–Scherrer formula, given by Equation (2),

$$D = \frac{0.9\lambda}{\beta cos\theta} \quad (2)$$

where λ is the wavelength of the X-ray radiation, β is the FWHM of the diffracted peaks, and θ is the glancing angle. The value of D is calculated to be ~24 nm which is concomitant with the result obtained from TEM analysis.

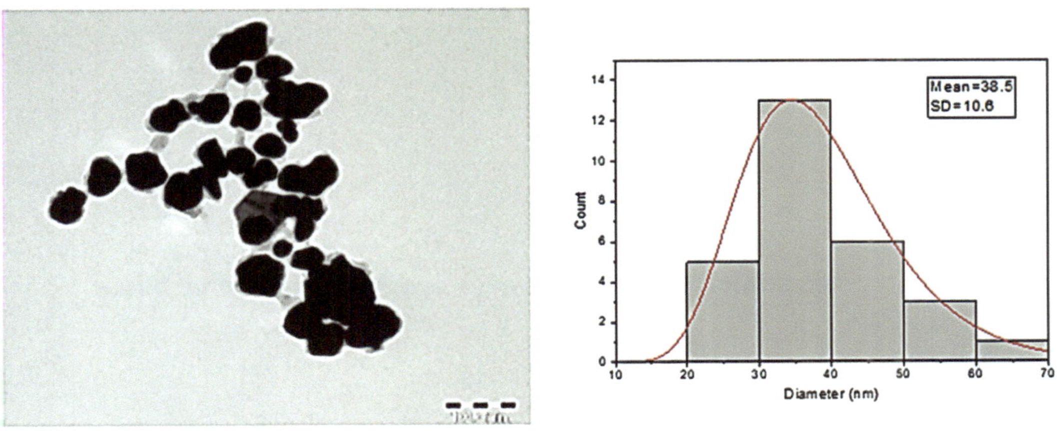

Figure 5. TEM images and diameter distribution of the AuNPs.

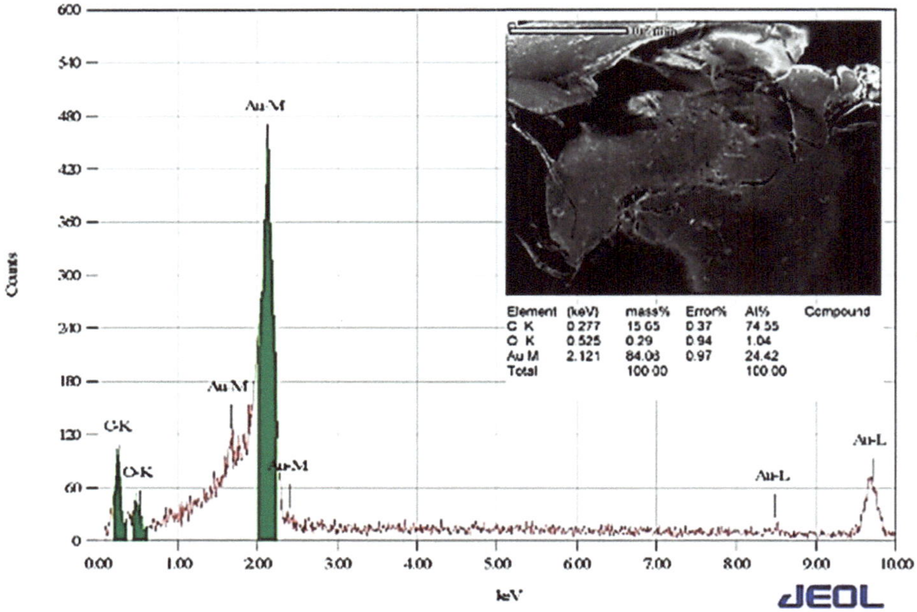

Figure 6. Analysis of elemental composition by EDAX.

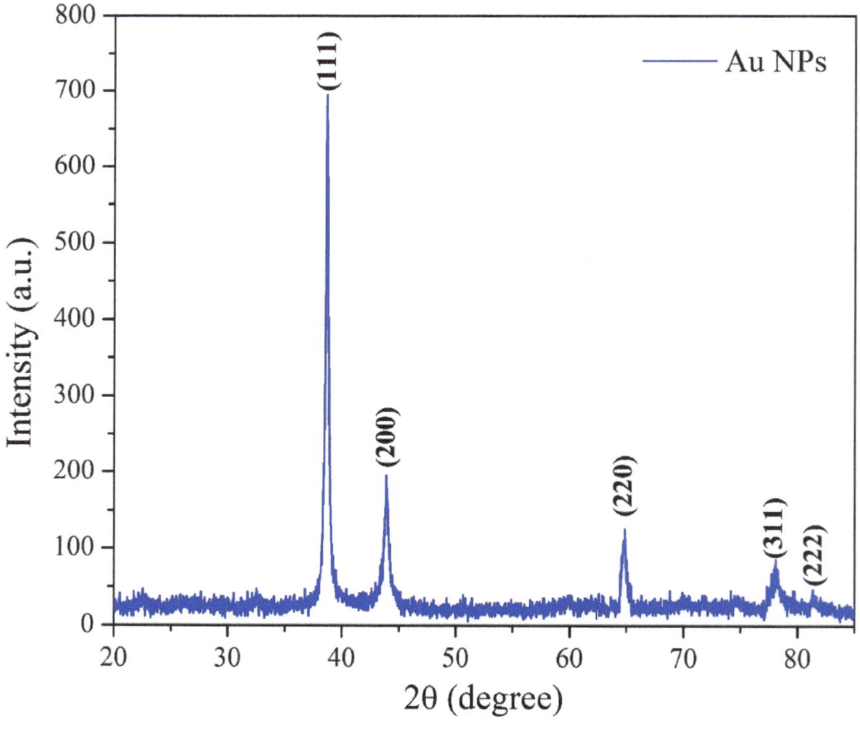

Figure 7. XRD pattern of the synthesized AuNPs.

3.4. Antifungal Activity of the AuNPs

A marked antifungal activity of AuNPs against five out of 18 *Aspergillus* isolates was observed. Three out of eleven (28%) *A. flavus* isolates (AM2, AM11, and AM15) and both *A. terreus* isolates (AM10 and 8) were inhibited (Table 8). The inhibition zones in the highest AuNP concentration (1000 ppm) varied between 6.3 mm for *A. terreus* and 9.3 mm against *A. flavus* (AM2). The rest of the 18 isolates were not inhibited by AuNPs, showing 0 mm inhibition zones. The broth microdilution method showed the inhibition of AuNPs for all five isolates found susceptible in the disc method. The five isolates were remarkably inhibited by the highest AuNP concentration while *A. flavus* (AM2) was inhibited already in the 200 ppm concentration (Table 9). MIC was 1000 ppm for A. flavus (AM15) while MEC was 25 ppm for (AM2) (Table 9).

Table 8. Efficacy of different concentrations of AuNPs against fungi as the zone of inhibition in disc method.

No.	Fungi	Inhibition Zone (mm)				
		25	50	100	200	1000
1	*A. flavus (AM1)*	0 ± 0	0 ± 0	0 ± 0	0 ± 0	0 ± 0
2	*A. flavus (AM2)*	0 ± 0	6.7 ± 0.6	7 ± 1	7.3 ± 0.6	9.3 ± 2.3
3	*A. flavus (AM3)*	0 ± 0	0 ± 0	0 ± 0	0 ± 0	0 ± 0
4	*A. flavus (AM4)*	0 ± 0	0 ± 0	0 ± 0	0 ± 0	0 ± 0
5	*A. flavus (AM5)*	0 ± 0	0 ± 0	0 ± 0	0 ± 0	0 ± 0
6	*A. fumigatus (AM6)*	0 ± 0	0 ± 0	0 ± 0	0 ± 0	0 ± 0
7	*A. fumigatus (AM7)*	0 ± 0	0 ± 0	0 ± 0	0 ± 0	0 ± 0
8	*A. niger (AM8)*	0 ± 0	0 ± 0	0 ± 0	0 ± 0	0 ± 0
9	*A. niger (AM9)*	0 ± 0	0 ± 0	0 ± 0	0 ± 0	0 ± 0
10	*A. terreus (AM10)*	6.7 ± 0.6	7 ± 0	7 ± 0	7 ± 0	7 ± 0
11	*A. flavus (AM11)*	0 ± 0	6.7 ± 0.6	6.7 ± 0.6	7 ± 0	8.7 ± 0.6
12	*A. flavus (AM12)*	0 ± 0	0 ± 0	0 ± 0	0 ± 0	0 ± 0
13	*A. flavus (AM13)*	0 ± 0	0 ± 0	0 ± 0	0 ± 0	0 ± 0
14	*A. terreus 8*	0 ± 0	0 ± 0	0 ± 0	0 ± 0	6.3 ± 0.6
15	*A. flavus (AM14)*	0 ± 0	0 ± 0	0 ± 0	0 ± 0	0 ± 0
16	*A. flavus (AM15)*	0 ± 0	0 ± 0	0 ± 0	0 ± 0	6.7 ± 0.6
17	*A. flavus (AM16)*	0 ± 0	0 ± 0	0 ± 0	0 ± 0	0 ± 0
18	*A. niger (AM17)*	0 ± 0	0 ± 0	0 ± 0	0 ± 0	0 ± 0

Table 9. Effect of different concentrations of AuNPs on fungal growth (growth rate score 0 = no growth, 4 = max growth) measured using broth microdilution technique.

No.	Fungi	Growth Rate (Scores/4)				
		25	50	100	200	1000
1	*A. flavus (AM2)*	3 ± 0 *	3 ± 0	3 ± 0	3 ± 0	2 ± 0
2	*A. terreus (AM10)*	4 ± 0	4 ± 0	4 ± 0	2 ± 0 *	2 ± 0
3	*A. flavus (AM11)*	4 ± 0	4 ± 0	4 ± 0	3 ± 0 *	1 ± 0
4	*A. terreus 8*	4 ± 0	4 ± 0	4 ± 0	3 ± 0 *	2 ± 0
5	*A. flavus (AM15)*	4 ± 0	3 ± 0 *	1 ± 0	1 ± 0	0 ± 0 **

*: MEC, **: MIC.

The AuNPs synthesized were not especially efficient against Aspergillus isolates. Only five out of eighteen isolates were inhibited by AuNPs. At the same time, the isolates were mostly resistant against the commercial drugs tested and needed the highest concentration to be inhibited. The relatively low inhibition efficiency of AuNPs is no surprise because it has been reported several times previously. AuNPs synthesized using the seed extract of Abelmoschus esculentus were not especially efficient against *A. flavus* and *A. niger* while they inhibited *Candida albicans* remarkably [62]. Elsewhere, AuNPs showed high activity against several Candida species [63]. The fungi Agaricus bisporus mediated AuNPs showed high antifungal activity against *A. flavus* but not against *A. terreus* [64]. When different synthesis and purification processes of the AuNPs were compared, the porification was observed to affect the efficiency of AuNPs against *C. albicans* [65]. Furthermore, AuNPs have been shown to inhibit *C. albicans* and *Sacharomyces cerevesiae* less than AgNPs [66]. AuNPs showed almost no activity against *C. albicans, C. tropicalis*, and *Fusarium oxysprorum*, while they were inhibited by NPs of Ag, Zn, and Cu synthesized using *A. kambarensis* extract [67]. Several other studies on varying and many times low efficiency against fungi have been published [55,68,69]. However, in some studies, high antifungal activity against *A. niger, A. flavus*, and *A. fumigatus* was also reported [70,71].

When thinking about the resistance against commercial drugs, the isolates that were inhibited by AuNPs behaved in various ways. For instance, *A. flavus* AM 15 needed the max concentration of 16 µg/mL Amphotericin B to be inhibited totally according to the microdilution method (Table 5). AM15 was inhibited to score 3 (out of 4) with 2 ug/mL concentration being relatively resistant. A. flavus AM11 was inhibited to score 3 only in the highest concentration. *Aspergillus terreus* AM10 was also resistant against Amphotericin B having a score of 3 at the highest concentration while a score of 4 up to 8 µg/mL concentration. *Aspergillus terreus* 8 was inhibited to score 2 only in the highest concentration. The least resistant isolate was *A. flavus* AM2 that was inhibited to score 3 already in the lowest concentration (1 µg/mL) and totally in 8 µg/mL.

3.5. *Effect of AuNPs on the Activity of Fungal Extracellular Enzymes*

The mechanism of antifungal activity of metallic nanoparticles has been attributed to several factors at molecular or physiological levels at membrane levels such as cell wall degradation or changes in the activity of extracellular enzymes [72,73]. Hence, we further tested the activity of extracellular enzymes before and after treatment with AuNPs.

Members of the *Aspergillus* genus are known to produce extracellular enzymes such as amylase, protease, deoxyribonuclease (DNase), lipase, elastase, and keratinase for growth, reproduction, and survival inside the host [74,75]. These extracellular enzymes are responsible for the formation of the extracellular matrix which can help in fungal evasion of killing by neutrophils and leading to a blockade in the production of reactive oxygen species. The components of the extracellular matrix of these pathogenic fungi (e.g., polysaccharides) are positively regulated by extracellular enzymes and can protect fungi from attacks by the host immune system [76]. Therefore, the extracellular components are outstanding targets for antifungal therapy [77].

With this background, the determination of the activity of extracellular enzymes after treatment with the AuNPs can be used as a tool to predict the possible survival mechanism. Hence, in this study, the effect of AuNPs on the extracellular enzymes was tested. *Aspergillus flavus* was selected in this report due to its high cytotoxic feature and its significant role in respiratory tract infections (invasive aspergillosis) and resultant complications in immunocompromised patients [78].

Profiling of extracellular enzymes produced by *A. flavus* (AM11, the most susceptible fungi) before treatment with AuNPs indicated that enzymes such as alkaline phosphatase, acid phosphatase, Naphthol-AS-BI-phosphohydrolase, α-galactosidase, ß-glucosidase, and α-mannosidase possessed higher activity ($20 \geq 40$ nmoles among 36.8% of enzymes). Low to moderate amounts of activity (5–10 nmoles among 63.2% of the enzymes) were noted predominantly among other enzymes identified. After treatment with AuNPs, the activities of enzymes such as acid phosphatase, Naphthol-AS-BI-phosphohydrolase, α-galactosidase, and ß-glucosidase decreased (82.4% of changes in the altered profile of enzymes). However, the activity profiles of enzymes such as ß-glucuronidase increased. The outcomes indicate that the activity of fungal extracellular enzymes diminished significantly after treatment with AuNPs (Figure 8).

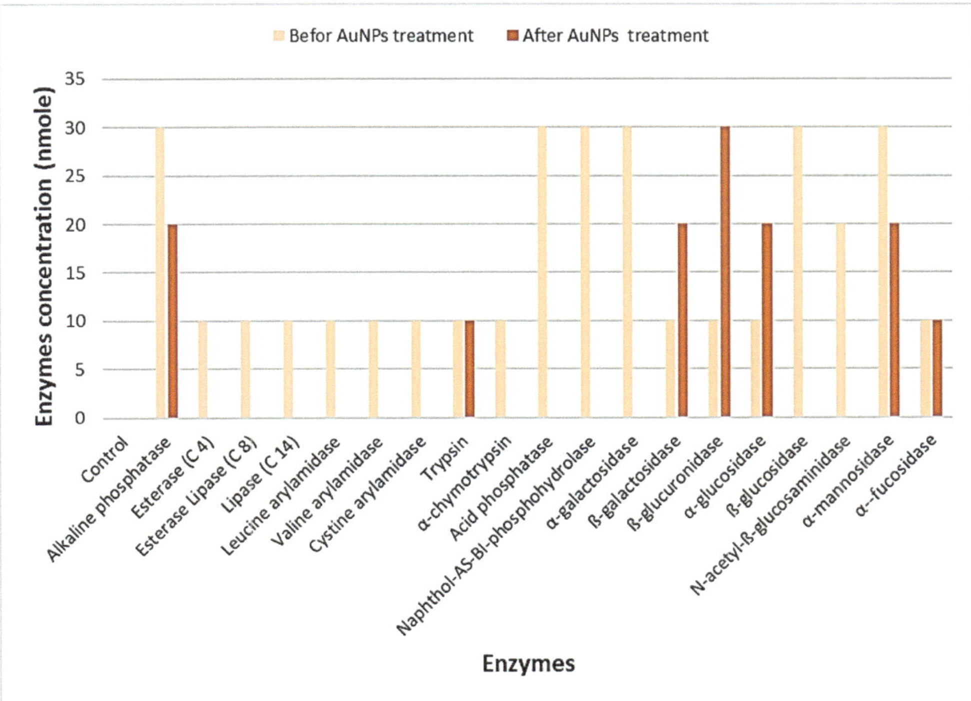

Figure 8. Effects of AuNPs on the activity of extracellular enzymes of *A. flavus* (AM11).

4. Conclusions

Mentha piperita represented a strong source for the synthesis and formation of AuNPs. The synthesized AuNPs were characterized structurally and morphologically. Further, the possible encapsulates were identified. In addition, this is the first-ever international report with regard to a few aspects related to antifungal effects, as mentioned at the end of the introduction section. Our findings related to the resistance pattern of these nosocomial isolates to common antifungals may indicate the emergence of serious acquired resistance to such agents, which might have arisen from the misuse of antifungals. This problem needs to be highlighted for proper maintenance of fungal infections in the future. Additionally, there are limited studies published in Saudi Arabia focusing on resistant *Aspergillus* and the worsening threat. Hence, this study would be effective for the use of nanomedicine in the management of nosocomial fungal infections of the respiratory tract. To conclude, the synthesized AuNPs were effective against five isolates of *Aspergillus* species that can cause invasive aspergillosis. Further studies are warranted to elucidate the interactions of these nanomaterials with the eukaryotic fungi at molecular levels.

Author Contributions: A.A.: Collection and drafting, A.H.B.: Supervision and editing and F.A.: Analysis, writing, revising the final version. Authors have read and agreed to the published version of the manuscript. All authors have read and agreed to the published version of the manuscript.

Funding: This research was funded by Researchers Supporting Project number (RSP-2021/364), King Saud University, Riyadh, Saudi Arabia.

Institutional Review Board Statement: Experimental protocols were approved by the institutional committee of the King Saud University, Research Project No. E-18-3066.

Informed Consent Statement: Not applicable.

Data Availability Statement: Not applicable.

Conflicts of Interest: All authors declare no conflict of interest.

References

1. Navale, V.; Vamkudoth, K.R.; Ajmera, S.; Dhuri, V. *Aspergillus* derived mycotoxins in food and the environment: Prevalence detection, and toxicity. *Toxicol. Rep.* **2021**, *8*, 1008–1030. [CrossRef] [PubMed]
2. Vonberg, R.P.; Gastmeier, P. Nosocomial aspergillosis in outbreak settings. *J. Hosp. Infect.* **2006**, *63*, 246–254. [CrossRef] [PubMed
3. Richardson, M.; Rautemaa-Richardson, R. Exposure to *Aspergillus* in Home and Healthcare Facilities' Water Environments: Focu on Biofilms. *Microorganisms* **2019**, *7*, 7. [CrossRef] [PubMed]
4. Nicolle, M.-C.; Benet, T.; Vanhems, P. Aspergillosis: Nosocomial or community-acquired? *Med. Mycol.* **2011**, *49*, S24–S2 [CrossRef]
5. Lozano, R.; Naghavi, M.; Foreman, K.; Lim, S.; Shibuya, K.; Aboyans, V.; Abraham, J.; Adair, T.; Aggarwal, R.; Ahn, S.; et a Global and regional mortality from 235 causes of death for 20 age groups in 1990 and 2010: A systematic analysis for the Globa Burden of Disease Study 2010. *Lancet* **2012**, *380*, 2095–2128. [CrossRef]
6. Soubani, A.O.; Khanchandani, G.; Ahmed, H.P. Clinical significance of lower respiratory tract *Aspergillus* culture in elderl hospitalized patients. *Eur. J. Clin. Microbiol. Infect. Dis. Off. Publ. Eur. Soc. Clin. Microbiol.* **2004**, *23*, 491–494. [CrossRef]
7. Donoghue, M.; Seibel, N.L.; Francis, P.S.; Walsh, T.J. Chapter 27—Fungal infections of the respiratory tract. In *Clinical Mycolog* 2nd ed.; Anaissie, E.J., McGinnis, M.R., Pfaller, M.A., Eds.; Churchill Livingstone: Edinburgh, UK, 2009; pp. 561–589.
8. Sharma, A.; Bisht, D.; Das, S.; Rai, G.; Dutt, S.; Arora, V.K. Molecular Detection of *Aspergillus* in Sputum of Patients with Lowe Respiratory Tract Infections. *Int. J. Appl. Basic Med. Res.* **2020**, *10*, 86–90.
9. Zachar, O. Formulations for COVID-19 early stage treatment via silver nanoparticles inhalation delivery at home and hospita *ScienceOpen Prepr.* **2020**. [CrossRef]
10. Jacob, J.A.; Salmani, J.M.M.; Chen, B. Magnetic nanoparticles: Mechanistic studies on the cancer cell interaction. *Nanotechnol. Re* **2016**, *5*, 481–488. [CrossRef]
11. Tran, N.T.T.; Wang, T.-H.; Lin, C.-Y.; Tai, Y. Synthesis of methotrexate-conjugated gold nanoparticles with enhanced cance therapeutic effect. *Biochem. Eng. J.* **2013**, *78*, 175–180. [CrossRef]
12. Sattarahmady, N.; Tondro, G.H.; Gholchin, M.; Heli, H. Gold nanoparticles biosensor of Brucella spp. genomic DNA: Visual an spectrophotometric detections. *Biochem. Eng. J.* **2015**, *97*, 1–7. [CrossRef]
13. Hu, W.; Yan, Z.; Li, H.; Qiu, J.; Zhang, D.; Li, P.; Pan, Y.; Guo, H. Development of a new colloidal gold immunochromatographi strip for rapid detecting subgroup A of avian leukosis virus using colloidal gold nanoparticles. *Biochem. Eng. J.* **2019**, *148*, 16–2 [CrossRef]
14. Jiang, C.; Jiang, Z.; Zhu, S.; Amulraj, J.; Deenadayalan, V.K.; Jacob, J.A.; Qian, J. Biosynthesis of silver nanoparticles and the identification of possible reductants for the assessment of *in vitro* cytotoxic and *in vivo* antitumor effects. *J. Drug Deliv. Sci. Techno* **2021**, *63*, 102444. [CrossRef]
15. Liu, X.; Shan, K.; Shao, X.; Shi, X.; He, Y.; Liu, Z.; Jacob, J.A.; Deng, L. Nanotoxic Effects of Silver Nanoparticles on Norma HEK-293 Cells in Comparison to Cancerous HeLa Cell Line. *Int. J. Nanomed.* **2021**, *16*, 753. [CrossRef] [PubMed]
16. Huang, H.; Shan, K.; Liu, J.; Tao, X.; Periyasamy, S.; Durairaj, S.; Jiang, Z.; Jacob, J.A. Synthesis, optimization and characterizatio of silver nanoparticles using the catkin extract of *Piper longum* for bactericidal effect against food-borne pathogens via conventiona and mathematical approaches. *Bioorganic Chem.* **2020**, *103*, 104230. [CrossRef] [PubMed]
17. Wen, X.; Wang, Q.; Dai, T.; Shao, J.; Wu, X.; Jiang, Z.; Jacob, J.A.; Jiang, C. Identification of possible reductants in the aqueou leaf extract of mangrove plant *Rhizophora apiculata* for the fabrication and cytotoxicity of silver nanoparticles against huma osteosarcoma MG-63 cells. *Mater. Sci. Eng. C* **2020**, *116*, 111252. [CrossRef] [PubMed]
18. Mishra, P.; Ray, S.; Sinha, S.; Das, B.; Khan, M.I.; Behera, S.K.; Yun, S.; Tripathy, S.; Mishra, A. Facile bio-synthesis of gol nanoparticles by using extract of *Hibiscus sabdariffa* and evaluation of its cytotoxicity against U87 glioblastoma cells unde hyperglycemic condition. *Biochem. Eng. J.* **2016**, *105*, 264–272. [CrossRef]
19. Mythili, R.; Selvankumar, T.; Srinivasan, P.; Sengottaiyan, A.; Sabastinraj, J.; Ameen, F.; Al-Sabri, A.; Kamala-Kannan, S. Govarthanan, M.; Kim, H. Biogenic synthesis, characterization and antibacterial activity of gold nanoparticles synthesised fro vegetable waste. *J. Mol. Liq.* **2018**, *262*, 318–321. [CrossRef]
20. Iram, S.; Zahera, M.; Khan, S.; Khan, I.; Syed, A.; Ansary, A.A.; Fuad, A.; Shair, O.; Khan, M. Gold nanoconjugates reinforce th potency of conjugated cisplatin and doxorubicin. *Colloids Surf. B Biointerfaces* **2017**, *160*, 254–264. [CrossRef]
21. Rahim, M.; Iram, S.; Syed, A.; Ameen, F.; Hodhod, M.S.; Khan, M.S. Nutratherapeutics approach against cancer: Tomato-mediate synthesised gold nanoparticles. *IET Nanobiotechnol.* **2018**, *12*, 1–5. [CrossRef]

22. Botteon, C.; Silva, L.; Ccana-Ccapatinta, G.; Silva, T.; Ambrosio, S.; Veneziani, R.; Bastos, J.K.; Marcato, P. Biosynthesis and characterization of gold nanoparticles using Brazilian red propolis and evaluation of its antimicrobial and anticancer activities. *Sci. Rep.* **2021**, *11*, 1–16.
23. Alsamhary, K.; Al-Enazi, N.; Alshehri, W.A.; Ameen, F. Gold nanoparticles synthesised by flavonoid tricetin as a potential antibacterial nanomedicine to treat respiratory infections causing opportunistic bacterial pathogens. *Microb. Pathog.* **2020**, *139*, 103928. [CrossRef] [PubMed]
24. Ameen, F.; AlYahya, S.A.; Bakhrebah, M.A.; Nassar, M.S.; Aljuraifani, A. Flavonoid dihydromyricetin-mediated silver nanoparticles as potential nanomedicine for biomedical treatment of infections caused by opportunistic fungal pathogens. *Res. Chem. Intermed.* **2018**, *44*, 5063–5073. [CrossRef]
25. Sathishkumar, P.; Preethi, J.; Vijayan, R.; Yusoff, A.R.M.; Ameen, F.; Suresh, S.; Balagurunathan, R.; Palvannan, T. Anti-acne, anti-dandruff and anti-breast cancer efficacy of green synthesised silver nanoparticles using *Coriandrum sativum* leaf extract. *J. Photochem. Photobiol. B Biol.* **2016**, *163*, 69–76. [CrossRef]
26. Ameen, F.; Abdullah, M.M.; Al-Homaidan, A.A.; Al-Lohedan, H.A.; Al-Ghanayem, A.A.; Almansob, A. Fabrication of silver nanoparticles employing the cyanobacterium *Spirulina platensis* and its bactericidal effect against opportunistic nosocomial pathogens of the respiratory tract. *J. Mol. Struct.* **2020**, *1217*, 128392. [CrossRef]
27. Ameen, F.; Al-Homaidan, A.A.; Al-Sabri, A.; Almansob, A.; AlNAdhari, S. Antioxidant, antifungal and cytotoxic effects of silver nanoparticles synthesized using marine fungus *Cladosporium halotolerans*. *Appl. Nanosci.* **2021**, 1–9. [CrossRef]
28. Manohar, M.; Marzinke, M.A. Chapter 3—Application of chromatography combined with mass spectrometry in therapeutic drug monitoring. In *Clinical Challenges in Therapeutic Drug Monitoring*; Clarke, W., Dasgupta, A., Eds.; Elsevier: Amsterdam, The Netherlands, 2016; pp. 45–70.
29. Cowen, L.E.; Sanglard, D.; Howard, S.J.; Rogers, P.D.; Perlin, D.S. Mechanisms of Antifungal Drug Resistance. *Cold Spring Harb. Perspect. Med.* **2014**, *5*, a019752-a. [CrossRef]
30. Ghannoum, M.A.; Rice, L.B. Antifungal agents: Mode of action, mechanisms of resistance, and correlation of these mechanisms with bacterial resistance. *Clin. Microbiol. Rev.* **1999**, *12*, 501–517. [CrossRef]
31. Espinel-Ingroff, A.; Arthington-Skaggs, B.; Iqbal, N.; Ellis, D.; Pfaller, M.; Messer, S.; Inaldi, M.; Fothergill, A.; Gibbs, D.L.; Wang, A. Multicenter evaluation of a new disk agar diffusion method for susceptibility testing of filamentous fungi with voriconazole, posaconazole, itraconazole, amphotericin B, and caspofungin. *J. Clin. Microbiol.* **2007**, *45*, 1811–1820. [CrossRef]
32. Hudzicki, J. Kirby-Bauer disk diffusion susceptibility test protocol. *Am. Soc. Microbiol.* **2009**, *15*, 55–63.
33. John, H. Reference method for broth dilution antifungal susceptibility testing of filamentous fungi, approved standard. M38-A2. *Clin. Lab. Stand. Inst.* **2008**, *28*, 1–35.
34. Pietrzak, K.; Twarużek, M.; Czyżowska, A.; Kosicki, R.; Gutarowska, B. Influence of silver nanoparticles on metabolism and toxicity of moulds. *Acta Biochim. Pol.* **2015**, *62*, 851–857. [CrossRef] [PubMed]
35. Kumar, S.; Stecher, G.; Li, M.; Knyaz, C.; Tamura, K. MEGA X: Molecular evolutionary genetics analysis across computing platforms. *Mol. Biol. Evol.* **2018**, *35*, 1547. [CrossRef] [PubMed]
36. Bullis, S.S.; Krywanczyk, A.; Hale, A.J. Aspergillosis myocarditis in the immunocompromised host. *IDCases* **2019**, *17*, e00567. [CrossRef]
37. Gnat, S.; Łagowski, D.; Nowakiewicz, A.; Dyląg, M. A global view on fungal infections in humans and animals: Opportunistic infections and microsporidioses. *J. Appl. Microbiol.* **2021**, *131*, 2095–2113. [CrossRef]
38. Alkfaji, F.; Hussaini, I.M.A. Biosynthesis of silver nanoparticles with Mentha spicata against *Aspergillus niger*. *Drug Invent. Today* **2020**, *14*, 806–811.
39. Rudramurthy, S.M.; Paul, R.A.; Chakrabarti, A.; Mouton, J.W.; Meis, J.F. Invasive aspergillosis by *Aspergillus flavus*: Epidemiology, diagnosis, antifungal resistance, and management. *J. Fungi* **2019**, *5*, 55. [CrossRef]
40. Badiee, P.; Alborzi, A.; Moeini, M.; Haddadi, P.; Farshad, S.; Japoni, A.; Ziyaeyan, M. Antifungal susceptibility of the *Aspergillus* species by Etest and CLSI reference methods. *Arch. Iran. Med.* **2012**, *15*, 429–432.
41. Reichert-Lima, F.; Lyra, L.; Pontes, L.; Moretti, M.L.; Pham, C.D.; Lockhart, S.R.; Schreiber, A. Surveillance for azoles resistance in *Aspergillus* spp. highlights a high number of amphotericin B-resistant isolates. *Mycoses* **2018**, *61*, 360–365. [CrossRef]
42. Rudramurthy, S.M.; Chakrabarti, A.; Geertsen, E.; Mouton, J.W.; Meis, J.F. In vitro activity of isavuconazole against 208 *Aspergillus flavus* isolates in comparison with 7 other antifungal agents: Assessment according to the methodology of the European Committee on Antimicrobial Susceptibility Testing. *Diagn. Microbiol. Infect. Dis.* **2011**, *71*, 370–377. [CrossRef]
43. Arendrup, M. Update on antifungal resistance in *Aspergillus* and *Candida*. *Clin. Microbiol. Infect.* **2014**, *20*, 42–48. [CrossRef] [PubMed]
44. Enoch, D.; Idris, S.; Aliyu, S.; Micallef, C.; Sule, O.; Karas, J. Micafungin for the treatment of invasive aspergillosis. *J. Infect.* **2014**, *68*, 507–526. [CrossRef] [PubMed]
45. Herbrecht, R.; Denning, D.W.; Patterson, T.F.; Bennett, J.E.; Greene, R.E.; Oestmann, J.-W.; Kern, W.; Marr, K.; Ribaud, P.; Lortholary, O.; et al. Voriconazole versus amphotericin B for primary therapy of invasive aspergillosis. *N. Engl. J. Med.* **2002**, *347*, 408–415. [CrossRef] [PubMed]
46. Maertens, J.A.; Rahav, G.; Lee, D.-G.; Ponce-de-León, A.; Sánchez, I.C.R.; Klimko, N.; Sonet, A.; Haider, S.; Velez, J.; Raad, I.; et al. Posaconazole versus voriconazole for primary treatment of invasive aspergillosis: A phase 3, randomised, controlled, non-inferiority trial. *Lancet* **2021**, *397*, 499–509. [CrossRef]

47. Xin Lee, K.; Shameli, K.; Miyake, M.; Kuwano, N.; Bt Ahmad Khairudin, N.B.; Bt Mohamad, S.E.; Yew, Y.P. Green synthesis of gold nanoparticles using aqueous extract of *Garcinia mangostana* fruit peels. *J. Nanomater.* **2016**, *2016*, 8489094. [CrossRef]
48. Le, V.T.; Ngu, N.N.Q.; Chau, T.P.; Nguyen, T.D.; Nguyen, V.T.; Nguyen, T.L.H.; Cao, X.T.; Doan, V.-D. Silver and Gold Nanoparticles from Limnophila rugosa Leaves: Biosynthesis, Characterization, and Catalytic Activity in Reduction of Nitrophenols. *J. Nanomater.* **2021**, *2021*, 8582165. [CrossRef]
49. Boruah, J.S.; Devi, C.; Hazarika, U.; Reddy, P.V.B.; Chowdhury, D.; Barthakur, M.; Kalita, P. Green synthesis of gold nanoparticles using an antiepileptic plant extract: In vitro biological and photo-catalytic activities. *RSC Adv.* **2021**, *11*, 28029–28041. [CrossRef]
50. Mariychuk, R.; Smolková, R.; Bartošová, V.; Eliašová, A.; Grishchenko, L.M.; Diyuk, V.E.; Lisnyak, V.V. The regularities of the *Mentha piperita* L. extract mediated synthesis of gold nanoparticles with a response in the infrared range. *Appl. Nanosci.* **2021**, *2021*, 1–13. [CrossRef]
51. Mathur, A.; Prasad, G.; Rao, N.; Babu, P.; Dua, V. Isolation and identification of antimicribial compound from *Mentha pipirita*. *Rasayan J.* **2011**, *4*, 36–42.
52. Das, S.; Das, A.; Thamarai Selvan, R.; Raj, S.A. The antibacterial and aroma finishing of cotton fabrics by *Mentha pipertia* extract. *J. Text. Inst.* **2020**, *112*, 1181–1190. [CrossRef]
53. Brahmi, F.; Khodir, M.; Mohamed, C.; Pierre, D. Chemical composition and biological activities of *Mentha* species. *Aromat. Med. Plants Back Nat.* **2017**, *10*, 47–79.
54. Salehi, B.; Stojanović-Radić, Z.; Matejić, J.; Sharopov, F.; Antolak, H.; Kręgiel, D.; Sen, S.; Sharifi-Rad, M.; Acharya, K.; Sharifi-Rad, R.; et al. Plants of Genus *Mentha*: From Farm to Food Factory. *Plants* **2018**, *7*, 70. [CrossRef] [PubMed]
55. Aljabali, A.A.A.; Akkam, Y.; Al Zoubi, M.S.; Al-Batayneh, K.M.; Al-Trad, B.; Abo Alrob, O.; Benamara, M.; Evans, D.J. Synthesis of Gold Nanoparticles Using Leaf Extract of *Ziziphus zizyphus* and their Antimicrobial Activity. *Nanomaterials* **2018**, *8*, 174. [CrossRef] [PubMed]
56. Jafarizad, A.; Safaee, K.; Gharibian, S.; Omidi, Y.; Ekinci, D. Biosynthesis and in-vitro study of gold nanoparticles using *Mentha* and *Pelargonium* extracts. *Procedia Mater. Sci.* **2015**, *11*, 224–230. [CrossRef]
57. Khademi-Azandehi, P.; Moghaddam, J. Green synthesis, characterization and physiological stability of gold nanoparticles from *Stachys lavandulifolia* Vahl extract. *Particuology* **2015**, *19*, 22–26. [CrossRef]
58. Lee, B.; Yoon, S.; Lee, J.W.; Kim, Y.; Chang, J.; Yun, J.; Ro, J.C.; Lee, J.S.; Lee, J.H. Statistical characterization of the morphologies of nanoparticles through machine learning based electron microscopy image analysis. *ACS Nano* **2020**, *14*, 17125–17133. [CrossRef] [PubMed]
59. Raliya, R.; Tarafdar, J. Biosynthesis of gold nanoparticles using *Rhizoctonia bataticola* TFR-6. *Adv. Sci. Eng. Med.* **2013**, *5*, 1073–1076. [CrossRef]
60. Aromal, S.A.; Babu, K.D.; Philip, D. Characterization and catalytic activity of gold nanoparticles synthesized using *Ayurvedic arishtams*. *Spectrochim. Acta Part A Mol. Biomol. Spectrosc.* **2012**, *96*, 1025–1030. [CrossRef]
61. Singh, H.; Du, J.; Singh, P.; Yi, T.H. Ecofriendly synthesis of silver and gold nanoparticles by *Euphrasia officinalis* leaf extract and its biomedical applications. *Artif. Cells Nanomed. Biotechnol.* **2018**, *46*, 1163–1170. [CrossRef]
62. Jayaseelan, C.; Ramkumar, R.; Rahuman, A.; Perumal, P. Green synthesis of gold nanoparticles using seed aqueous extract of *Abelmoschus esculentus* and its antifungal activity. *Ind. Crop. Prod.* **2013**, *45*, 423–429. [CrossRef]
63. Wani, I.A.; Ahmad, T. Size and shape dependant antifungal activity of gold nanoparticles: A case study of *Candida*. *Colloids Surf. B Biointerfaces* **2013**, *101*, 162–170. [CrossRef] [PubMed]
64. Eskandari-Nojedehi, M.; Jafarizadeh-Malmiri, H.; Rahbar-Shahrouzi, J. Hydrothermal green synthesis of gold nanoparticles using mushroom (*Agaricus bisporus*) extract: Physico-chemical characteristics and antifungal activity studies. *Green Process. Synth.* **2018**, *7*, 38–47. [CrossRef]
65. López-Lorente, Á.I.; Cárdenas, S.; González-Sánchez, Z.I. Effect of synthesis, purification and growth determination methods on the antibacterial and antifungal activity of gold nanoparticles. *Mater. Sci. Eng. C* **2019**, *103*, 109805. [CrossRef] [PubMed]
66. Khatoon, U.T.; Rao, G.V.S.N.; Mohan, M.K.; Ramanaviciene, A.; Ramanavicius, A. Comparative study of antifungal activity of silver and gold nanoparticles synthesized by facile chemical approach. *J. Environ. Chem. Eng.* **2018**, *6*, 5837–5844. [CrossRef]
67. Gholami-Shabani, M.; Sotoodehnejadnematalahi, F.; Shams-Ghahfarokhi, M.; Eslamifar, A.; Razzaghi-Abyaneh, M. Physicochemical properties, anticancer and antimicrobial activities of metallic nanoparticles green synthesized by *Aspergillus kambarensis*. *IET Nanobiotechnol.* **2022**, *16*, 1. [CrossRef]
68. Fatima, F.; Bajpai, P.; Pathak, N.; Singh, S.; Priya, S.; Verma, S.R. Antimicrobial and immunomodulatory efficacy of extracellularly synthesized silver and gold nanoparticles by a novel phosphate solubilizing fungus *Bipolaris tetramera*. *BMC Microbiol.* **2015**, *15*, 1–10. [CrossRef]
69. BalaKumaran, M.D.; Ramachandran, R.; Balashanmugam, P.; Jagadeeswari, S.; Kalaichelvan, P.T. Comparative analysis of antifungal, antioxidant and cytotoxic activities of mycosynthesized silver nanoparticles and gold nanoparticles. *Mater. Technol.* **2020**, 1–11. [CrossRef]
70. Elegbede, J.A.; Lateef, A.; Azeez, M.A.; Asafa, T.B.; Yekeen, T.A.; Oladipo, I.C.; Aina, D.A.; Beukes, L.S.; Gueguim-Kana, E.B. Biofabrication of gold nanoparticles using xylanases through valorization of corncob by *Aspergillus niger* and *Trichoderma longibrachiatum*: Antimicrobial, antioxidant, anticoagulant and thrombolytic activities. *Waste Biomass Valorization* **2020**, *11*, 781–791. [CrossRef]

71. Ojo, S.A.; Lateef, A.; Azeez, M.A.; Oladejo, S.M.; Akinwale, A.S.; Asafa, T.B.; A Yekeen, T.; Akinboro, A.; Oladipo, I.C.; Gueguim-Kana, E.B.; et al. Biomedical and catalytic applications of gold and silver-gold alloy nanoparticles biosynthesized using cell-free extract of *Bacillus safensis* LAU 13: Antifungal, dye degradation, anti-coagulant and thrombolytic activities. *IEEE Trans. Nanobioscience* **2016**, *15*, 433–442. [CrossRef]
72. Ameen, F.; Alsamhary, K.; Alabdullatif, J.A.; Alnadhari, S. A review on metal-based nanoparticles and their toxicity to beneficial soil bacteria and fungi. *Ecotoxicol. Environ. Saf.* **2021**, *213*, 112027. [CrossRef]
73. Du, J.; Zhang, Y.; Yin, Y.; Zhang, J.; Ma, H.; Li, K.; Wan, N. Do environmental concentrations of zinc oxide nanoparticle pose ecotoxicological risk to aquatic fungi associated with leaf litter decomposition? *Water Res.* **2020**, *178*, 115840. [CrossRef] [PubMed]
74. Duran, R.M.; Gregersen, S.; Smith, T.D.; Bhetariya, P.J.; Cary, J.W.; Harris-Coward, P.Y.; Mattison, C.P.; Grimm, C.; Calvo, A.M. The role of *Aspergillus flavus* veA in the production of extracellular proteins during growth on starch substrates. *Appl. Microbiol. Biotechnol.* **2014**, *98*, 5081–5094. [CrossRef] [PubMed]
75. Balakrishnan Sangeetha, A.; Abdel-hadi, A.; Hassan, A.S.; Shobana, C.S.; Suresh, S.; Abirami, B.; Selvam, K.P.; Al-Baradie, R.S.; Banawas, S.; Alaidarous, M.; et al. Evaluation of in vitro activities of extracellular enzymes from *Aspergillus* species isolated from corneal ulcer/keratitis. *Saudi J. Biol. Sci.* **2020**, *27*, 701–705. [CrossRef] [PubMed]
76. Mayer, F.L.; Wilson, D.; Hube, B. *Candida albicans* pathogenicity mechanisms. *Virulence* **2013**, *4*, 119–128. [CrossRef]
77. Garcia-Rubio, R.; de Oliveira, H.C.; Rivera, J.; Trevijano-Contador, N. The fungal cell wall: *Candida*, *Cryptococcus*, and *Aspergillus* species. *Front. Microbiol.* **2020**, *10*, 2993. [CrossRef]
78. Lee, R.J.; Workman, A.D.; Carey, R.M.; Chen, B.; Rosen, P.L.; Doghramji, L.; Adappa, N.D.; Palmer, J.N.; Kennedy, D.W.; Cohen, N.A. Fungal Aflatoxins Reduce Respiratory Mucosal Ciliary Function. *Sci. Rep.* **2016**, *6*, 33221. [CrossRef]

nanomaterials

MDPI

Article

Near-Infrared Light-Triggered Nitric Oxide Nanogenerators for NO-Photothermal Synergistic Cancer Therapy

Weiwei Liu [1,2], Farouk Semcheddine [1], Zengchao Guo [1], Hui Jiang [1,*] and Xuemei Wang [1,*]

1 State Key Laboratory of Bioelectronics, National Demonstration Center for Experimental Biomedical Engineering Education, School of Biological Science and Medical Engineering, Southeast University, Nanjing 210096, China; liuw182@163.com (W.L.); fou_semch@yahoo.fr (F.S.); rmcqh242526@163.com (Z.G.)

2 School of Chemistry and Chemical Engineering, Southeast University, Nanjing 210096, China

* Correspondence: sungi@seu.edu.cn (H.J.); xuewang@seu.edu.cn (X.W.)

Abstract: Cancer is still one of the major health issues faced by human beings today. Various nanomaterials have been designed to treat tumors and have made great progress. Herein, we used amino-functionalized metal organic framework (UiO-66-NH_2) as superior templates and successfully synthesized the UiO-66-NH_2@Au_{shell} composite nanoparticles (UA) with high loading capacity and excellent photothermal properties through a simple and gentle method. In addition, due to the rich pore structure and excellent biocompatibility of the as-prepared composite nanoparticles, the hydrophobic NO donor BNN6 (N,N′-Di-sec-butyl-N,N′-dinitroso-1, 4-phenylenediamine) molecule was efficiently delivered. Based on the phenomenon where BNN6 molecules can decompose and release NO at high temperature, when UiO-66-NH_2@Au_{shell}-BNN6 composite nanoparticles (UA-BNN6) entered tumor cells and were irradiated by NIR, the porous gold nanoshells on the surface of composite nanoparticles induced an increase in temperature through the photothermal conversion process and promoted the decomposition of BNN6 molecules, releasing high concentration of NO, thus efficiently killing HeLa cells through the synergistic effect of NO-photothermal therapy. This effective, precise and safe treatment strategy controlled by NIR laser irradiation represents a promising alternative in the field of cancer treatment.

Keywords: MOF; BNN6; gold nanoshells; NO-photothermal therapy

1. Introduction

Since the discovery of NO as an important biological signal molecule, it has attracted considerable attention from researchers [1,2]. As a gas transmitter, NO has been found to play an important role in regulating various cellular events, such as vascular dilation, platelet aggregation and adhesion, inflammatory response, immune response, and neurotransmission. Due to its specific properties, NO has great clinical value in cardiovascular diseases, wound healing, and antibacterial and tumor therapy [3–9]. In particular, NO gas therapy provides an efficient and green therapy method for the treatment of malignant tumors that cause great harm to human health. Due to the constant efforts of scientific researchers, we have a more profound understanding of the mechanism of NO-killing tumor cells. It was found that a high concentration of NO (1 > uM) could induce the apoptosis of tumor cells [10]. The mechanisms behind this apoptotic effect are the ability of NO to induce an oxidative and nitrifying stress, damage mitochondria and DNA, inhibit DNA synthesis and repair, deaminate DNA, and inhibit cell respiration. Although NO has a strong anti-tumor effect, its short life cycle and sensitivity to biological substances limit its clinical applications. Therefore, a major challenge faced by researchers is to develop an effective NO donor and NO-encapsulated cargo carrier to control the release of NO in space, time, and measurement [11,12].

In view of the above problems, various exogenous NO donors have been reported, such as diazeniumdiolates [13], s-nitrosothiols [14], metal-nitrosyl complexes [15], and nitrobenzene derivatives [16]. However, they often have the disadvantage of spontaneous uncontrolled release, which seriously hinders their clinical application. Usually these NO donors are attached or immobilized on small drug molecules, organic polymer materials, or inorganic materials [17–20]. At the same time, in order to efficiently release NO in tumor tissues, such NO delivery platforms can generally release NO in response to specific stimuli, such as light, heat, or pH [21–24]. Among several methods, photoexcitation stands out for its ease of operation and safety. Currently, the main light source applied is ultraviolet and visible light. However, this kind of light source has the disadvantages of shallow tissue penetration and significant side effects on surrounding tissues, whereas near-infrared light represents the best choice due to its deep tissue penetration and low phototoxicity [25,26]. In recent years, the NO release platform based on near infrared light excitation has been developed, and relatively ideal experimental results have been obtained.

MOF-NH_2@Au_{shell} composite nanomaterials not only have the advantages of good biocompatibility, easy degradation, and easy chemical modification of MOF materials [27,28] but also possess excellent photothermal properties and highly porous gold nanoshells [29,30]. Compared with SiO_2@Au_{shell} materials coated with a dense gold nanoshell [31,32], MOF-NH_2@Au_{shell} composite nanomaterials can be easily functionalized due to their richness of material selection, the controllability of their size and structure, and the porosity of their gold shell. MOF-NH_2@Au_{shell} also has great advantages in terms of drug delivery and controlled release. Here, we used the UA as a carrier to immobilize the NO donor reagent, BNN6, obtaining a multifunctional UA-BNN6. The surface plasmon resonance properties of the gold nanoshell surface layer of the UA endow the material with excellent photothermal properties, whereas its rich hierarchical pore structure provides space for the loading of BNN6. The BNN6 molecule has low cytotoxicity and is easily degraded to release NO at high temperatures. It is considered to be an ideal peripheral NO donor. By combining the photothermal conversion performance of UA with the decomposition and release of NO by BNN6 at high temperature, the synergistic effect of NO-photothermal therapy on tumor cells initiated by infrared NIR response was achieved [33,34]. After the composite nanomaterial entered cancer cells, the NO donor molecule was successfully transported inside the cells. Under NIR irradiation, the nanomaterial generated a high temperature and BNN6 was degraded to a released high concentration of NO. In addition, the temperature increase induced by the composite nanomaterials also caused damage to cancer cells (Scheme 1). Compared with photothermal therapy alone, the synergistic effect of NO-photothermal therapy has the advantages of using low dosage, being simple to operate, and having few side effects. This green and efficient cancer treatment strategy has great clinical application value.

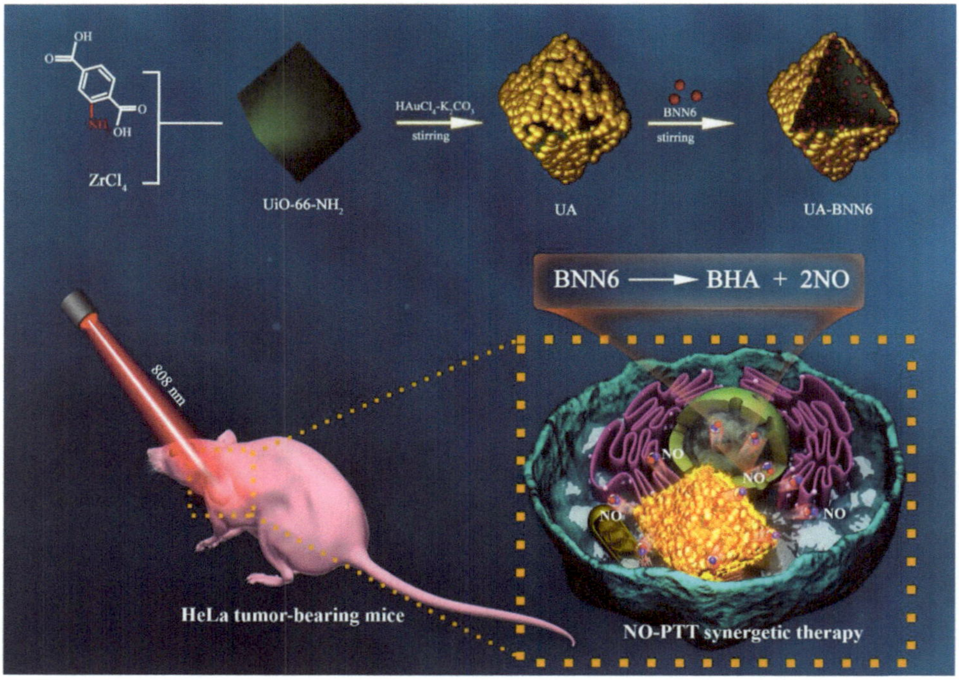

Scheme 1. A schematic illustration of the synthesis of UA-BNN6 and its use as an NIR laser-mediated NO release nanovehicle for tumor therapy.

2. Materials and Methods

2.1. Reagents and Instruments

Thiazolyl blue (MTT) and N,N′-Bis(1-methylpropyl)-1,4-phenylenediamine (BPA) (95%) were purchased from Sigma-Aldrich (St. Louis, MO, USA). Zircomiun tetrachloride (98%), 2-aminoterephthalic acid (98%), N,N-Dimethylformamide(DMF), potassium carbonate, hydrochloric acid (HCl), chloroauric acid ($HAuCl_4$), polyvinyl pyrrolidone (PVP, Kr = 40,000), sodium nitrite (≥99%), and other reagents were of analytical grade and were purchased from Sinoreagent (Shanghai, China). The Hoechst 33342, nitric oxide assay kit, and NO fluorescent probe DAF-FM DA were purchased from Beyotime Biotechnology Co. Ltd. (Nanjing, China). Phosphate buffer saline (PBS) and all other solutions were prepared throughout using ultrapure water (18.2 MΩ cm^{-1}, Millipore, MA, USA).

The scanning electron microscope images, energy dispersive spectrometer (EDS), and EDS mapping images were acquired with a Zeiss Ultra Plus SEM (Zeiss, Oberkochen, Germany). The fluorescence and UV−vis spectra were recorded using a Shimadzu RF-5301PC fluoremeter (Kyoto, Japan) and a Biomate 3S spectrophotometer (Thermo Fisher, MA, USA), respectively. The FT-IR spectrum was performed on a Nicolet iS5 IR spectrometer (Thermo Fisher, Waltham, MA, USA) in the range of 400–4000 cm^{-1}. The MTT assays were carried out using a MK3 microplate reader (Thermo Fisher, Waltham, MA, USA). The confocal microscopic images were acquired with a Nikon Ti-E confocal microscopy (Nikon, Tokyo, Japan). The NIR laser (λ = 808 nm, maximum power of 1 W/cm^2) was purchased from Lei Rui Company, Changchun, China. The infrared mappings were imaged by an FLIR C3 infrared camera (FLIR Systems Inc, Wilsonville, OR, USA).

2.2. Synthesis of UiO-66-NH_2 Nanoparticles

UiO-66-NH_2 nanoparticles were synthesized based on previous reports [29]. Briefly, 40 mg $ZrCl_4$ and 31 mg of 2-aminoterephthalic acid were dispersed in 5 mL DMF and then added to 28 μL deionized water. The final mixture was hydrothermally treated at 120 °C for 24 h. The obtained product was washed with DMF and ethanol, respectively. The final product was dispersed in 10 mL of deionized water.

2.3. Synthesis of UA

The preparation of growth solution was as follows: 50 mg K_2CO_3 was dispersed in 100 mL water, then 1.5 mL of 50 mM $HAuCl_4$ aqueous solution was added and stirred at 900 rpm for 10 min. Finally, the mixture was placed in a dark environment for 12 h.

The synthesis of UA was as follows: 5 mL of the prepared UiO-66-NH_2 mixture was added to 50 mL of the growth solution. Then, 5 mL of 1 mg/mL PVP (MW = 40,000) aqueous solution and 1 mL of formaldehyde were added and stirred at room temperature for 10 min. The reaction solution changed from colorless to dark blue. The final product was washed with dionized water, centrifuged 3 times, and then dispersed in 1 mL of deionized water.

2.4. Synthesis of BNN6

Next, 2.34 mL (10 mmol) of N, N′-Bis(1-methylpropyl)-1,4-phenylenediamine (BPA) was diluted in 18 mL of ethanol, and 20 mL (6 M) of degassed $NaNO_2$ aqueous solution was added dropwise to the above solution with a separating funnel and stirred for 30 min under nitrogen atmosphere. Subsequently, 20 mL (6 M) of HCl was added dropwise through a separating funnel. As the reaction progressed, the color of the reaction solution gradually changed from red to orange, and off-white precipitate appeared. After stirring for 4 h, the precipitate was collected by centrifugation and then centrifuged and washed with 50% ethanol several times. The final product was freeze-dried and stored at −20 °C, away from light.

2.5. Adsorption of BNN6 on UA

Next, 10 mg UA and 10 mg BNN6 were dispersed in 10 mL of DMSO. After stirring at room temperature for 24 h, the obtained solution was washed with DMSO and centrifuged several times to remove free BNN6. The final product was then washed with dionized water and freeze-dried, and the adsorption capacity of BNN6 to the nanoparticles was measured using an ultraviolet-visible spectrometer.

2.6. Cytotoxicity test of UA

The cytotoxicity test of UA was evaluated by MTT. Briefly, 4000 Hela cells per well were inoculated into 96-well plates and cultured in DMEM medium. After incubating for 12 h at 37 °C, the medium was removed, and 100 μL of fresh medium containing different concentrations of samples was added to each well, respectively. After another 24 h of incubation, medium was discarded, and 100 μL serum-free medium containing MTT (0.5 mg/mL) was added. After a further incubation of 4 h at 37 °C, medium was removed, and 150 μL DMSO was added to each well. After shaking on a shaker for 10 min, the absorption was detected at 490 nm using a microplate reader.

2.7. Cellular Uptake Study

The synthesis UA–FIT (15 mg of UA and 5 mg of FITC) was dissolved in 5 mL DMSO. After the mixture was stirred at room temperature for 24 h, the obtained product was centrifuged and washed several times using DMSO and deionized water and then dispersed in deionized water.

Cell uptake of composite nanoparticles: to monitor the uptake of nanomaterials by cells, we used UA–FITC to observe the cells under a laser confocal microscope. First, HeLa cells were inoculated in a confocal dish and cultured with DMEM at 37 °C for 12 h.

Subsequently, 100 μL of UA–FITC in DMEM solution (100 μg/mL) was added, and cells were incubated for another 6 h. The medium was then removed, and cells were washed with PBS (pH 7.4) three times to remove the non-uptaken nanoparticles. Finally, the cell nucleus was labeled with Hoechst 33342 dye, and cells were visualized using a confocal laser scanning microscope (Nikon, Tokyo, Japan) under a 60-fold oil immersion lens.

2.8. Photothermal Properties of UA

The photothermal properties of nanoparticles were evaluated by infrared photothermal imager (FLIR Systems Inc, Wilsonville, OR, USA), and 1 mL of a certain amount of UA dissolved in PBS was placed in a centrifuge tube. The solution was then irradiated with a near-infrared laser (Lei Rui, Changchun, China) (1 W. cm^{-2}) with a wavelength of 808 nm, and the temperature was recorded every 30 s with an infrared thermal imaging camera (FLIR Systems Inc, Wilsonville, OR, USA).

2.9. Photothermal Stability of UA

Using the same above-mentioned procedure, 1 mL of PBS containing 100 μg of UA was placed in a centrifuge tube and irradiated with NIR laser (Lei Rui, Changchun, China) (808 nm, 1 W/cm^2) for 5 min, followed by natural cooling for 15 min. Temperature changes were recorded by infrared thermal imager every 30 s. The above procedure was repeated 5 times.

2.10. NO Release Performance of UA-BNN6

Based on the phenomenon that BNN6 molecules decompose and release NO at high temperatures, the photothermal conversion performance of the gold nanoshells present on the surface of the composite nanoparticles under NIR radiation provides conditions for the release of NO, thus ensuring this NO release process is controlled by NIR switch on/off. The experimental procedures for the evaluation of the NO-releasing performance of our nanomaterials are as follows: the PBS solution and a certain concentration of different samples were prepared and exposed to NIR (808 nm) with a certain energy density for a certain period of time. Subsequently, the NO concentration and fluorescence intensity of the as-prepared solutions in each group were measured using a nitric oxide detection kit (Beyotime Biotechnology, Nanjing, China) according to the manufacturer's instructions. DAF-FM DA was used as a NO fluorescence probe.

2.11. Intracellular NO Fluorescence Detection

The presence of NO in HeLa cells was observed using DAF-FM DA as a NO fluorescent probe. HeLa cells were inoculated into 6-well plates and incubated with DMEM medium at 37 °C for 12 h. The medium was then removed and DMEM medium containing UA-BNN6 or UA (100 μg/mL) was added. After further incubation for 6 h, an appropriate volume of diluted DAF-FM DA was added and incubated for 20min for staining. Finally, the cells were washed with PBS (pH 7.4) three times to remove the non-uptaken DAF-FM DA, and irradiation was continued with NIR for a certain time (0–30 min). Cell fluorescence was observed by confocal laser scanning microscopy (LSCM) under excitation wavelength of 488 nm.

2.12. Determination of Intracellular NO Concentration

The concentration of NO in HeLa cells was determined using a nitric oxide detection kit purchased from Biyuntian. The specific steps were as follows: first, HeLa cells with a concentration of 1×10^5 cells/well were seeded in 6-well plates and incubated at 37 °C for 12 h with DMEM medium. The medium was then removed, and DMEM medium containing UA-BNN6 or UA (100 μg/mL) was added. After further incubation for 6 h, cells were irradiated with NIR for a certain period of time (0–30 min) and then collected. Finally, nitric oxide assay kit was used to determine the concentration of NO in cells as per the manufacturer's instructions.

2.13. In Vitro NO-Photothermal Therapy Synergistic Effect against Cancer Cells

The NO-photothermal synergistic treatment efficiency of composite nanoparticles on tumor cells was evaluated using MTT assay. HeLa cells were seeded in a 96-well plate (100 μL/well) at a density of 1×10^4 cells/well. After incubating at 37 °C for 24 h, the medium was removed. Then, 100 μL of fresh medium containing different samples was added to each well. After another 4 h incubation, some of the wells were irradiated with 808 nm NIR laser (1 $W \cdot cm^{-2}$) for 20 min. After another 24 h of incubation, the medium was removed, 100 μL of serum-free medium containing MTT (0.5 mg/mL) was added, and plates were placed it in a 37 °C incubator for 2 h. The liquid was then discarded, and 150 μL of DMSO was added to each well and placed on a shaker for 10 min, under dark conditions. Finally, the absorbance was measured at 490 nm with a microplate reader.

2.14. Animal Model

BALB/c female nude mice (5–6 weeks old) were purchased from Nanjing Yunqiao Purui Biotechnology Co., Ltd. (Nanjing, china) and were given *ad libitum* access to food and water. All animal experiments were conducted in accordance with the guidelines of the Southeast University Animal Research and Ethics Committee and approved by the National Institute of Biological Sciences and the Southeast University Animal Care Research Advisory Committee. HeLa cells were inoculated subcutaneously into the right side of each nude mouse to obtain HeLa xenograft-bearing mice.

2.15. In Vivo Antitumor Efficacy Study

The in vivo antitumor efficacy of UA-BNN6 was evaluated using HeLa tumor-bearing mice (23–28 g) as animal model. When tumors reached a size of 100 mm^3, nude mice were randomly divided into five groups: PBS (control group), UA, UA-BNN6, UA +NIR group and UA-BNN6+NIR group. PBS (150 uL), UA and UA-BNN6 (2 mg ml^{-1}, 150 uL) groups received an intratumoral injection. NIR laser irradiation (808 nm) was set at 1.0 W cm^{-2} for 15 min. A second treatment was repeated 6 days after the first one. The weight and tumor size of nude mice were recorded every two days. The volume was calculated using the following formula: $V = d^2 \times D/2$ (d is the shortest diameter of the tumor and D is the longest diameter of the tumor). After 14 days of treatment, nude mice were euthanized and tumors were removed and weighed. Finally, the main organs of nude mice were excised and stained with hematoxylin and eosin (H&E) for histological analysis.

3. Results and Discussion

In this study, UA were synthesized by mixing amino group-rich UiO-66-NH_2 nanoparticles with $HAuCl_4$ and K_2CO_3.The deposition of gold seeds on the surface of the composite nanoparticles takes place through electrostatic adsorption with the amino groups part, whereas the gold nanoshells with porous structure is generated under the reduction action of formaldehyde [29]. UiO-66-NH_2 nanoparticles were obtained by the coordination of $ZrCl_4$ with the organic ligand amino terephthalic acid. As shown in Figure 1a, the TEM image revealed the relatively uniform size (particle size of about 60 nm) of the synthesized nanoparticles, with a typical octahedral crystal morphology of UiO-66 nanocrystals. In this study, we used UiO-66-NH_2 nanocrystals as a template to further synthesize UA. TEM images (Figure 1b,c) clearly show that the synthesized nanoparticles were homogenous, with a particle size of about 80 nm, indicating that the thickness of the gold nanoshells of the as-prepared nanoparticles is about 10 nm. The obtained nanoparticles still retained the crystalline morphology of UiO-66-NH_2. In addition, it can be clearly observed that the composite nanomaterials have a hierarchical porous structure, composed of the pore structure in the center of UiO-66-NH_2 and the large pore channel structure of the gold nanoshells on the surface. Furthermore, the EDS spectra (Figure 1d) of the nanoparticles showed obvious peaks of the Au element. EDS mapping image (Figure 1e) of the same sample also confirms that the material contained a high amount of the Au element, as high as 73.46%. The above results fully demonstrate that UA with a hierarchical pore structure have been successfully synthesized.

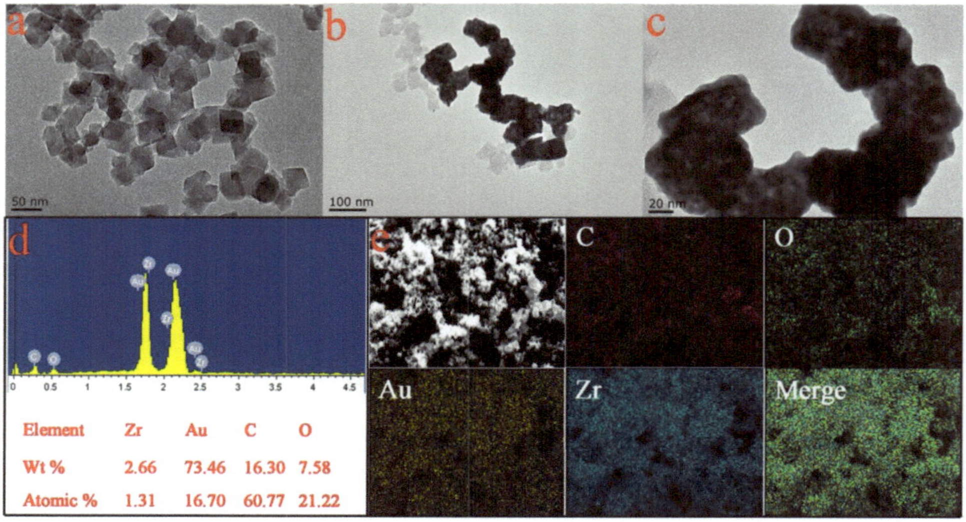

Figure 1. (**a**) TEM images of UiO-66-NH_2. (**b**,**c**) TEM image, (**d**) EDS spectra and (**e**) EDS mapping image of UA.

Benefiting from the advantages of the abundant hierarchical pore structure and the high specific surface area, UA become an ideal carrier with efficient loading capacity. In this study, BNN6 was efficiently loaded into the pores of UA through π-π force and hydrogen bonding. The synthesis of the NO donor BNN6 was obtained by reacting BPA and $NaNO_2$ in an acidic environment based on existing reports, and its synthesis path is shown in Figure S1. Subsequently, UA were placed in a DMSO solution of BNN6 and stirred overnight to obtain UA-BNN6. The infrared spectroscopy of each sample (Figure 2a) shows that UA-BNN6 had an obvious absorption peak at 1377 cm^{-1}, which is attributed to the stretching vibration peak of N-N = O, whereas no obvious absorption peak was seen in the infrared spectrum of UA. These results indicate that BNN6 was successfully adsorbed on UA-BNN6. In addition, it was calculated from the UV-vis spectra (Figure S2) of the supernatant that when the concentration of BNN6 was 1 mg/mL, the loading amount of BNN6 into the nanoparticles was 2.8 mg/5 mg (Figure 2b). Finally, the SEM image (Figure 2c) of UA-BNN6 showed that the crystal morphology of the material did not change after adsorption of BNN6 molecules.

It is well known that gold nanomaterials have excellent photothermal conversion performance due to their surface plasmon resonance properties. In addition, their excellent photothermal stability and good biocompatibility render them a great thermal sensitizer, which has been widely used in the domain of photothermal therapy of cancer treatment. Similarly, UA have excellent photothermal properties due to the presence of gold nanoshells on their surface. According to the UV-vis spectrum of UA (Figure 3b), it can be seen that it has a strong oscillation absorption peak around 808nm, which indicates that our synthesized nanomaterials would have an excellent photothermal conversion performance under laser irradiation in the near-infrared region. Subsequently, we used an infrared thermal imager to evaluate the photothermal conversion performance of UA under 808 nm NIR laser irradiation. As shown in Figure 3c, when a PBS solution of UA (100 μg/mL) is irradiated by 808 nm NIR laser (1.0 W·cm^{-2}), the temperature increases rapidly with the increase of irradiation time. The temperature increased up to 64 °C after 300 s of irradiation. In contrast, the temperature of the free PBS solution in the control group remained at about 25 °C under the same conditions of laser irradiation. In addition, PBS solutions containing different concentrations of UA were irradiated by NIR laser under the same conditions (Figure 3a), and temperature changes were observed by infrared thermal imager. It can be seen that the change in temperature significantly depend on the concentration and irradiation time. The photothermal stability of the material was evaluated by controlling the heating/cooling cycle of the material by switching the laser on and off. The results show (Figure 3d) that the photothermal conversion performance of our synthesized nanomaterials hardly decreases even after the sample has been subjected to the repeated heating and cooling process five times. The above results fully demonstrate that UA have excellent photothermal properties and are promising candidates as photothermal agents.

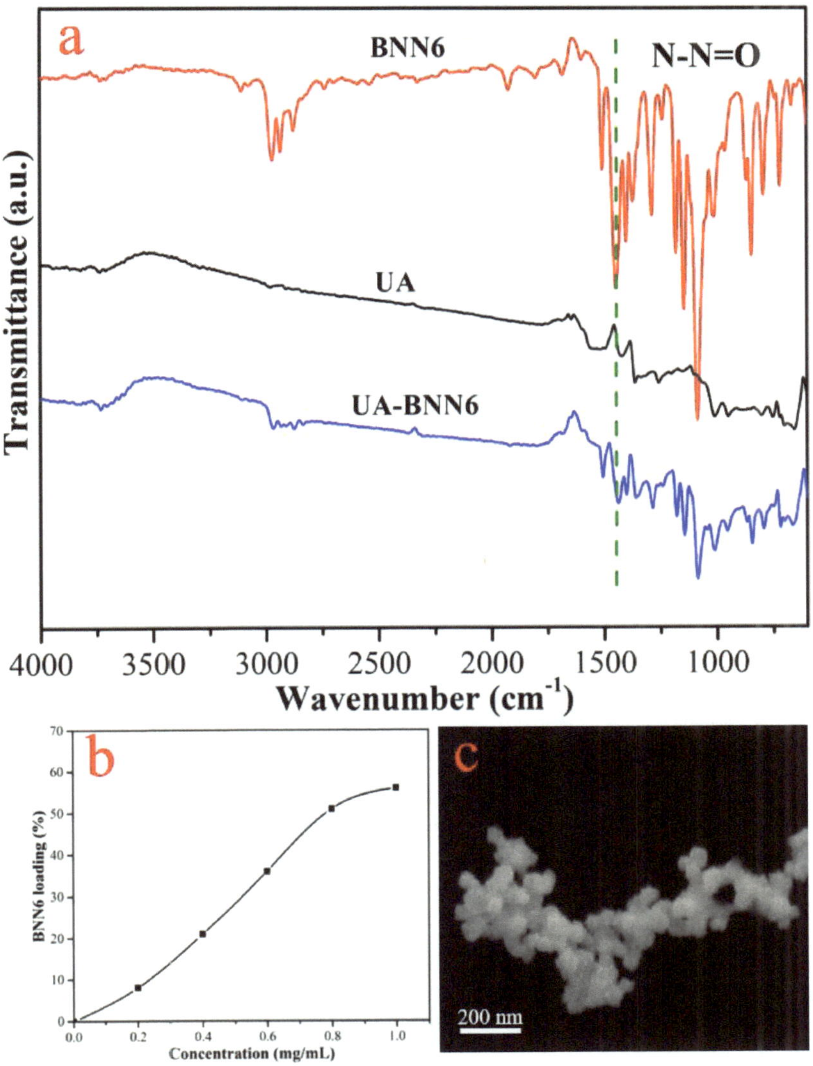

Figure 2. (**a**) The FTIR spectra of BNN6, UA and UA−BNN6. (**b**) The adsorption curve of UA to BNN6. (**c**) The SEM image of UA−BNN6.

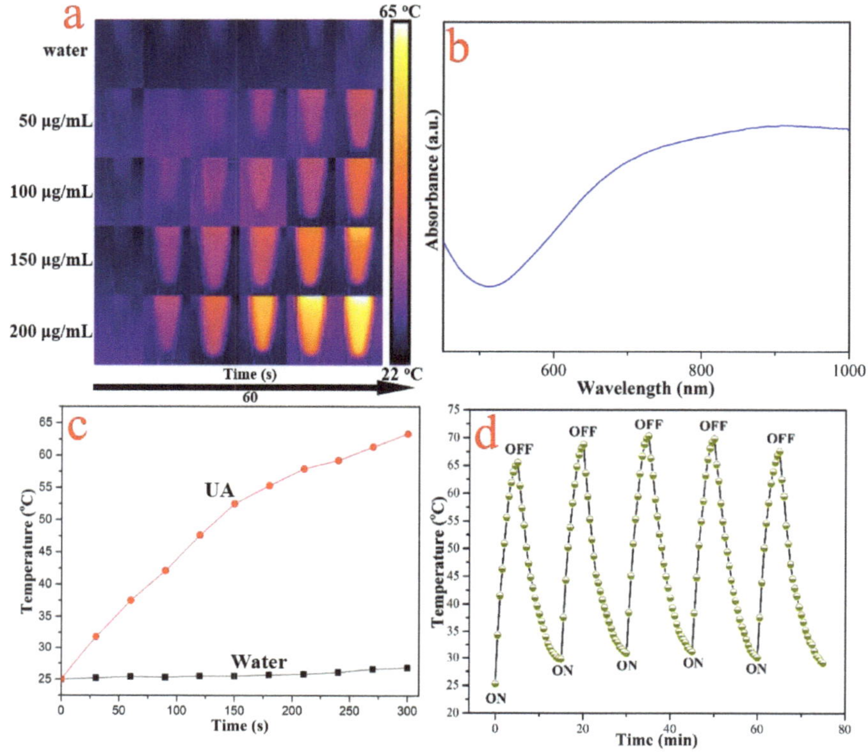

Figure 3. (**a**) Thermal images of PBS and different concentrations of UA under NIR laser irradiation. (**b**) UV-vis spectra of UA. (**c**) Temperature curve of water and of UA under NIR irradiation (100 μg mL^{-1}, 1.0 W cm^{-2}). (**d**) The temperature variation of UA-BNN6 aqueous solution over five on/off cycles upon 808 nm NIR laser irradiation (100 μg mL^{-1}, 1.0 W cm^{-2}).

Studies have reported that BNN6 can stimulate decomposition and release NO under ultraviolet light or a high-temperature environment. In addition, its low cytotoxicity makes it widely used as a NO donor reagent for NO gas therapy. The gold nanoshell layer on the surface of UA-BNN6 has high photothermal conversion performance and generates local high temperature under the irradiation of NIR laser. This provides favorable conditions for the decomposition of BNN6 molecules loaded on the material to release NO. This strategy of releasing NO in response to NIR light provides the possibility for the precise and controllable release of NO. In vitro NO release performance of our nanomaterials was evaluated using a nitric oxide detection kit and NO fluorescence probe DAF-FM DA. Figure 4a shows that when the concentration of composite nanoparticles is 100 μg/mL, the release rate of NO showed an obvious dependence on the NIR laser energy density, with most of NO is released after a 15 min NIR laser irradiation time. As expected, a release of NO can be hardly detected in the control group under the same conditions, which indicates that the NIR laser irradiation does not cause the decomposition of BNN6 molecules. This is consistent with the results of the UV-vis spectra (Figure S2), which showed an obvious absorption oscillation peak in the 365 nm region and a very weak absorption oscillation peak in the near-infrared region. Furthermore, to monitor the release of NO, UA-BNN6 were heated directly at 50 °C or continuously irradiated with an 808 nm near-infrared laser (1.0 W/cm^2). The results showed (Figure 4b) that the release of NO reached 4.62 μM when the NIR laser irradiation lasted for 20 min, while only a very small amount of NO was detected when the sample was directly heated. This indicates that photothermal agents can greatly improve the decomposition efficiency of BNN6 and release NO because the temperature on the material surface is much higher than the average temperature of the medium [35]. In addition, Figure 4c shows that the NO release behavior of UA-BNN6 is controlled by switching the NIR laser on or off. Upon turning on the NIR illumination, NO release from our nanomaterial is initiated. Conversely, when the NIR light source is turned off, the NO release from the nanomaterial stops immediately. Correspondingly, the release of NO was detected by NO fluorescence probe DAF-FM DA. Figure S3a shows that only the fluorescence intensity of the PBS solution of UA-BNN6 was significantly enhanced after NIR laser irradiation, showing obvious concentration Figure S3b and time Figure S3c dependence. The fluorescence intensity of the sample was basically stable after laser irradiation for 20 min, indicating that BNN6 molecule was basically decomposed completely. In the corresponding photographs (Figure S3c), it can be observed that with the increase of NIR laser irradiation time, the color of the mixture gradually changes from white to wine red, which is due to the decomposition of the beige BNN6 molecule into the wine red BHA molecule and NO at high temperatures.

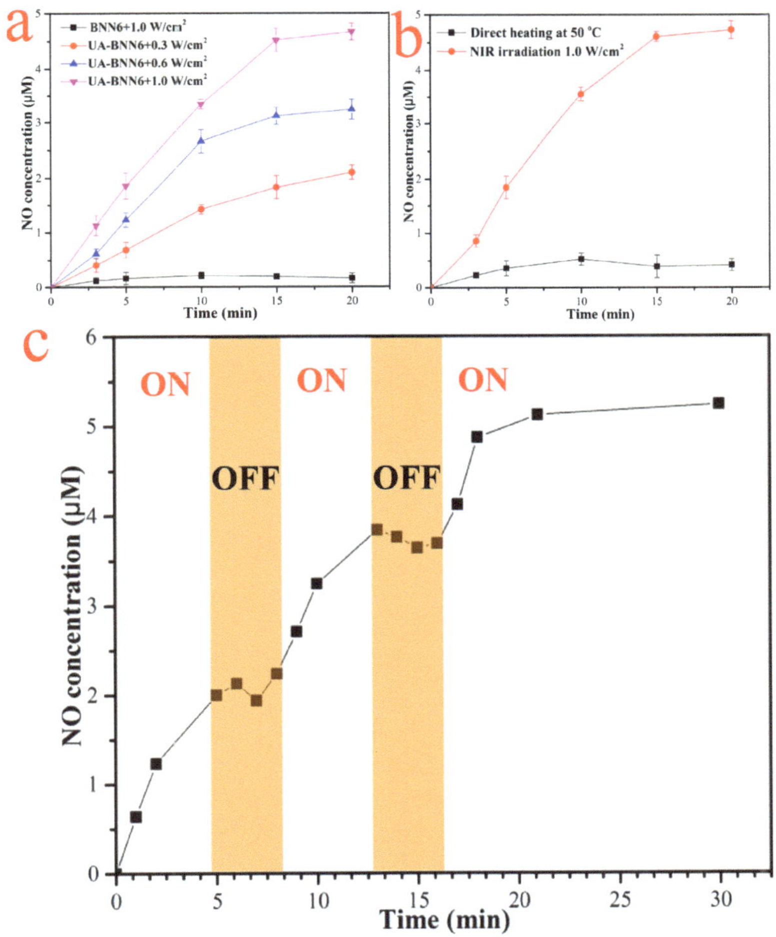

Figure 4. (**a**) The NO release curves of UA-BNN6 under various 808 nm NIR laser power densities. (**b**) The effects of direct heating and 808 nm NIR laser irradiation on NO release from UA-BNN6. (**c**) The on/off generation of NO upon NIR laser irradiation (808 nm, 1.0 W cm^{-2}).

The biocompatibility of UA was evaluated using the MTT assay. The results (Figure 5a) show that even if UA concentration reaches 50 μg/mL, the cell survival rate of HeLa cells is still as high as 87.7%, indicating that our nanomaterial has low cytotoxicity. Subsequently, UA–FITC obtained by replacing BNN6 molecules with FITC molecules was co-incubated with HeLa cells, and the nuclei were labeled with Hoechst 33342 dye. Finally, a laser scanning confocal microscope (LSCM) was used to observe the uptake or our nanomaterial by HeLa cells with a 60× oil-immersion objective. As displayed in Figure 5b, strong green fluorescence appears in the cytoplasm, while only blue fluorescence appears in the nucleus, indicating that the material mainly enters the cytoplasm but not the nucleus after being uptaken by cells. These results fully indicate that nanoparticles have good biocompatibility and cell uptake.

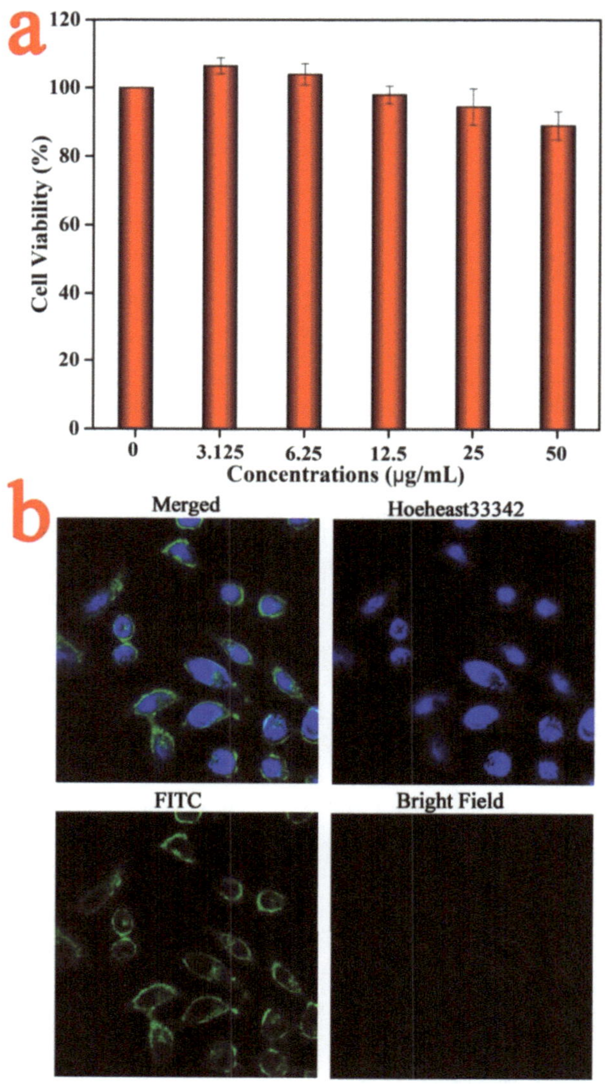

Figure 5. (**a**) The viabilities of HeLa cells after 24 h of incubation with different concentrations of UA. (**b**) Confocal laser scanning microscopic (CLSM) imaging of HeLa cells after co-incubation with UA–FITC (the nucleus was stained with Hoechst 33342).

Similarly, we used HeLa cells to investigate the intracellular NO release performance of UA-BNN6. The NO concentration in the cells was measured using a nitric oxide detection kit, and the cells were observed with the NO fluorescent probe DAF-FM DA. The fluorescence images (Figure S4) of each cell group showed that the cells only showed weak fluorescence after co-incubation with the sample UA-BNN6 or UA. Similarly, the cells incubated with the UA and irradiated with an NIR laser also showed weak fluorescence. In contrast, when the cells were incubated with the sample UA-BNN6 and irradiated with a NIR laser for 20 min, the fluorescence image showed strong fluorescence. Correspondingly, the NO concentration in the cells of each experimental group was determined using a nitric oxide detection kit. The results (Figure 6a) showed that after the cells were co-incubated with UA-BNN6 and irradiated with NIR laser, the concentration of NO was 3.5 times that of the control group. The concentration of NO in the other experimental groups was almost unchanged compared with the control group. Furthermore, the intracellular concentration of NO was determined under different NIR laser irradiation times. The results (Figure 6b) showed that the intracellular concentration of NO basically remained stable after continuous NIR laser irradiation for more than 20 min with (808 nm, 1.0 W/cm^2). This shows that our composite nanoparticles can promptly respond to NIR laser stimulation to release most of the NO. Correspondingly, the cells labeled with the NO fluorescent probe DAF-FM DA (Figure 6c) showed that the fluorescence intensity of the cells was obviously dependent on the duration of NIR laser irradiation, and the fluorescence intensity was basically stable after 20 min. These results further indicate that UA-BNN6 taken up by cells can respond to the stimulation of NIR light and release a large amount of NO in a precise and controllable manner.

Given the fact that BNN6 molecules degrade and release NO at high temperatures, and with the excellent photothermal properties of the gold nanoshells present on the surface of UA-BNN6 upon exposition to NIR laser irradiation, our synthesized nanomaterial can not only decompose and create a high-temperature environment for BNN6 but also lead to a local temperature increase in the cancer cells, finally achieving a synergistic NO-photothermal anti-cancer effect. Herein, using HeLa cells as an in vitro model, an MTT assay was used to evaluate the NO-photothermal synergistic anticancer properties of UA-BNN6. The results (Figure 6d) showed that the tumor cells maintained a high survival rate even when high concentrations of UA or UA-BNN6 were used, indicating the low cytotoxicity of the composite nanoparticles. On the contrary, cells co-incubated with the sample UA or UA-BNN6 and irradiated with a 808 nm NIR laser (1.0 W/cm^2) showed significant cytotoxicity. At the concentration of 50 μg/mL, each group showed different cytotoxicity results. Among them, only tumor cells incubated with UA or UA-BNN6 and irradiated with NIR laser exhibited significant cytotoxicity, where UA-BNN6 showed the most severe cytotoxicity. The former showed strong cytotoxicity, mainly due to the photothermal effect of gold nanoshells, while the latter benefited from the NO-photothermal synergistic antitumor effect of the material. Using composite nanomaterials that can respond to NIR stimulation and quickly initiate the process of NO-photothermal synergistic killing of cancer cells is an efficient and green anti-tumor method.

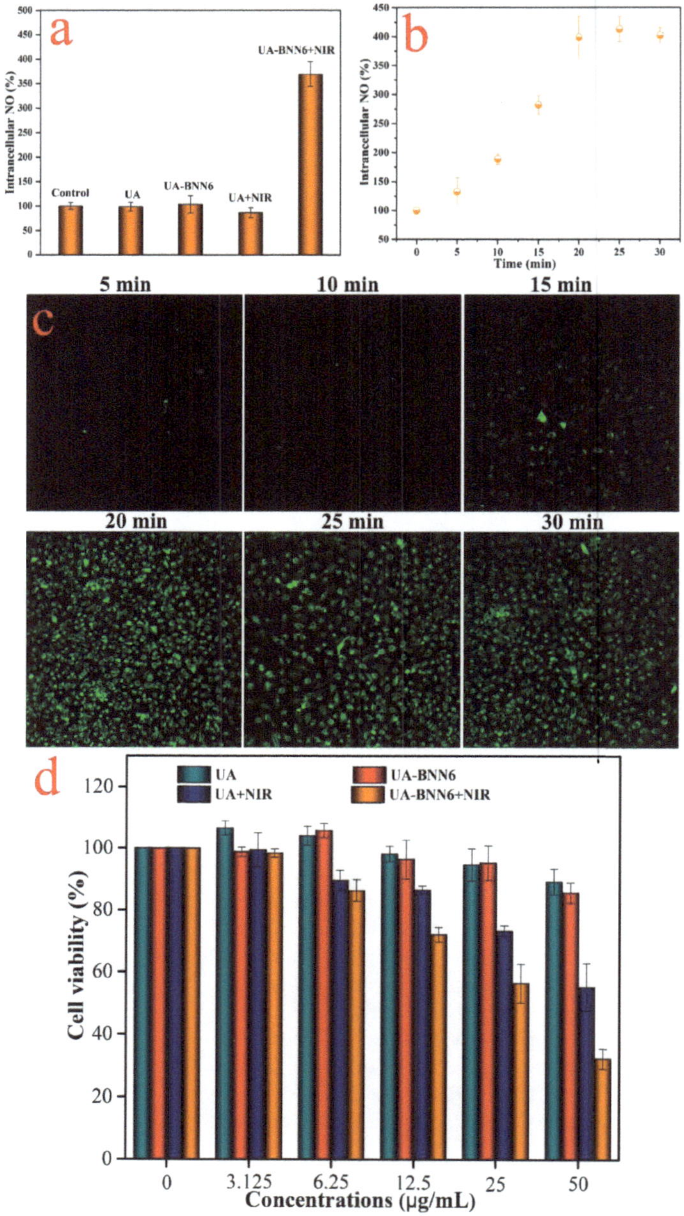

Figure 6. (**a**) Intracellular NO generation in HeLa cells after being treated with UA, UA-BNN6, UA+NIR, and UA-BNN6+NIR. The untreated cells were used as the control. (**b**) Intracellular NO generation in HeLa cells after being treated with UA-BNN6+NIR at different time intervals. (**c**) Fluorescence confocal images of the intracellular generation of NO in HeLa cells after treatment with UA-BNN6+NIR at different time intervals. DAF-FM DA was used as the NO fluorescent probe. (**d**) The viabilities of HeLa cells after being treated with UA, UA-BNN6, UA+NIR, and UA-BNN6+NIR (808 nm, 1.0 W cm^{-2}).

In view of the satisfactory synergistic killing effect of UA-BNN6 composite nanoparticles on cancer cells, HeLa tumor-bearing mice were used as animal models to evaluate the antitumor properties of the composite nanoparticles in vivo. Firstly, nude mice were injected with UA-BNN6 or PBS and irradiated with near-infrared laser (808 nm, 1.0 W/cm^2) for 5 min. To evaluate the photothermal efficiency of the material in nude mice, infrared thermal imaging system was used to observe the changes of surface temperature. As displayed in Figure 7a,b, the temperature at the tumor site of the nude mice injected with UA-BNN6 increased significantly upon laser irradiation, reaching 44.9 °C just after 1 min of irradiation and further rising to 54.3 °C after 5 min. In contrast, nude mice injected with PBS showed only a slight increase in temperature after laser irradiation for 5 min. These results indicate that the UA-BNN6 can be used as a thermosensitive agent for in vivo tumor photothermal therapy. Subsequently, to further verify the therapeutic performance of the composite nanoparticles on tumors, we randomly divided HeLa xenograft-bearing mice (with an initial tumor volume of 100 mm^3) into five groups: PBS group (control group), UA group, UA-BNN6 group, UA+NIR group, and UA-BNN6+NIR group. Nude mice in each group received corresponding treatment every 6 days. The results showed (Figure 7e) that the tumors of nude mice in the PBS, UA, and UA-BNN6 groups increased significantly within the first 14 days, mainly due to the low tumor-growth inhibition effect of the administrated samples. In contrast, tumors in nude mice of the UA+NIR and UA-BNN6+NIR groups were significantly shrunken and practically disappeared. However, it was evident that the wound healing of nude mice treated with UA+NIR was significantly slower than that of the UA-BNN6+NIR group. The former could inhibit tumor growth mainly through photothermal effect upon NIR irradiation. Indeed, the NIR laser irradiation of nude mice in the UA-BNN6+NIR group caused an increase in temperature in the tumor area and promoted the decomposition of the NO donor, BNN6, thus releasing a large amount of NO. Compared with the single PTT treatment, UA-BNN6+NIR achieved a remarkable NO-photothermal synergistic effect, resulting in a more efficient anti-tumor effect. Indeed, various studies have reported that NO can promote wound healing [36]. The tumor volume growth (Figure 7c) recorded in nude mice during the treatment period also showed that the most effective tumor growth inhibitory response was obtained in the UA+NIR and UA-BNN6+NIR groups. After 14 days of treatment, the tumors of the nude mice were removed and weighed, and the results (Figure 7e) showed that the final tumor weights were consistent with the tumor growth behavior in each group. In addition, no significant decrease in body weight of nude mice during different treatments was observed (Figure 7d), indicating the high biocompatibility of the composite nanoparticles. Finally, major organs (heart, liver, spleen, lung, and kidney) of nude mice were excised and stained with hematoxylin and eosin (H&E) for histological analysis. The results showed (Figure S5) that, compared with the PBS group, no obvious damage was seen in all major organs in each group, further confirming the excellent biocompatibility of the nanomaterial. The above results fully demonstrate that UA-BNN6 are highly efficient, biocompatible, and safe nanomaterials for synergistic NO-photothermal therapy of cancer, hence having outstanding application prospects as anti-tumor agents to realize targeted cancer clearance.

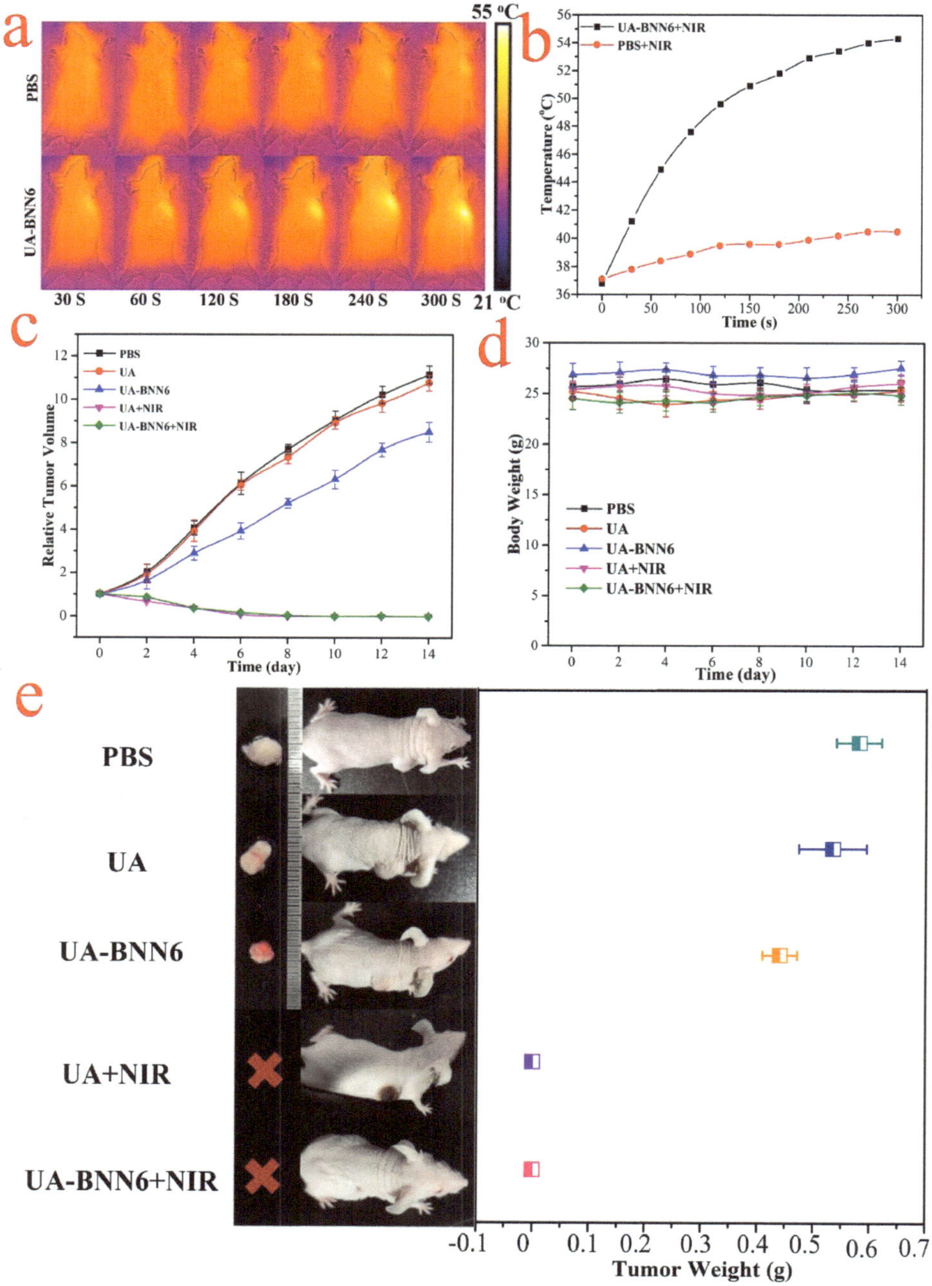

Figure 7. NO-photothermal synergistic tumor therapy in vivo. (**a**) Photothermal imaging of nude mice 6 h after injection of PBS or UA-BNN6 and 808 nm NIR laser treatment and (**b**) temperature change curve of tumor site; (**c**) changes of tumor volume in nude mice during different treatments; (**d**) changes in body weight of nude mice during treatment; (**e**) photographs of nude mice and tumors after different treatments (left), and the weight of removed tumors (right).

4. Conclusions

In summary, in this contribution UA with excellent NIR absorption properties were successfully synthesized by a simple and gentle method. Owing to the rich porous structure of the synthesized composite nanoparticles, UA can efficiently deliver the hydrophobic NO donor BNN6 into cancer cells/tissues. Based on the phenomenon that BNN6 molecules can decompose and release NO at high temperatures, UA-BNN6 that enter cancer cells can generate local high temperatures under NIR irradiation and promote the decomposition of BNN6 molecules to release relatively high concentrations of NO, so as to achieve a NO-photothermal therapy synergistic effect against cancer cells. This NIR-controlled treatment strategy has the advantages of high anti-cancer efficacy, precision, and good biocompatibility, paving a way for a future green and efficient cancer therapy approach for targeted cancer clearance.

Supplementary Materials: The following supporting information can be downloaded at: https://www.mdpi.com/article/10.3390/nano12081348/s1, Figure S1: The synthesis and decomposition route of BNN6, Figure S2: UV-vis absorption spectra of BNN6, Figure S3: Fluorescence curves of PBS solutions of UA-BNN6 after irradiation by NIR (808 nm, 1.0 W cm^{-2}), Figure S4: Confocal images of intracellular generation of NO in HeLa cells after treatment with UA, UA-BNN6, UA+NIR and UA-BNN6+NIR, Figure S5: H&E staining of tissue sections of main organs in nude mice.

Author Contributions: W.L. and X.W. conceived and designed the research. carried out the experiments with help from Z.G., and H.J. and X.W. supervised the whole process. W.L. and F.S. wrote the manuscript, and X.W. revised the manuscript. W.L. and X.W. analyzed the data. All authors have read and agreed to the published version of the manuscript.

Funding: This work is supported by the National Science Foundation of China (82061148012, 82027806, and 21974019), the National High-tech R&D Program and National Key Research and Development Program of China (2017YFA0205301), the Primary Research and Development Plan of Jiangsu Province (BE2019716), and the ISF-NSFC Joint Research Program (grant No.3258/20).

Institutional Review Board Statement: Not applicable.

Informed Consent Statement: Not applicable.

Data Availability Statement: Not applicable.

Conflicts of Interest: The authors declare no conflict of interest.

References

1. Ignarro, L.J.; Buga, G.M.; Wood, K.S.; Chaudhuri, G. Endothelium-derived relaxing factor produced and released from artery and vein is nitric oxide. *Proc. Natl. Acad. Sci. USA* **1987**, *84*, 9265–9269. [CrossRef] [PubMed]
2. Tejero, J.; Shiva, S.; Gladwin, M.T. Sources of vascular nitric oxide and reactive oxygen species and their regulation. *Physiol. Rev.* **2019**, *99*, 311–379. [CrossRef] [PubMed]
3. Paul, S.; Pan, S.; Mukherjee, A.; De, P. Nitric Oxide Releasing Delivery Platforms: Design, Detection, Biomedical Applications, and Future Possibilities. *Mol. Pharm.* **2021**, *18*, 3181–3205. [CrossRef] [PubMed]
4. Carpenter, A.W.; Schoenfisch, M.H. Nitric oxide release: Part II. Therapeutic applications. *Chem. Soc. Rev.* **2012**, *41*, 3742–3752. [CrossRef]
5. Liu, S.; Cai, X.; Xue, W.; Ma, D.; Zhang, W. Chitosan derivatives co-delivering nitric oxide and methicillin for the effective therapy to the methicillin-resistant S. aureus infection. *Carbohydr. Polym.* **2020**, *234*, 115928. [CrossRef]
6. Palmieri, E.M.; Gonzalez-Cotto, M.; Baseler, W.A.; Davies, L.C.; Ghesquière, B.; Maio, N.; Rice, C.M.; Rouault, T.A.; Cassel, T.; Higashi, R.M.; et al. Nitric oxide orchestrates metabolic rewiring in M1 macrophages by targeting aconitase 2 and pyruvate dehydrogenase. *Nat. Commun.* **2020**, *11*, 1–17. [CrossRef]
7. Wang, Y.; Yang, T.; He, Q. Strategies for engineering advanced nanomedicines for gas therapy of cancer. *Natl. Sci. Rev.* **2020**, *7*, 1485–1512. [CrossRef]
8. Qi, M.; Ren, X.; Li, W.; Sun, Y.; Sun, X.; Li, C.; Yu, S.; Xu, L.; Zhou, Y.; Song, S.; et al. NIR responsive nitric oxide nanogenerator for enhanced biofilm eradication and inflammation immunotherapy against periodontal diseases. *Nano Today* **2022**, *43*, 101447. [CrossRef]
9. Sestito, L.F.; Thomas, S.N. Lymph-directed nitric oxide increases immune cell access to lymph-borne nanoscale solutes. *Biomaterials* **2021**, *265*, 120411. [CrossRef]
10. Vannini, F.; Kashfi, K.; Nath, N. The dual role of iNOS in cancer. *Redox Biol.* **2015**, *6*, 334–343. [CrossRef]

11. Ghalei, S.; Mondal, A.; Hopkins, S.; Singha, P.; Devine, R.; Handa, H. Silk Nanoparticles: A natural polymeric platform for nitri oxide delivery in biomedical applications. *ACS Appl. Mater. Int.* **2020**, *12*, 53615–53623. [CrossRef]
12. Vong, L.B.; Nagasaki, Y. Nitric oxide nano-delivery systems for cancer therapeutics: Advances and challenges. *Antioxidant* **2020**, *9*, 791. [CrossRef]
13. Kauser, N.I.; Weisel, M.; Zhong, Y.L.; Lo, M.M.; Ali, A. Calcium dialkylamine diazeniumdiolates: Synthesis, stability, and nitri oxide generation. *J. Org. Chem.* **2020**, *85*, 4807–4812. [CrossRef]
14. Deng, Y.; Jia, F.; Chen, S.; Shen, Z.; Jin, Q.; Fu, G.; Ji, J. Nitric oxide as an all-rounder for enhanced photodynamic therapy: Hypoxi relief, glutathione depletion and reactive nitrogen species generation. *Biomaterials* **2018**, *187*, 55–65. [CrossRef]
15. Mascharak, P.K. Nitric oxide delivery platforms derived from a photoactivatable Mn (II) nitrosyl complex: Entry to photopharma cology. *J. Inorg. Biochem.* **2022**, 111804. [CrossRef]
16. Yang, Y.; Huang, Z.; Li, L. Advanced nitric oxide donors: Chemical structure of NO drugs, NO nanomedicines and biomedica applications. *Nanoscale* **2021**, *13*, 444–459. [CrossRef]
17. Riccio, D.A.; Schoenfisch, M.H. Nitric oxide release: Part I. Macromolecular scaffolds. *Chem. Soc. Rev.* **2012**, *41*, 3731–374 [CrossRef]
18. Li, M.; Li, J.; Chen, J.; Liu, Y.; Cheng, X.; Yang, F.; Gu, N. Platelet membrane biomimetic magnetic nanocarriers for targete delivery and in situ generation of nitric oxide in early ischemic stroke. *ACS Nano* **2020**, *14*, 2024–2035. [CrossRef]
19. Dong, X.; Liu, H.; Feng, H.; Yang, S.; Liu, X.; Lai, X.; Lu, Q.; Lovell, J.F.; Chen, H.; Fang, C. Enhanced drug delivery by nanoscal integration of a nitric oxide donor to induce tumor collagen depletion. *Nano Lett.* **2019**, *19*, 997–1008. [CrossRef]
20. Soren, O.; Rineh, A.; Silva, D.G.; Cai, Y.; Howlin, R.P.; Allan, R.N.; Feelisch, M.; Davies, J.C.; Connett, G.J.; Faust, S.N.; et a Cephalosporin nitric oxide-donor prodrug DEA-C3D disperses biofilms formed by clinical cystic fibrosis isolates of Pseudomona aeruginosa. *J. Antimicrob. Chemother.* **2020**, *75*, 117–125. [CrossRef]
21. Duan, Y.; Wang, Y.; Li, X.; Zhang, G.; Zhang, G.; Hu, J. Light-triggered nitric oxide (NO) release from photoresponsiv polymersomes for corneal wound healing. *Chem. Sci.* **2020**, *11*, 186–194. [CrossRef] [PubMed]
22. Zhou, X.; Meng, Z.; She, J.; Zhang, Y.; Yi, X.; Zhou, H.; Zhong, J.; Dong, Z.; Han, X.; Chen, M.; et al. Near-infrared light-responsiv nitric oxide delivery platform for enhanced radioimmunotherapy. *Nano Micro Lett.* **2020**, *12*, 1–14. [CrossRef] [PubMed]
23. Hu, D.; Deng, Y.; Jia, F.; Ji, J. Surface charge switchable supramolecular nanocarriers for nitric oxide synergistic photodynami eradication of biofilms. *ACS Nano* **2019**, *14*, 347–359. [CrossRef] [PubMed]
24. You, C.; Li, Y.; Dong, Y.; Ning, L.; Zhang, Y.; Yao, L.; Wang, F. Low-temperature trigger nitric oxide nanogenerators for enhance mild photothermal therapy. *ACS Biomater. Sci. Eng.* **2020**, *6*, 1535–1542. [CrossRef]
25. Tang, Y.; Wang, G. NIR light-responsive nanocarriers for controlled release. *J. Photochem. Photobiol. C* **2021**, *47*, 100420. [CrossRef
26. Zhang, X.; Wang, S.; Cheng, G.; Yu, P.; Chang, J. Light-responsive nanomaterials for cancer therapy. *Engineering* **2021**, in press [CrossRef]
27. Lai, X.; Jiang, H.; Wang, X. Biodegradable metal organic frameworks for multimodal imaging and targeting theranostics. *Biosensor* **2021**, *11*, 299. [CrossRef]
28. Yang, J.; Yang, Y. Metal-organic frameworks for biomedical applications. *Small* **2020**, *16*, 1906846. [CrossRef]
29. Liu, C.; Luo, L.; Zeng, L.; Xing, J.; Xia, Y.; Sun, S.; Zhang, L.; Yu, Z.; Yao, J.; Yu, Z.; et al. Porous Gold Nanoshells on Functiona NH_2-MOFs: Facile Synthesis and Designable Platforms for Cancer Multiple Therapy. *Small* **2018**, *14*, 1801851. [CrossRef]
30. Song, J.; Yang, X.; Yang, Z.; Lin, L.; Liu, Y.; Zhou, Z.; Shen, Z.; Yu, G.; Dai, Y.; Jacobson, O.; et al. Rational design of branche nanoporous gold nanoshells with enhanced physico-optical properties for optical imaging and cancer therapy. *ACS Nano* **2017** *11*, 6102–6113. [CrossRef]
31. Hu, Y.; Chi, C.; Wang, S.; Wang, L.; Liang, P.; Liu, F.; Shang, W.; Wang, W.; Zhang, F.; Li, S.; et al. A comparative study of clinica intervention and interventional photothermal therapy for pancreatic cancer. *Adv. Mater.* **2017**, *29*, 1700448. [CrossRef]
32. Liu, H.; Chen, D.; Li, L.; Liu, T.; Tan, L.; Wu, X.; Tang, F. Multifunctional gold nanoshells on silica nanorattles: A platform for the combination of photothermal therapy and chemotherapy with low systemic toxicity. *Angew. Chem. Int. Ed.* **2011**, *50*, 891–895 [CrossRef]
33. Liu, L.; Pan, X.; Liu, S.; Hu, Y.; Ma, D. Near-infrared light-triggered nitric oxide release combined with low-temperature photothermal therapy for synergetic antibacterial and antifungal. *Smart Mater. Med.* **2021**, *2*, 302–313. [CrossRef]
34. Huang, S.; Liu, H.; Liao, K.; Hu, Q.; Guo, R.; Deng, K. Functionalized GO nanovehicles with nitric oxide release and phototherma activity-based hydrogels for bacteria-infected wound healing. *ACS Appl. Mater. Int.* **2020**, *12*, 28952–28964. [CrossRef]
35. Nguyen, S.C.; Zhang, Q.; Manthiram, K.; Ye, X.; Lomont, J.P.; Harris, C.B.; Weller, H.; Alivisatos, A.P. Study of Heat Transfe Dynamics from Gold Nanorods to the Environment via Time-Resolved Infrared Spectroscopy. *ACS Nano* **2016**, *10*, 2144–2151 [CrossRef]
36. Shende, P.; Sahu, P. Synergistic effect of nitric oxide for wound healing using etherification of cotton. *J. Pharm. Innov.* **2021**, *98*, 1–7 [CrossRef]

nanomaterials

MDPI

Article

Highly Stable, Graphene-Wrapped, Petal-like, Gap-Enhanced Raman Tags

Ming Chen [1], Bin Wang [2], Jingfan Wang [1], Hongliang Liu [1], Zhixiang Chen [1], Xiaoxuan Xu [2,*] and Xing Zhao [1,*]

1 Tianjin Key Laboratory of Micro-Scale Optical Information Science and Technology, Institute of Modern Optics, Nankai University, Tianjin 300350, China; dreamchenming@163.com (M.C.); wjf9816@163.com (J.W.); drliuhl@nankai.edu.cn (H.L.); dtchzx333@163.com (Z.C.)

2 College of Artificial Intelligence, Nankai University, Tianjin 300350, China; wb@nankai.edu.cn

* Correspondence: xuxx@nankai.edu.cn (X.X.); zhaoxingtjnk@nankai.edu.cn (X.Z.)

Abstract: Gap-enhanced Raman tags (GERTs) were widely used in cell or biological tissue imaging due to their narrow spectral linewidth, weak photobleaching effect, and low biological matrix interference. Here, we reported a new kind of graphene-wrapped, petal-like, gap-enhanced Raman tags (GP-GERTs). The 4-Nitrobenzenethiol (4-NBT) Raman reporters were embedded in the petal-like nanogap, and graphene was wrapped on the surface of the petal-like, gap-enhanced Raman tags. Finite-difference time-domain (FDTD) simulations and Raman experimental studies jointly reveal the Raman enhancement mechanism of graphene. The *SERS* enhancement of GP-GERTs is jointly determined by the petal-like "interstitial hotspots" and electron transfer between graphene and 4-NBT molecules, and the total Raman enhancement factor (*EF*) can reach 10^{10}. Mesoporous silica was grown on the surface of GP-GERTs by tetraethyl orthosilicate hydrolysis to obtain Raman tags of MS-GP-GERTs. Raman tag stability experiments showed that: MS-GP-GERTs not only can maintain the signal stability in aqueous solutions of different pH values (from 3 to 12) and simulated the physiological environment (up to 72 h), but it can also stably enhance the signal of different Raman molecules. These highly stable, high-signal-intensity nanotags show great potential for *SERS*-based bioimaging and multicolor imaging.

Keywords: gap-enhanced Raman tags; graphene; Raman-enhanced mechanism; stability

1. Introduction

Surface-enhanced Raman scattering (*SERS*) is an ultrasensitive vibrational spectroscopy technique that is widely used in various fields, such as chemistry, physics, biology, and medicine [1–5]. The nanoprobes designed based on *SERS* technology are called "*SERS* tags", which are usually composed of metal nanoparticles and Raman reporter molecules [6]. The strong characteristic Raman signal can be generated by "*SERS* tags"; thus, it has a similar optical-labeling function with fluorescent dyes and quantum dots, showing great potential in the field of biological analysis and imaging [7–9]. However, this simple metal nanoparticle-Raman reporter molecule structure lacks stability, and the Raman signal is easily disturbed by the environment. Usually, a layer of material needs to be wrapped around the structure to protect the Raman signal molecule. When necessary, specific target molecules need to be modified outside the protective layer to obtain *SERS* probes with various biological functions.

Gap-enhanced Raman tags (GERTs) are an emerging class of *SERS* tags. In GERTs, Raman reporter molecules are embedded between metal core shells, which reduces the influence of the external environment and nanoparticle aggregation on Raman signals [10–13]. Recently, Zhang et al. [14]. proposed a petal-like, gap-enhanced Raman tag (P-GERTs), which enables single-particle detection due to the strong electromagnetic field hot spots generated by both the inner gap and the outer petaloid structure. However, the Raman reporters adsorbed on the surface of the outer petal structure of P-GERTs are still affected by the external environment, resulting in unstable *SERS* signals and poor controllability.

Graphene is one of the most widely studied and mature two-dimensional (2D) atomic materials [15–17]. It has the advantages of single atomic layer thickness, unique phononic structure, high-electron mobility, chemical inertness, and biocompatibility [18–20]. Moreover, graphene can also be used as a shell-separating nanoparticle extension material for the fabrication of ultrathin shells at the atomic layer scale, while also providing additional chemical enhancement [21]. Numerous studies have shown that the combination of graphene and metal nanoparticles can enhance the *SERS* performance of substrates [22–24]. Qiu et al. [25]. used the electrostatic interaction between graphene oxide and gold nanorods to synthesize graphene oxide-wrapped gold nanorods and reduce the biological toxicity of gold nanorods. Zhang's research group [26,27] combined graphene with silver nanoparticles to obtain stable and highly reproducible *SERS* substrates. Li et al. [28]. reported the enhanced Raman spectra of graphene-wrapped gold nanoparticles, which exhibited good pH stability and high-temperature stability.

In this work, we developed new graphene-wrapped, petal-like, gap-enhanced Raman tags (GP-GERTs for brevity). The 4-NBT Raman reporters were embedded between the petal-like nano-gaps, and graphene was wrapped on the surface of the petal-like gold nanoparticles by electrostatic interaction. We explained the Raman enhancement mechanism of GP-GERTs through FDTD simulations and Raman experimental studies. The "interstitial hot spots" between petals and the "charge transfer" between graphene and 4-NBT jointly enhanced the *SERS* signal, and the total Raman *EF* can reach 1.75×10^{10}. Mesoporous silica was grown on the surface of GP-GERTs by hydrolysis of tetraethyl orthosilicate to obtain MS-GP-GERTs. Due to the special structure of Raman tags and the chemical inertness of graphene, MS-GP-GERTs showed excellent stability in serum environment and aqueous solutions with different pH values. The Raman signal of MS-GP-GERTs remained stable, whether immersed in an aqueous solution with a pH ranging from 3 to 12 or immersed in a simulated physiological environment for a long time (up to 72 h). In addition, the MS-GP-GERTs structure can also achieve stable *SERS* enhancement for a variety of different Raman reporters. This highly stable, high-signal intensity Raman tag has great potential for applications in different types of biomedical imaging and multicolor Raman imaging.

2. Materials and Methods

2.1. Materials

Chloroauric acid ($HAuCl_4 \cdot 4H_2O$)) and cetyltrimethylammonium chloride (CTAC, 99%) were purchased from Shanghai Meryer Chemical Reagent Co., Ltd (Meryer, Shanghai, China). The 4-Nitrobenzenethiol (4-NBT, 95%), 1,4-benzenedithiol (1,4-BDT, 98%), 4-mercaptobenzonitrile (4-MBN, 95%), Biphenyl-4,4′-dithiol (B-4,4′-D, 98%), and 2-naphthalene thiophenol (2-NT, 98%) were obtained from Shanghai Macklin Biochemical Technology Co., Ltd (Macklin, Shanghai, China). Ascorbic acid (AR, 99%), sodium hydroxide solid powder (NaOH, AR, 96%), hydrochloric acid (ACS, 37%), sodium chloride (AR, 99.5%), glucose monohydrate (98%), and ethyl orthosilicate (TEOS) were purchased from Shanghai Aladdin Biochemical Technology Co., Ltd (Aladdin, Shanghai, China). Bovine serum albumin (98%) was purchased from Hefei Qiansheng Biotechnology Company (Qiansheng, Hefei, China). Mechanically exfoliated single-layer graphene (99%) was purchased from Shenzhen Suiheng Graphene Technology Co., Ltd (Suiheng, Shenzhen, China). Anhydrous ethanol and isopropanol were purchased from Tianjin Huaxun Pharmaceutical Technology Co., Ltd. (Huaxun, Tianjin, China). Deionized water (18.25 MΩ) was used for all experiments.

2.2. Synthesis of GP-GERTs and MS-GP-GERTs

Gold nanocores were synthesized by referring to the work of Zhang et al. [14]. The synthesis process of GP-GERTs and MS-GP-GERTs are shown in Figure 1.

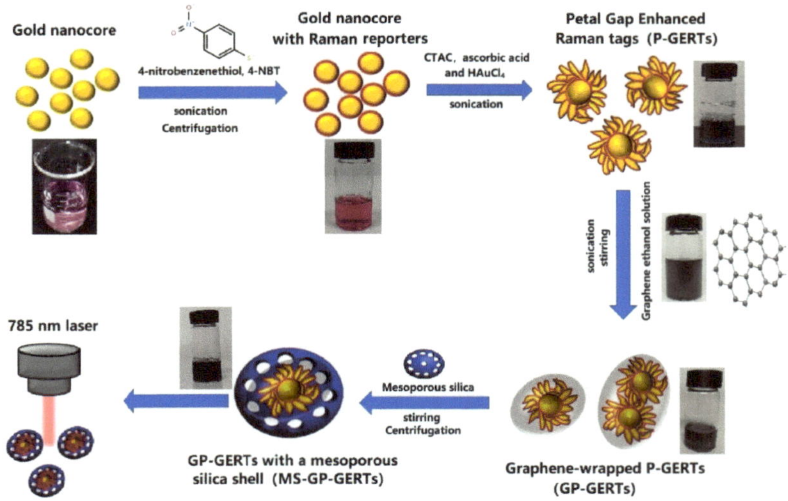

Figure 1. Schematic illustration of the synthesis process of the GP-GERTs and MS-GP-GERTs and *SERS* measurement by Raman system.

Firstly, Raman reporter molecules were modified on the surface of gold nanocores. The 4-NBT ethanol solution (500 µL, 10 mM) was added to 10 mL of gold nucleus solution (1 nM) and sonicated for 30 min; the obtained sol particles were washed with CTAC solution (50 mM) by centrifugation and then re-dispersed in CTAC (50 mM) solution for later use. In this process, the ethanol solution of 4-NBT was replaced by the ethanol solution of different Raman molecules, and the gold nanocores modified with different Raman molecules can be obtained.

Secondly, gold nanopetals were grown on the surface of gold nanocores modified with Raman reporters. A total of 0.5 mL of 4-NBT molecule-modified gold nanonucleus solution was added to a mixed solution of CTAC (8 mL, 0.05 M), ascorbic acid (250 µL, 0.04 M) and chloroauric acid (500 µL, 5 mM). The solution changed from colorless to pink, purple, and blue, and it continued to be sonicated for 30 min. P-GERTs with 4-NBT Raman reports were obtained.

Thirdly, petal-gap enhanced Raman tags was wrapped by graphene. A total of 4 mL of the prepared P-GERTs were mixed with 2 mL of graphene ethanol solution (20 mg/L), and NaOH (0.05 M) was added to adjust the pH of the solution to 12.5. The mixture was stirred at 40 °C for 8 h; graphene could be bonded on the surface of P-GERTs via electrostatic interaction. After the dispersion was cooled to room temperature, it was centrifuged at 14,000 rcf for 15 min to remove the supernatant and was re-dispersed in CTAC to obtain GP-GERTs.

Finally, mesoporous silica shells were grown on the surface of GP-GERTs through TEOS hydrolysis. An amount of 1.5 mL of deionized water was added to 2 mL of GP-GERTs, and sodium hydroxide solution (0.05 M) was added dropwise to adjust the pH to 12.5. Next, 50 µL of 5% TEOS in isopropanol was added slowly to the solution under stirring. This procedure was repeated three times at 30-min intervals, and then the mixture was reacted for 24 h at 40 °C. The resulting solution was then centrifuged at 14,000 rcf for 15 min, the supernatant was removed, and the particles were redispersed in 2 mL of ethanol to obtain MS-GP-GERTs.

2.3. Characterization of GERTs

Scanning electron microscope (SEM) images were acquired on a Merlin Compact (Zeiss, Oberkochen, Germany) with an accelerating voltage of 10 kV. Transmission electron microscopy (TEM) images were acquired on a FEI Talos F200c (ThermoFisher, Waltham, MA, USA) at 120 kV. UV-Vis spectra were acquired using a Lambda 45 UV-Vis spectrophotometer (PerkinElmer, Waltham, MA, USA). Fourier transform infrared spectra (FTIR) were acquired using a VERTEX V70 Fourier infrared spectrometer (Bruker, Karlsruhe, Germany). Raman spectra were recorded with a laser confocal Raman spectrometer (XperRam200, Nanobase, Seoul, Korea) equipped with an Olympus upright microscope with a 40× objective, numerical aperture (NA) of 0.65, and an Andor scientific grade TE-cooled CCD. The Raman excitation wavelength was set to 785 nm, the power was 60 mW, the integration time was set to 1 s, and the spectral resolution was 2.5 cm^{-1}. All Raman-testing experiments were performed on silicon substrates. The baseline of spectral data was removed by the LabSpec software (Horiba, Paris, France).

2.4. Stability Experiments of MS-GP-GERTs

For the experiment of PH stability, 2 mL of the prepared MS-GP-GERTs solution was centrifuged and dispersed in aqueous solutions of different pH values (pH values ranged from 3 to 12). After soaking for 30 min, it was centrifuged again and redispersed in 2 mL of ethanol. A small amount of the sample was drawn dropwise on the silicon substrate using a pipette, and the MS-GP-GERTs sample to be tested was obtained after natural drying.

Using 10% bovine serum albumin, 10% glucose solution, and 0.9% normal saline to simulate a physiological environment, the time stability of Raman tags in a physiological environment was tested. The specific test method is as follows: 2 mL of the prepared MS-GP-GERTs solution was centrifuged and dispersed in 2 mL of bovine serum albumin solution (10%). The mixing time of MS-GP-GERTs and 10% bovine serum albumin solution was controlled as 2 h, 12 h, 24 h, 36 h, and 72 h, respectively. It was then washed twice by centrifugation with absolute ethanol and redispersed in 2 mL of ethanol. A small amount of sample was drawn dropwise on the silicon substrate using a pipette, and the MS-GP-GERTs sample to be tested was obtained after natural drying. Two additional tests can be performed by replacing the 10% bovine serum albumin solution with 10% glucose solution or 0.9% normal saline.

3. Results

3.1. UV-Vis Absorption Spectroscopy

According to the synthesis process of Raman tags, we tested three sets of UV-Vis absorption spectroscopy for comparison. The red and black curves of Figure 2a represent the absorption spectra of gold nanocores and gold nanocores modified 4-NBT molecules, respectively. As shown in Figure 2a, the absorption peak of gold nanocores is around 524 nm, which is consistent with the absorption peak of 20 nm gold particles [29]. After the adsorption of 4-NBT molecules on the surface of the gold nanocores, the absorption peak has a red shift of 3 nm. The blue and green curves in Figure 2a represent the absorption spectra of P-GERTs and (MS)P-GERTs, respectively. Different from the absorption peak of gold nanocores, the absorption peak of P-GERTs is around 638 nm, which can be explained by the larger diameter of Raman tags and the generation of a large number of electromagnetic hot spots. In addition, compared with P-GERTs, the absorption peak of (MS)P-GERTs has a 5 nm red shift, which indicates that mesoporous silica can grow on the surface of P-GERTs [30].

Figure 2b is the UV-Vis absorption spectroscopy of graphene, GP-GERTs, and MS-GP-GERTs. As shown in Figure 2b, in the range of 300–900 nm, the absorption curve of graphene shows a decreasing trend, which is the same as that reported in the previous literature [31]. In addition, after the Raman tags were combined with graphene, the absorption peak of the Raman tags in the visible light range disappeared, which was mainly affected by graphene. It was also confirmed that the Raman tags were encapsulated by graphene, and GP-GERTs and MS-GP-GERTs were successfully synthesized.

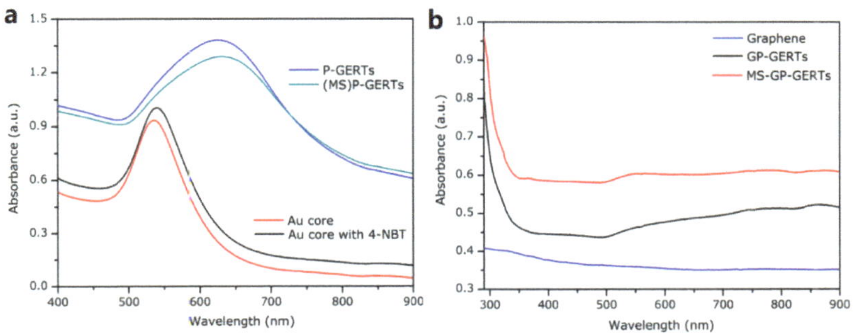

Figure 2. UV-Vis Absorption Spectroscopy: (**a**) absorption spectroscopy of gold core, gold core modified Raman molecule, P-GERTs, and (MS) P-GERTs; (**b**) UV-Vis absorption spectroscopy of graphene, GP-GERTs and MS-GP-GERTs.

3.2. Fourier Transform Infrared Spectroscopy (FTIR)

Figure 3 shows the FTIR spectra of graphene, GP-GERTs, and MS-GP-GERTs. As shown in Figure 3, graphene has no obvious characteristic peaks in the infrared region, which is because the single-layer graphene we use is prepared by mechanical exfoliation. During the measurement of FTIR spectra, graphene is compressed into graphite flakes. Therefore, none of the peaks were measured. GP-GERTs also showed the same trend, with no obvious peaks in the infrared region, which also indicated that graphene was wrapped on the surface of petal-like gold nanostructures. The FTIR spectra of MS-GP-GERTs showed characteristic peaks of mesoporous silica. The peak around 790 cm^{-1} is derived from the stretching vibration of Si-O bond, and the peak around 1075 cm^{-1} is derived from the Si-O-Si group [32]. Furthermore, since our mesoporous silica was obtained by hydrolysis of TEOS (isopropanol as solvent), and the composite structure was not calcined when measuring the FTIR spectrum. Therefore, the vibration peaks of C-H bond appeared at 2850 cm^{-1} and 2920 cm^{-1}, and the vibration peak of the Si-OH bond appeared near 3014 cm^{-1} [33].

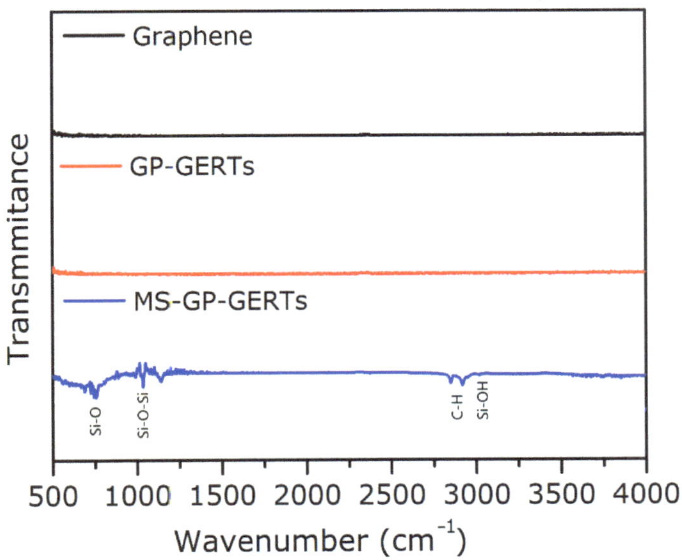

Figure 3. FTIR spectra of graphene, GP-GERTs, and MS-GP-GERTs.

3.3. SEM and TEM

Figure 4 shows the SEM images of P-GERTs. Figure 4a is a 100 KX magnification of P-GERTs, and the scale bar is 100 nm. The shape of P-GERTs nanoparticles is similar to that of petals, and the agglomeration between particles is not obvious, and they are relatively independent of each other. Figure 4b is the P-GERTs magnified by 200 KX, and the scale bar is 20 nm. According to Figure 4b, the morphology of P-GERTs nanoparticles can be better observed, and the diameter of the particles can be estimated. The SEM images confirmed that the P-GERTs had a petal-like shape with particle diameters ranging from about 60 nm to 80 nm.

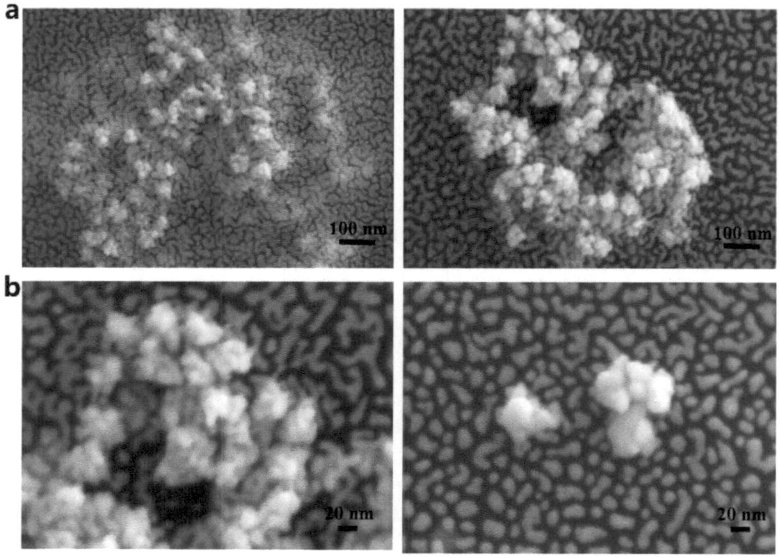

Figure 4. SEM images of P-GERTs: (**a**) 100 KX magnification; (**b**) 200 KX magnification.

Figure 5a shows the TEM image of P-GERTs, and the scale bar is 20 nm. The microstructure of P-GERTs can be more clearly observed from Figure 5a, which consists of an inner gold core, middle nano gaps (marked by a red arrow in Figure 5a), and outer nano petals. The total diameter of P-GERTs is about 70 nm, which is mutually confirmed with the previous SEM images. Figure 5b,c are TEM images of monolayer graphene and GP-GERTs, respectively, and the scale bar is 2 μm. The complete morphology of the monolayer graphene can be seen from Figure 5b, and combined with Figure 5c, it can be clearly seen that most of the P-GERTs nanoparticles are covered by the mesh-like monolayer graphene. Figure 5d shows a TEM image of (MS)P-GERTs with a scale bar of 200 nm, and the upper right inset shows a further magnified single (MS)P-GERTs particle with a scale bar of 20 nm. As shown in Figure 5d, the mesoporous silica layer wraps around the P-GERTs with a thickness of about 15–25 nm, which also shows a clear mesoporous structure in the enlarged inset. Figure 5e shows the TEM picture of MS-GP-GERTs, and the scale bar is 200 nm. From Figure 5e, it can be seen that most of the MS-GP-GERTs nanoparticles are dispersed independently in the solution, and the individual nanoparticles are marked by red arrows, and the diameter of the particles is about 80–95 nm. It can be observed that a mesoporous silica layer grows on the surface of P-GERTs while being covered by excess graphene in solution. It should be pointed out that there is also a very thin graphene layer (about 0.5–1 nm) between the mesoporous silica layer and the P-GERTs particle. This is because the excess monolayer graphene was well mixed with the P-GERTs before the mesoporous silica cladding is grown. Unfortunately, such thin graphene is difficult to observe together with mesoporous silica and P-GERTs in TEM characterization. However, in our later Raman experiments and calculations of the enhancement mechanism, the graphene interlayer was proven to exist.

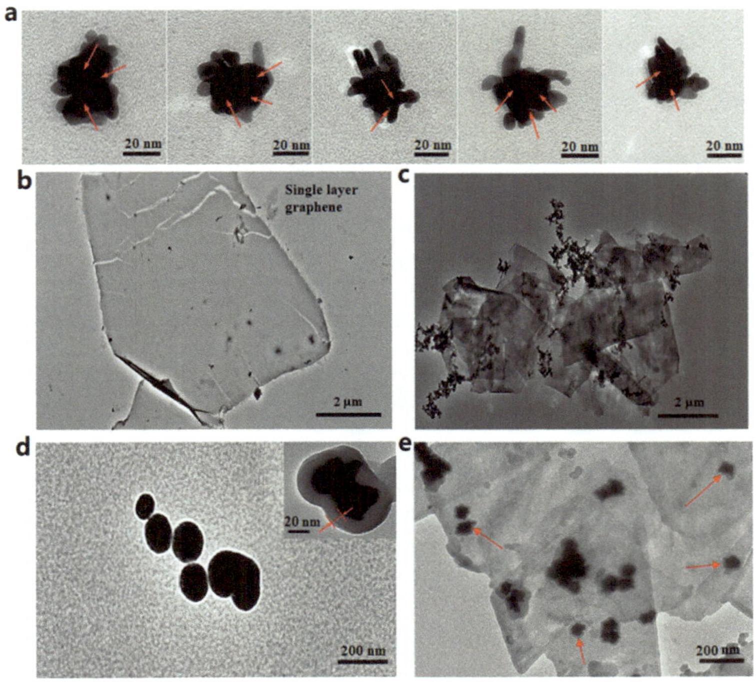

Figure 5. (**a**) TEM image of P-GERTs; (**b**) TEM image of monolayer graphene; (**c**) TEM image of GP-GERTs; (**d**) TEM image of (MS) P-GERTs; (**e**) TEM image of MS-GP-GERTs.

3.4. Simulation

FDTD solutions (Lumerical, Vancouver, BC, Canada) simulation software was used to simulate the spatial distribution of the electromagnetic field strength of the Raman tags. Figure 6a shows a simulation model for simulating the electric field effect of a Raman tag. As shown in Figure 6a, the laser source is total-field scattered-field, and the laser is polarized along the X direction and propagates in the -Z direction. The real structures of P-GERTs are very complex with random petal-like structures, and in the FDTD simulations, we used small gold nanospheres to simulate the petal structures. Figure 6c shows the 3D model of the P-GERTs we used, which consists of an inner gold core (radius 15 nm), a molecular layer of 4-NBT in the middle (thickness 1 nm), and outer petals (small gold spheres with a radius of 9 nm). A total of 26 small golden spheres were used to simulate the petals of P-GERTs. A total of 18 of them were evenly distributed on three circles perpendicular to the X, Y, and Z axes, and the distance between the center of the small gold sphere and the center of the inner gold core is 25 nm. The remaining eight were distributed in the center of the remaining blank area, and the distance between the centers of the spheres is also 25 nm. The total particle size of the P-GERTs is about 68 nm, and the specific size is given in Figure 6b. As shown in Figure 6d–f, the frequency domain filed and power monitor are placed in the XY plane, and the 4-NBT molecular layer refractive index was set to one, according to other literatures [34,35]. The graphene layer thickness was set to 1 nm, and the refractive index of graphene was set to 2.63 + 1.28 i [36]. The thickness of the SiO_2 layer was set to 20 nm, and the refractive index of SiO_2 and Au were both derived from the parameters in the software. The calculation priority is from inside to outside, and the refractive index of the medium surrounding the Raman label was set to 1.33 (simulated water environment).

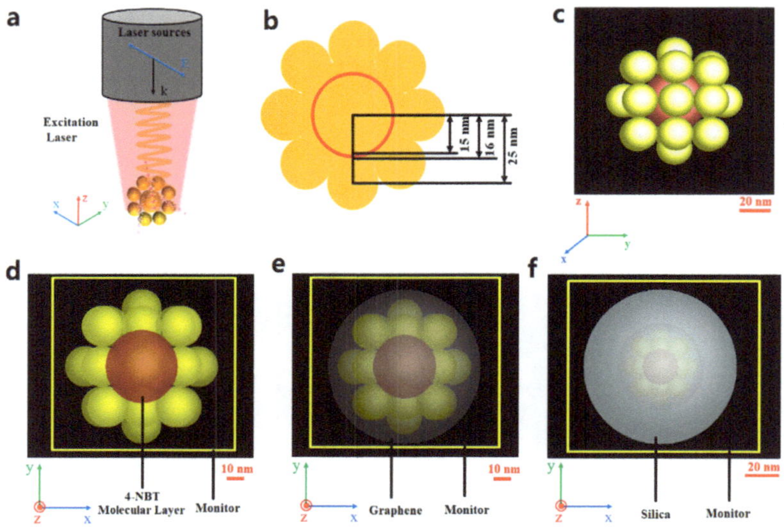

Figure 6. (**a**) Schematic diagram of the Raman tag electric field effects simulation model; (**b**) 2D dimension drawing of P-GERTs simulation model; (**c**) 3D simulation model of P-GERTs, the scale bar is 20 nm; (**d–f**) Monitor setup planes for P-GERTs, GP-GERTs, and MS-GP-GERTs, the scale bars are 10 nm, 10 nm and 20 nm, respectively.

To investigate the effect of different laser wavelengths on the electric field enhancement effect of P-GERTs, we calculated the electromagnetic field strengths of P-GERTs when excited at three different laser wavelengths (532, 638, and 785 nm). As shown in Figure 7a–c, the electric field intensity distributions of single P-GERTs were similar under the excitation of three laser wavelengths, and the electromagnetic hot spots were mainly concentrated in the gap between the gold core and the petals. When excited by 638 nm laser, the maximum ratio of electromagnetic field strength is 38.6, when excited by 785 nm laser, the maximum ratio is 33, and when excited by 532 nm laser, the maximum ratio is 28.7. According to the previous absorption spectrum of P-GERTs, the resonance absorption peak of P-GERTs is around 638 nm; thus, 638 nm belongs to resonance excitation, which is also the reason for the strongest electric field enhancement effect. However, resonance excitation can lead to a significant endothermic effect of metal nanoparticles, and the temperature around the nanoparticles increases rapidly, which in turn affects the stability of Raman tags [37]. Both 785 nm and 532 nm belong to non-resonant excitation, and the thermal effect is not obvious, which can ensure the long-term stability of the Raman tags. Compared with 532 nm, the electric field enhancement effect is stronger at 785 nm excitation; thus, the excitation light of 785 nm wavelength is used in the subsequent simulation and Raman experiments.

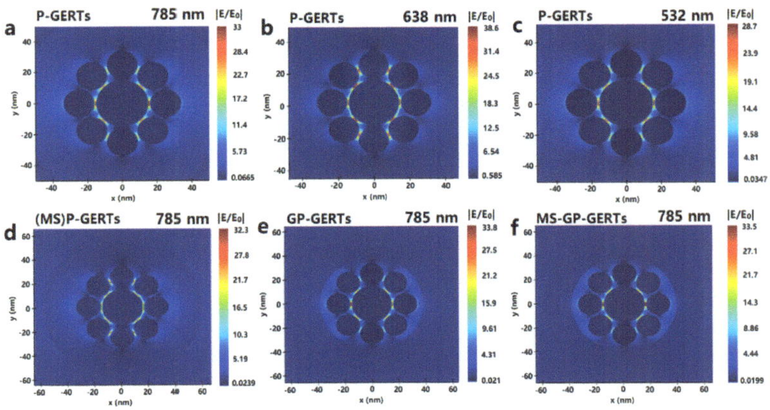

Figure 7. (**a–c**) Electric field enhancement distributions of single P-GERTs calculated by FDTD at excitation wavelengths of 785, 638, and 532 nm; (**d–f**) Electric field enhancement distributions of single (MS)P-GERTs, GP-GERTs, and MS-GP-GERTs, all with excitation wavelengths of 785 nm.

Figure 7d–f shows the electric field enhancement distributions for single (MS)P-GERTs, GP-GERTs, and MS-GP-GERTs. As shown in Figure 7d–f, the electromagnetic hot spots of single (MS)P-GERTs, GP-GERTs, and MS-GP-GERTs are also concentrated in the gap between the gold core and the petals. Silica layer and graphene layer have little effect on the electromagnetic field strength of P-GERTs. The maximum ratio of electromagnetic field strength of single (MS)P-GERTs, GP-GERTs, and MS-GP-GERTs were 32.3, 33.8, and 33.5, respectively. Several studies have shown that graphene has the effect of enhancing the Raman signal, but the enhancement mechanism is chemical enhancement, and the enhancement effect of graphene on the electromagnetic field is not obvious [38–40]. Our simulation results also provide support for this claim.

3.5. Raman Spectra and Enhancement Mechanism of GERTs

Figure 8a shows the Raman spectrum of 4-NBT molecule (10^{-3} mol/L) on silicon substrate. As shown in Figure 8a, there are five main characteristic peaks in the Raman spectrum of 4-NBT molecule, which are 1335 cm^{-1}, 722 cm^{-1}, 854 cm^{-1}, 1080 cm^{-1}, and 1570 cm^{-1}, respectively. The peak of 1335 cm^{-1} corresponds to the strong mode $\nu(NO_2)$, the remaining four peaks correspond to the four weak modes, π(CH) + π(CS) + π(CC), π(CH), ν(CS), and ν(CC), respectively. Figure 8b shows the Raman spectrum of graphene (10^{-3} mol/L) on silicon substrate. Within 1200–2800 cm^{-1}, there are three Raman characteristic peaks of graphene, which are D peak at 1330 cm^{-1}, G peak at 1575 cm^{-1}, and 2D peak at 2670 cm^{-1}. Figure 8c shows the Raman spectra of four Raman tags of P-GERTs, (MS)P-GERTs, GP-GERTs, and MS-GP-GERTs. As shown in Figure 8c, the Raman peaks of all Raman tags are similar to those of the 4-NBT molecule. We found that the Raman signal intensities of P-GERTs and (MS)P-GERTs were comparable, and the Raman signal intensities of GP-GERTs and MS-GP-GERTs were comparable. We can clearly see that after graphene-wrapped P-GERTs, the Raman signal of GP-GERTs is significantly enhanced. The same conclusion can be drawn by comparing the Raman signal intensities of MS-GP-GERTs and (MS)P-GERTs. To determine the concentration detection limit of MS-GP-GERTs Raman tags, the prepared MS-GP-GERTs Raman tag solutions were diluted with ethanol. We prepared samples with Raman tags of MS-GP-GERTs at concentrations of 100 pM, 10 pM, 1 pM, 100 fM, and 10 fM for Raman measurement, and the experimental results obtained are shown in Figure 8e. From Figure 8e, it can be seen that even at a concentration of 100 fM of MS-GP-GERTs, a 4-NBT Raman curve with a good signal-to-noise ratio can be obtained. At further dilution to 10 fM, the Raman signal of 4-NBT was hardly detected.

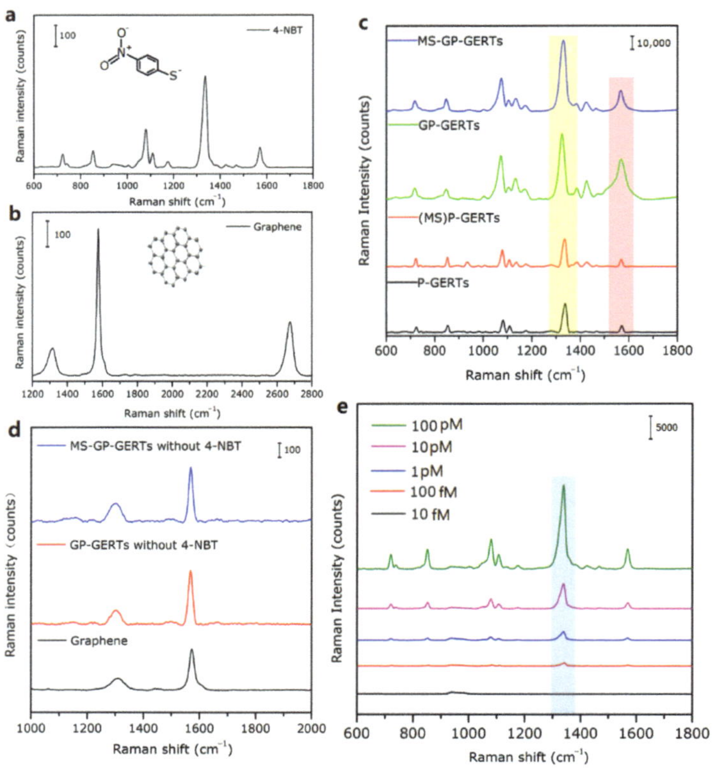

Figure 8. (**a**) Raman spectrum of 4-NBT on silicon substrate; (**b**) Raman spectrum of graphene on silicon substrate; (**c**) Raman spectra of P-GERTs, (MS)P-GERTs, GP-GERTs, and MS-GP-GERTs; (**d**) Raman spectra of composite structures MS-GP-GERT, GP-GERT, and graphene without 4-NBT molecules; (**e**) Raman spectra of different MS-GP-GERTs Raman tag concentrations.

It is generally reported that there are two enhancement mechanisms of *SERS* effect, namely physical enhancement mechanism and chemical enhancement mechanism [41–43]. The physical enhancement mechanism can be explained as the "interstitial hot spots" of rough metal surfaces or nanostructures are excited by the laser, causing localized surface plasmon resonance to generate electromagnetic field enhancement (EM). The chemical enhancement mechanism can be regarded as the signal enhancement caused by the charge transfer between the substrate and the molecule, namely the charge transfer mechanism (CT). It is worth mentioning that the graphene D peak (1330 cm^{-1}) used in our experiments is very close to the strong mode $\nu(NO_2)$ peak (1335 cm^{-1}) of the 4-NBT molecule, and the graphene G peak (1575 cm^{-1}) is very close to the $\nu(CC)$ peak (1570 cm^{-1}) of the 4-NBT molecule. Coincidentally, in our experiments, the most obvious Raman signal enhancements are also the two peaks at 1335 cm^{-1} (the yellow-shaded area in Figure 8c) and 1570 cm^{-1} (the red-shaded area in Figure 8c). According to Zhu et al. [44], the intensity ratio of the D peak and the G peak in the Raman spectra will change during the process of converting graphene to graphene oxide. Due to the increased defects in graphene oxide, the D peak will eventually become stronger. To determine the source of the Raman peak enhancement in Figure 8c, we replaced the 4-NBT ethanol solution with ethanol solution during the synthesis of MS-GP-GERTs and GP-GERTs so that the final composite structure does not contain 4-NBT molecules. Figure 8d shows the Raman spectra of the composite structures, MS-GP-GERTs, GP-GERTs, and graphene. As shown in Figure 8d, although the signal intensities of graphene D peak (1330 cm^{-1}) and G peak (1575 cm^{-1}) in MS-GP-GERTs and GP-GERTs were slightly enhanced, the ratio of the two hardly changed. This indicates that the graphene in the composite structure has not changed. It was further confirmed that the enhancement of Raman peak in Figure 8c originated from the *SERS* signal of 4-NBT molecules.

Therefore, the Raman signal enhancement mechanism of GP-GERTs tags may have three situations: one can be explained as CT mechanism, the electron transfer between graphene and 4-NBT molecules causes the Raman signal enhancement of GP-GERTs; the other one can be explained as the EM mechanism, the "interstitial hotspot" around the P-GERTs tag enhances the Raman signal of graphene, leading to the enhancement of the Raman signal of GP-GERTs; the third possibility is that these two enhancement mechanisms coexist.

According to the principle of electromagnetic field enhancement, the approximate electromagnetic enhancement factor (EF_{EM}) can usually be calculated with the following formula [45]:

$$EF_{EM} = \frac{|E_{out}(\omega_0)|^2 |E_{out}(\omega_s)|^2}{|E_0|^4} \approx \frac{|E_{out}(\omega_0)|^4}{|E_0|^4} \tag{1}$$

where E_0 is the incident electric field strength, which is usually set to 1 V/m. $E_{out}(\omega_0)$ is the local electric field strength of incident light (frequency ω_0), and $E_{out}(\omega_s)$ is the local electric field strength of Raman scattered light (frequency ω_s).

According to the electric field simulation results in Section 3.3, the EF_{EM} of P-GERTs, (MS)P-GERTs, GP-GERTs, and MS-GP-GERTs Raman tags can be calculated to be 1.18×10^6, 1.09×10^6, 1.31×10^6, and 1.26×10^6, respectively.

On the other hand, in the actual Raman experiment, the following formula can be used to calculate the enhancement factor (*EF*) of the experimental results [46]:

$$EF = \frac{I_{SERS}/N_{SERS}}{I_{\text{Raman}}} / N_{\text{Raman}} \tag{2}$$

Among them, *EF* represents the enhancement factor obtained by analyzing the experimental results. I_{SERS} and I_{Raman} represent the Raman intensity of *SERS* and the Raman intensity of the molecule itself, respectively. N_{SERS} and N_{Raman} represent the number of molecules in the *SERS* experiment and the number of molecules in the ordinary Raman experiment, respectively.

For normal Raman measurements, the concentration of the 4-NBT Raman molecule is 1 mM. For the calculation of the number of molecules in the *SERS* experiment, according to literature, the surface area of gold nanospheres with a diameter of 20 nm is about 1256 nm^2, and it can be assumed that the adsorption area of each Raman molecule is 0.2 nm^2 [47,48]. We estimate the number of Raman molecules adsorbed on each P-GERT to be 6280, and multiplying the concentration of gold nanocores can calculate the number of 4-NBT Raman molecules in the *SERS* experiment. The enhancement factors of P-GERTs, (MS)P-GERTs, GP-GERTs, and MS-GP-GERTs can be calculated according to the signal intensity of 4-NBT molecule at 1335 cm^{-1} in the Raman experiment and Equation (2), respectively, $EF = 7.56 \times 10^9$, $EF = 7.95 \times 10^9$, $EF = 1.67 \times 10^{10}$, and $EF = 1.75 \times 10^{10}$. According to the signal intensity of graphene at 1575 cm^{-1} in the Raman experiment and Equation (2), the enhancement factors of GP-GERTs and MS-GP-GERTs can be calculated to be $EF = 5.3 \times 10^9$ and $EF = 4.34 \times 10^9$, respectively. Compared with the previous calculation results of the electromagnetic enhancement factor (EF_{EM}), EF_{EM} is 10^6 order of magnitude, while the actual *EF* in the experiment is 10^9~10^{10} order of magnitude, and the difference between EF_{EM} and actual *EF* is 10^3~10^4 order of magnitude. This indicates that the Raman signal enhancement of GP-GERTs comes not only from EM enhancement, but also from chemical enhancement (CT mechanism), and the Raman signal enhanced by CT mechanism is in the order of 10^3~10^4.

Based on the experimental and theoretical calculation results, we propose a schematic diagram of the possible Raman enhancement mechanism of GP-GERTs tags. As shown in

Figure 9, the EM mechanism and the CT mechanism work together on 4-NBT molecules and graphene. The EM mechanism originates from "electromagnetic hot spots" inside and on the surface of the petal-like nanotags. The CT mechanism mainly comes from the following three aspects: Firstly, the gold nanoparticles are excited by light to generate hot electrons, which are transferred to 4-NBT molecules, and the 4-NBT molecules are excited to generate *SERS* signals. Secondly, the hot electrons on the gold nanoparticles may also be transferred to the graphene to enhance the Raman signal of the graphene or transferred to the 4-NBT molecule after passing through the graphene. Thirdly, the hot electrons generated by the photo-excited graphene are transferred to the 4-NBT molecule, which in turn generates the *SERS* signal.

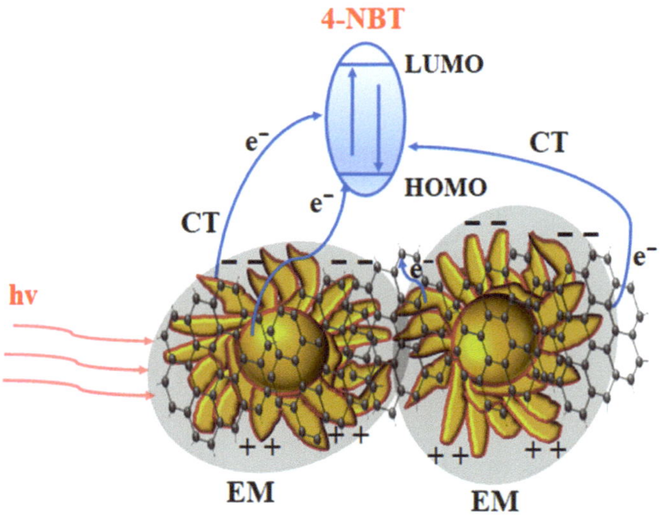

Figure 9. Schematic of Raman enhancement mechanism of GP-GERTs.

3.6. Stability of MS-GP-GERTs

The stability of the Raman signal is an important indicator for evaluating Raman tags. A stable Raman signal for a long time is an important basis for the practical application value of Raman tags. For Raman tags applied in the field of bioimaging, it is usually necessary to maintain stability under various storage conditions and physiological environments (such as different pH and serum solutions). Figure 10a shows the experimental data of the stability study of MS-GP-GERTs in aqueous solutions with different pH values. As shown in Figure 10a, the Raman spectra and signal intensities of MS-GP-GERTs were hardly affected despite the wide range of pH changes (from pH = 3 to pH = 12). This indicates that MS-GP-GERTs have strong pH stability, especially in acidic environments, and the Raman tags reported previously are difficult to maintain in acidic solutions (pH < 5) [14,30] since the Raman 2D peak of graphene is more easily affected by the pH of the solution [49,50]. To evaluate the stability of graphene quality, similar to Figure 8d, we investigated the changes of the 2D peaks of graphene in the composite structure MS-GP-GERTs without 4-NBT molecules in aqueous solutions with different pH values. The experimental results are shown in Figure 10b. It can be seen from Figure 10b that the 2D peaks of graphene were less affected by the pH values of the solution, and the intensity of the 2D peaks hardly changes. Only when the pH value was 3, the 2D characteristic peak was shifted slightly to lower wave numbers. Therefore, we believe that the quality of graphene in the composite structure MS-GP-GERTs is stable in solutions of different pH values.

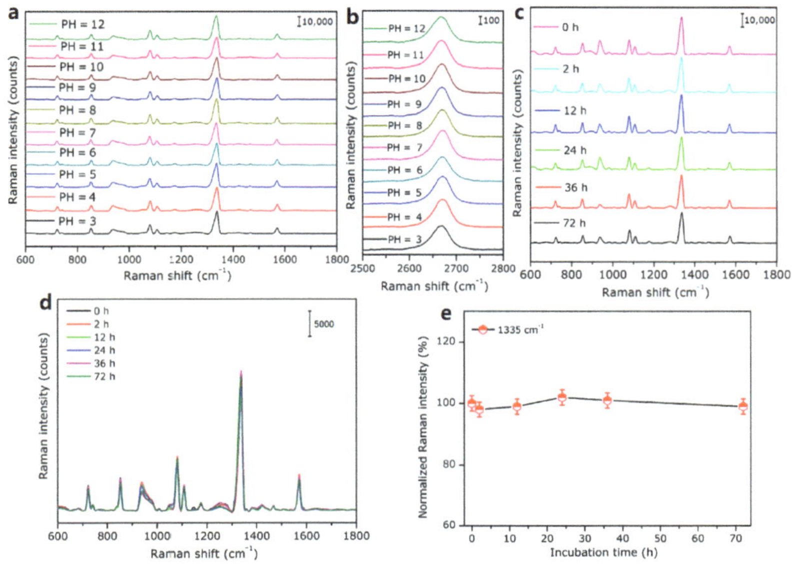

Figure 10. (**a**) *SERS* spectra of MS-GP-GERTs in aqueous solutions with different pH values; (**b**) 2D peak Raman curves of graphene in composite MS-GP-GERTs in aqueous solutions with different pH values; (**c**) *SERS* spectra of MS-GP-GERTs in 10% bovine serum albumin solution at different incubation times; (**d**) *SERS* spectra of MS-GP-GERTs in 10% glucose solution at different incubation times; (**e**) variation curve of normalized Raman intensity at 1335 cm^{-1} during incubation of MS-GP-GERTs with normal saline.

Subsequently, we carried out a temporal stability test of the Raman signal of MS-GP-GERTs in 10% bovine serum albumin solution. As shown in Figure 10c, MS-GP-GERTs were stable for a long time in 10% bovine serum albumin solution. After 72 h, the Raman spectral curves and signal intensities of MS-GP-GERTs remained basically unchanged, showing excellent temporal stability in a simulated serum physiological environment. Considering the actual physiological environment, we further investigated the stability of MS-GP-GERTs in glucose solution and normal saline. The concentration of glucose solution was set to 10%, and 0.9% normal saline was prepared with sodium chloride and deionized water. Similar to 10% bovine serum albumin, the incubation time of MS-GP-GERTs with glucose solution (10%) and normal saline (0.9%) was varied from 0 h to 72 h, and the obtained experimental results are shown in Figure 10d,e. From Figure 10d, it can be seen that the Raman spectra of MS-GP-GERTs are very stable during the incubation time, and the Raman tags show excellent temporal stability in glucose solution. Figure 10e shows the change of the normalized Raman intensity of the 1335 cm^{-1} Raman peak during the incubation of MS-GP-GERTs with normal saline. The Raman intensity of MS-GP-GERTs was also stable during the incubation time, and saline had little effect on the stability of the Raman signal.

For most *SERS* tags, due to the direct adsorption of Raman reporters on the surface of metal substrates, it is difficult to maintain stable performance under harsh conditions, such as long-term laser irradiation and strong acid conditions [51–53]. This limits the application of *SERS* tags to a certain extent, especially in in vivo imaging applications. In our experiments, MS-GP-GERTs showed excellent stability in the serum environment and aqueous solutions with different pH values. The main reasons are analyzed as follows: First, our Raman reporters are not simply adsorbed on the surface of the metal substrate but embedded in the inner gap of the petal-shaped gold nanoparticles, which makes the Raman reporters less disturbed by the external environment. Secondly, graphene is wrapped on the surface of P-GERTs, and the chemical inertness of graphene can protect Raman reporters from harsh environments. In addition, the protective layers of the mesoporous silica grow on the surface of the graphene-wrapped, petal-like tags, which further improve the stability and biocompatibility of the Raman tags.

Finally, we changed the Raman reporters inside the MS-GP-GERTs to study the stability of the graphene-wrapped, petal-like, gap-enhanced Raman tag structure to the *SERS* enhancement of different Raman reporters. Figure 11a shows the schematic structure of MS-GP-GERTs with different Raman reporters. Different colors correspond to different Raman reporters, and red, green, blue, and purple correspond to 1,4-BDT molecules, 4-MBN molecules, B-4,4′-D, and 2-NT molecules, respectively. Figure 11b shows the *SERS* spectra corresponding to Figure 11a. As can be seen in Figure 11b, the graphene-wrapped, petal-like, gap-enhanced Raman tag structure can achieve stable *SERS* enhancement for all four Raman reporter molecules. Similar to Figure 8e, we further examined the concentration detection limit of MS-GP-GERTs tags with different Raman reporters. Figure 11c,d show the Raman signals of MS-GP-GERTs tags with different Raman reporters at concentrations of 10 pM and 100 fM, respectively. Raman spectra of different reporter molecules with good signal-to-noise ratio can also be obtained when the Raman tags concentration is 100 fM. This means that *SERS* tags with different Raman characteristic peaks can be easily prepared by only replacing the internal Raman reporters of MS-GP-GERTs. It shows the great potential of MS-GP-GERTs in different types of biomedical imaging and multicolor Raman imaging.

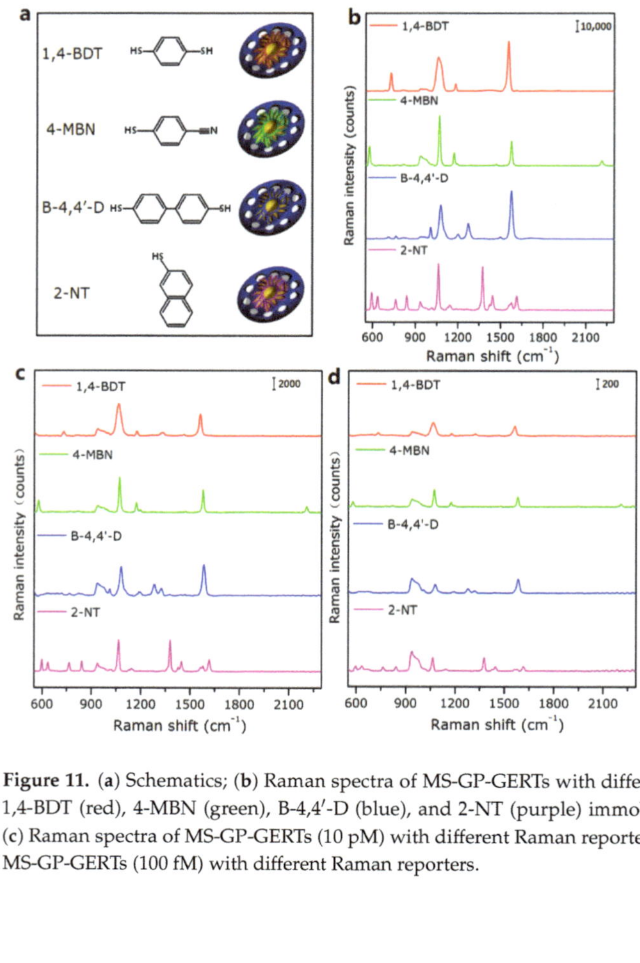

Figure 11. (**a**) Schematics; (**b**) Raman spectra of MS-GP-GERTs with different Raman reporters of 1,4-BDT (red), 4-MBN (green), B-4,4′-D (blue), and 2-NT (purple) immobilized in the nanogaps; (**c**) Raman spectra of MS-GP-GERTs (10 pM) with different Raman reporters; (**d**) Raman spectra of MS-GP-GERTs (100 fM) with different Raman reporters.

4. Conclusions

In summary, we developed new graphene-wrapped, petal-like, gap-enhanced Raman tags and demonstrated the detailed synthesis process. We explained the specific role of graphene in Raman tags through FDTD simulations and Raman spectroscopy experimental studies. The Raman enhancement of GP-GERTs is determined by both CT mechanism and EM mechanism, and the total Raman *EF* can reach 10^{10}. The stability experiments show that MS-GP-GERTs not only have excellent stability in different PH values (from 3 to 12) aqueous solutions, but also maintain signal stability in the simulated physiological environment for a long time (up to 72 h). In addition, the MS-GP-GERTs structure can also achieve stable enhancement of different Raman reporters, showing a bright application prospect in the field of biomedical imaging.

Author Contributions: M.C. and X.Z. conceived the original ideas. M.C. conducted the experiments and analyzed the data. B.W. and J.W. participated in the simulation and part of the data analysis. H.L. and Z.C. participated in Raman spectroscopy measurements and data analysis. M.C., X.Z. and X.X. participated in writing the paper. B.W. provided experimental conditions and methodological guidance. X.X. and X.Z. supervised the project. All authors have read and agreed to the published version of the manuscript.

Funding: This research was funded by Tianjin Key R & D Program (No. 19YFZCSY00250).

Institutional Review Board Statement: Not applicable.

Informed Consent Statement: Not applicable.

Data Availability Statement: The data presented in this article are available on request from the corresponding author.

Acknowledgments: We would like to thank Feng Min from the School of Physical Sciences of Nankai University for his help in SEM characterization. We would like to thank Li Ang from the State Key Laboratory of Medicinal Chemical Biology, Nankai University, for his assistance in TEM characterization. We also want to thank Xianghui Wang and Lu Sun from the Institute of Modern Optics for their help in UV-Vis absorption spectroscopy and Fourier Near Infrared Spectroscopy measurements.

Conflicts of Interest: The authors declare no conflict of interest.

References

1. Ding, S.Y.; You, E.M.; Tian, Z.Q.; Martin, M. Electromagnetic theories of surface-enhanced Raman spectroscopy. *Chem. Soc. Rev.* **2017**, *46*, 4042–4076. [CrossRef] [PubMed]
2. Muehlethaler, C.; Leona, M.; Lombardi, J.R. Review of surface enhanced Raman scattering applications in forensic science. *Anal. Chem.* **2016**, *88*, 152–169. [CrossRef] [PubMed]
3. Langer, J.; Jimenez, A.; Berasturi, D.; Aizpurua, J.; Alvarez-Puebla, R.A.; Auguie, B.; Baumberg, J.J.; Bazan, G.C.; Bell, S.E.J.; Boisen, A.; et al. Present and future of surface-enhanced Raman scattering. *ACS. Nano* **2019**, *14*, 28–117. [CrossRef] [PubMed]
4. Li, J.F.; Huang, Y.F.; Ding, Y.; Yang, Z.L.; Li, S.B.; Zhou, X.S.; Fan, F.R.; Zhang, W.; Zhou, Z.Y.; Wu, D.Y.; et al. Shell-isolated nanoparticle-enhanced Raman spectroscopy. *Nature* **2010**, *464*, 392–395. [CrossRef] [PubMed]
5. Kneipp, J.; Kneipp, H.; Kneipp, K. SERS—A single-molecule and nanoscale tool for bioanalytics. *Chem. Soc. Rev.* **2008**, *37*, 1052–1060. [CrossRef] [PubMed]
6. Wang, Y.; Yan, B.; Chen, L. SERS tags: Novel optical nanoprobes for bioanalysis. *Chem. Rev.* **2013**, *113*, 1391–1428. [CrossRef]
7. Wang, Z.; Zong, S.; Wu, L.; Zhu, D.; Cui, Y.P. SERS-activated platforms for immunoassay: Probes, encoding methods, and applications. *Chem. Rev.* **2017**, *117*, 7910–7963. [CrossRef]
8. Gandra, N.; Singamaneni, S. Bilayered raman-intense gold nanostructures with hidden tags (brights) for high-resolution bioimaging. *Adv. Mater.* **2013**, *25*, 1022–1027. [CrossRef]
9. Kang, J.W.; So, P.T.C.; Dasari, R.R.; Lim, D.K. High resolution live cell Raman imaging using subcellular organelle-targeting SERS-sensitive gold nanoparticles with highly narrow intra-nanogap. *Nano Lett.* **2015**, *15*, 1766–1772. [CrossRef]
10. Bao, Z.; Zhang, Y.; Tan, Z.; Yin, X.; Di, W.; Ye, J. Gap-enhanced Raman tags for high-contrast sentinel lymph node imaging. *Biomaterials* **2018**, *163*, 105–115. [CrossRef]
11. Khlebtsov, B.N.; Bratashov, D.N.; Byzova, N.A.; Dzantiev, B.B.; Khlebtsov, N.G. SERS-based lateral flow immunoassay of troponin I by using gap-enhanced Raman tags. *Nano Res.* **2019**, *12*, 413–420. [CrossRef]

12. Gu, Y.; He, C.; Zhang, Y.; Lin, L.; Thackray, B.D.; Ye, J. Gap-enhanced Raman tags for physically unclonable anticounterfeiting labels. *Nat. Commun.* **2020**, *11*, 516. [CrossRef] [PubMed]
13. Khlebtsov, B.; Pylaev, T.; Khanadeev, V.; Bratashov, D.; Khlebtsov, N. Quantitative and multiplex dot-immunoassay using gap-enhanced Raman tags. *RSC Adv.* **2017**, *7*, 40834–40841. [CrossRef]
14. Zhang, Y.; Gu, Y.; He, J.; Thackray, B.D.; Ye, J. Ultrabright gap-enhanced Raman tags for high-speed bioimaging. *Nat. Commun* **2019**, *10*, 3905. [CrossRef] [PubMed]
15. Geim, A.K. Graphene: Status and prospects. *Science.* **2009**, *324*, 1530–1534. [CrossRef]
16. Tiwari, S.K.; Sahoo, S.; Wang, N.; Huczko, A. Graphene research and their outputs: Status and prospect. *J. Sci.-Adv. Mater. Device* **2020**, *5*, 10–29. [CrossRef]
17. Ren, W.; Cheng, H.M. The global growth of graphene. *Nat. Nanotechnol.* **2014**, *9*, 726–730. [CrossRef]
18. Zhang, X.; Jing, Q.; Ao, S.; Schneider, G.F.; Kireev, D.; Zhang, Z.J.; Fu, W.Y. Ultrasensitive field-effect biosensors enabled by the unique electronic properties of graphene. *Small* **2020**, *16*, 1902820. [CrossRef]
19. Guo, S.; Zhang, Y.; Ge, Y.; Zhang, S.L.; Zeng, H.B.; Zhang, H. 2D V-V binary materials: Status and challenges. *Adv. Mater.* **2019**, *31*, 1902352. [CrossRef]
20. Zheng, T.T.; Zhou, Y.; Feng, E.D.; Tian, Y. Surface-enhanced Raman scattering on 2D nanomaterials: Recent developments and applications. *Chin. J. Chem.* **2021**, *39*, 745–756. [CrossRef]
21. Huh, S.; Park, J.; Kim, Y.S.; Kim, K.S.; Hong, B.H.; Nam, J.M. UV/ozone-oxidized large-scale graphene platform with large chemical enhancement in surface-enhanced Raman scattering. *ACS Nano* **2011**, *5*, 9799–9806. [CrossRef] [PubMed]
22. Jiang, T.; Wang, X.L.; Tang, S.S.; Zhou, J.; Gu, C.J.; Tang, J. Seed-mediated synthesis and SERS performance of graphene oxide-wrapped Ag nanomushroom. *Sci. Rep.* **2017**, *7*, 1–9. [CrossRef] [PubMed]
23. Yang, W.; Li, Z.; Lu, Z.; Yu, J.; Huo, Y.Y.; Man, B.Y.; Pan, J.; Si, H.P.; Jiang, S.Z.; Zhang, C. Graphene-Ag nanoparticles-cicada wings hybrid system for obvious SERS performance and DNA molecular detection. *Opt. Express* **2019**, *27*, 3000–3013. [CrossRef] [PubMed]
24. Wang, J.C.; Qiu, C.C.; Pang, H.; Wu, J.Y.; Sun, M.T.; Liu, D.M. High-performance SERS substrate based on perovskite quantum dot–graphene/nano-Au composites for ultrasensitive detection of rhodamine 6G and p-nitrophenol. *J. Mater. Chem. C* **2021**, *9*, 9011–9020. [CrossRef]
25. Qiu, X.J.; You, X.R.; Chen, X.; Chen, H.L.; Dhinakar, A.; Liu, S.H.; Guo, Z.Y.; Wu, J.; Liu, Z.M. Development of graphene oxide-wrapped gold nanorods as robust nanoplatform for ultrafast near-infrared SERS bioimaging. *Int. J. Nanomed.* **2017**, *12*, 43–49. [CrossRef]
26. Gong, T.C.; Zhang, J.; Zhu, Y.; Wang, X.Y.; Zhang, X.L.; Zhang, J. Optical properties and surface-enhanced Raman scattering of hybrid structures with Ag nanoparticles and graphene. *Carbon* **2016**, *102*, 245–254. [CrossRef]
27. Quan, J.M.; Zhang, J.; Li, J.Y.; Zhang, X.L.; Wang, M.; Wang, N.; Zhu, Y. Three-dimensional AgNPs-graphene-AgNPs sandwiched hybrid nanostructures with sub-nanometer gaps for ultrasensitive surface-enhanced Raman spectroscopy. *Carbon* **2019**, *147*, 105–111. [CrossRef]
28. Zhang, Y.J.; Chen, Q.Q.; Chen, X.; Wang, A.; Tian, Z.Q.; Li, J.F. Graphene-coated Au nanoparticle-enhanced Raman spectroscopy. *J. Raman Spectrosc.* **2021**, *52*, 439–445. [CrossRef]
29. Haiss, W.; Thanh, N.T.K.; Aveyard, J.; Fernig, D.G. Determination of size and concentration of gold nanoparticles from UV−Vis spectra. *Anal. Chem.* **2007**, *79*, 4215–4221. [CrossRef]
30. Zhang, Y.Q.; Qiu, Y.Y.; Lin, L.; Gu, H.C.; Xiao, Z.Y.; Ye, J. Ultraphotostable mesoporous silica-coated gap-enhanced Raman tags (GERTs) for high-speed bioimaging. *ACS Appl. Mater. Interfaces* **2017**, *9*, 3995–4005. [CrossRef]
31. Bepete, G.; Anglaret, E.; Ortolani, L.; Morandi, V.; Huang, K.; Penicaud, A.; Drummond, C. Surfactant-free single-layer graphene in water. *Nat. Chem.* **2017**, *9*, 347–352. [CrossRef] [PubMed]
32. Baumgartner, B.; Hayden, J.; Loizillon, J.; Steinbacher, S.; Grosso, D.; Lendl, B. Pore size-dependent structure of confined water in mesoporous silica films from water adsorption/desorption using ATR–FTIR spectroscopy. *Langmuir* **2019**, *35*, 11986–11994. [CrossRef] [PubMed]
33. Shadmani, S.; Gabriel, L.; Hornyak, D.C.; Htet, K.; Tanujjal, B. Morphology and Visible Photoluminescence Modulation in Dye-free Mesoporous Silica Nanoparticles using a Simple Calcination Step. *Mater. Res. Bull.* **2022**, *152*, 111842. [CrossRef]
34. Zhuang, X.; Miranda, P.B.; Kim, D.; Shen, Y.R. Mapping molecular orientation and conformation at interfaces by surface nonlinear optics. *Phys. Rev. B* **1999**, *59*, 12632–12640. [CrossRef]
35. Cecchet, F.; Lis, D.; Guthmuller, J.; Champagne, B.; Caudano, Y.; Silien, C.; Mani, A.A.; Thiry, P.A.; Peremans, A. Orientational Analysis of Dodecanethiol and p-Nitrothiophenol SAMs on Metals with Polarisation-Dependent SFG Spectroscopy. *Chemphyschem* **2010**, *11*, 607–615. [CrossRef]
36. Cheon, S.; Kihm, K.D.; Kim, H.G.; Lim, G.; Park, J.S.; Lee, J.S. How to reliably determine the complex refractive index (RI) of graphene by using two independent measurement constraints. *Sci. Rep.* **2014**, *4*, 6364. [CrossRef]
37. Lin, D.H.; Qin, T.Q.; Wang, Y.Q.; Sun, X.Y.; Chen, L.X. Graphene oxide wrapped SERS tags: Multifunctional platforms toward optical labeling, photothermal ablation of bacteria, and the monitoring of killing effect. *ACS Appl. Mater. Interfaces* **2014**, *6*, 1320–1329. [CrossRef]
38. Zhang, X.L.; Wang, N.; Liu, R.J.; Wang, X.Y.; Zhu, Y.; Zhang, J. SERS and the photo-catalytic performance of Ag/TiO_2/graphene composites. *Opt. Mater. Express* **2018**, *8*, 704. [CrossRef]

39. Xu, W.; Mao, N.; Zhang, J. Graphene: A platform for surface-enhanced Raman spectroscopy. *Small* **2013**, *9*, 1206–1224. [CrossRef]
40. Kaushik, V.; Kagdada, H.L.; Singh, D.K.; Pathak, S. Enhancement of SERS effect in Graphene-Silver hybrids. *Appl. Surf. Sci.* **2022**, *574*, 151724. [CrossRef]
41. Wang, Y.; Du, X.C.; Hu, A.J.; Hu, Y.; Gong, C.H.; Wang, X.F.; Xiong, J. Coupling enhancement mechanisms, materials, and strategies for surface-enhanced Raman scattering devices. *Analyst* **2021**, *146*, 5008–5032. [CrossRef] [PubMed]
42. Trivedi, D.J.; Barrow, B.; Schatz, G.C. Understanding the chemical contribution to the enhancement mechanism in SERS: Connection with Hammett parameters. *J. Chem. Phys.* **2020**, *153*, 124706. [CrossRef] [PubMed]
43. Valley, N.; Jensen, L.; Autschbach, J.; Schatz, G.C. Theoretical studies of surface enhanced hyper-Raman spectroscopy: The chemical enhancement mechanism. *J. Chem. Phys.* **2010**, *133*, 054103. [CrossRef] [PubMed]
44. Zhu, B.; Wang, F.; Wang, G.; Gu, Y.Z. Oxygen-containing-defect-induced synergistic nonlinear optical enhancement of graphene/CdS nanohybrids under single pulse laser irradiation. *Photonics Res.* **2018**, *6*, 1158–1169. [CrossRef]
45. Jiang, M.; Wang, Z.; Zhang, J. TiO_2/AgNPs SERS substrate for the detection of multi-molecules with a self-cleaning and high enhancement factor using the UV-induced method. *Opt. Mater. Express* **2022**, *12*, 1010–1018. [CrossRef]
46. Le Ru, E.C.; Blackie, E.; Meyer, M.; Etchegoin, P.G. Surface enhanced Raman scattering enhancement factors: A comprehensive study. *J. Phys. Chem. C* **2007**, *111*, 13794–13803. [CrossRef]
47. Sergiienko, S.; Moor, K.; Gudun, K.; Yelemessova, Z.; Bukasov, R. Nanoparticle–nanoparticle vs. nanoparticle–substrate hot spot contributions to the SERS signal: Studying Raman labelled monomers, dimers and trimers. *Phys. Chem. Chem. Phys.* **2017**, *19*, 4478–4487. [CrossRef]
48. Pugmire, D.L.; Tarlov, M.J.; Van Zee, R.D.; Naciri, J. Structure of 1, 4-benzenedimethanethiol self-assembled monolayers on gold grown by solution and vapor techniques. *Langmuir* **2003**, *19*, 3720–3726. [CrossRef]
49. Camerlingo, C.; Verde, A.; Manti, L.; Meschini, R.; Delfino, I.; Lepore, M. Graphene-based Raman spectroscopy for pH sensing of X-rays exposed and unexposed culture media and cells. *Sensors* **2018**, *18*, 2242. [CrossRef]
50. Zou, J.; Chen, Y.; Shu, X.; Li, X.; Song, Y.; Zhao, Z. Proper pH value enhances giant magneto-impedance effect of FINEMET/rGO composite ribbons by electroless plating. *Mater. Sci. Eng. B* **2021**, *265*, 115004. [CrossRef]
51. Chen, H.Y.; Guo, D.; Gan, Z.F.; Jiang, L.; Chang, S.; Li, D.W. A phenylboronate-based SERS nanoprobe for detection and imaging of intracellular peroxynitrite. *Microchim. Acta* **2019**, *186*, 1–8. [CrossRef]
52. Li, M.; Wu, J.; Ma, M.; Feng, Z.; Mi, Z.; Rong, P.; Liu, D. Alkyne-and nitrile-anchored gold nanoparticles for multiplex SERS imaging of biomarkers in cancer cells and tissues. *Nanotheranostics* **2019**, *3*, 113–119. [CrossRef]
53. Zhang, Y.; Qian, J.; Wang, D.; Wang, Y.L.; He, S.L. Multifunctional gold nanorods with ultrahigh stability and tunability for in vivo fluorescence imaging, SERS detection, and photodynamic therapy. *Angew. Chem. Int. Edit.* **2013**, *52*, 1148–1151. [CrossRef]

nanomaterials

MDPI

Article

Magnetic-Core–Shell–Satellite Fe_3O_4-Au@Ag@(Au@Ag) Nanocomposites for Determination of Trace Bisphenol A Based on Surface-Enhanced Resonance Raman Scattering (SERRS)

Jie Huang [1], Tianxiang Zhou [1], Wenshi Zhao [1,2], Min Zhang [1], Zhibo Zhang [1], Wangsheng Lai [1], Naveen Reddy Kadasala [3], Huilian Liu [1,*] and Yang Liu [1,*]

1 Key Laboratory of Functional Materials Physics and Chemistry of the Ministry of Education, Jilin Normal University, Changchun 130103, China
2 Changchun Institute of Optics, Fine Mechanics and Physics, Chinese Academy of Sciences, Changchun 130033, China
3 Department of Chemistry, Towson University, Towson, MD 21252, USA
* Correspondence: lhl541@126.com (H.L.); liuyang@jlnu.edu.cn (Y.L.)

Abstract: As a typical representative of endocrine-disrupting chemicals (EDCs), bisphenol A (BPA) is a common persistent organic pollutant in the environment that can induce various diseases even at low concentrations. Herein, the magnetic Fe_3O_4-Au@Ag@(Au@Ag) nanocomposites (CSSN NCs) have been prepared by self-assembly method and applied for ultra-sensitive surface-enhanced resonance Raman scattering (SERRS) detection of BPA. A simple and rapid coupling reaction of Pauly's reagents and BPA not only solved the problem of poor affinity between BPA and noble metals, but also provided the SERRS activity of BPA azo products. The distribution of hot spots and the influence of incremental introduction of noble metals on the performance of SERRS were analyzed by a finite-difference time-domain (FDTD) algorithm. The abundance of hot spots generated by core–shell–satellite structure and outstanding SERRS performance of Au@Ag nanocrystals were responsible for excellent SERRS sensitivity of CSSN NCs in the results. The limit of detection (LOD) of CSSN NCs for BPA azo products was as low as 10^{-10} M. In addition, the saturation magnetization (Ms) value of CSSN NCs was 53.6 emu·g^{-1}, which could be rapidly enriched and collected under the condition of external magnetic field. These magnetic core–shell–satellite NCs provide inspiration idea for the tailored design of ultra-sensitive SERRS substrates, and thus exhibit limitless application prospects in terms of pollutant detection, environmental monitoring, and food safety.

Keywords: magnetic; core–shell–satellite nanocomposites; surface-enhanced resonance Raman scattering; coupling reaction; FDTD; bisphenol A

1. Introduction

Endocrine-disrupting chemicals (EDCs), also known as environmental hormones, are exogenous substances that affect mammalian reproduction by interfering with the endocrine system of organisms [1–3]. Bisphenol A (BPA), as a typical representative of EDCs, has been used principally in manufacturing polycarbonates and epoxy resins which are the main raw materials of plasticizers, plastic bottles and cups, food storage and packaging materials, and other commonly used industrial products [4,5]. Unfortunately, a large amount of evidence shows that the persistence, bioaccumulation, and biomagnification of BPA can bring about serious negative effects on human health and the ecosystem [6,7]. Although BPA is a low-toxicity chemical, it can induce a variety of diseases even at very low concentrations, such as congenital disabilities, diabetes, cardiovascular diseases, multiple cancers, and especially reproductive system diseases [8–10]. Hence, it is urgent to explore a rapid, sensitive, efficient, and low-cost detection method for BPA.

At present, direct detection and indirect detection are two commonly used methods for BPA detection. The direct detection method of BPA mainly includes high-performance liquid chromatography (HPLC), gas chromatography/mass spectrometry (GC-MS) and liquid chromatography/mass spectrometry (LC-MS) [11–13]. Nevertheless, these methods ordinarily need complex sample pretreatment, long-term testing, professional testers, and expensive equipment, which limit their practical application. By comparison, indirect detection methods, such as plasmon resonance biosensors and enzyme-linked immunosorbent assay, have high sensitivity [14,15]. However, the complex antibody preparation and the blurry recognition boundaries between specificity and non-specificity still need to be solved [16]. In recent years, as a highly sensitive spectral analysis technology, surface-enhanced Raman scattering (SERS) has stimulated considerable research enthusiasm due to its high sensitivity, easy operation, multiplexing ability, and applicability, and it is extensively used in the detection of chemical and biological molecules [17–20]. Both electromagnetic enhancement (EM) and chemical enhancement (CE) are widely recognized SERS enhancement mechanisms [21,22]. A large number of experimental results have proven that the EM mechanism shows an enhancement of 10^4 to 10^{11}, while the CE mechanism only contributes to 10 to 10^3 enhancement [23]. Therefore, it is widely acknowledged that EM plays a key role in SERS enhancement [24]. Given that the enhancement of EM caused by localized surface plasmon resonance (LSPR) excitation usually occurs in nanogaps between metal nanoparticles (named "hot spots"), it is highly desirable to increase the number of "hot spots" for obtaining high performance SERS substrates [25–27].

Inducing aggregation and constructing interlayers are two commonly used methods for introducing high- density hot spots. Nonetheless, the introduction of aggregation will inevitably bring about non-uniform distribution of noble metals, which is not conducive to the purpose of increasing hot spots [28]. In comparison, the use of polymers to achieve the adsorption or self-assembly of hot spots on surface of composite SERS substrates has attracted more and more attention [29,30]. Polyethyleneimine dithiocarbamate (PEI-DTC) polymer with excellent bonding strength is an ideal interlayer, which can significantly improve the affinity of Au and Ag nanocrystals to the SERS surfaces due to the presence of the bidentate ligands with two chelating sulfur groups [31]. Unfortunately, single-element Au nanocrystals or Ag nanocrystals have limited plasmonic absorption [32]. The bimetallic Au@Ag nanocrystals can not only have richer plasmonic modes and tunable LSPR but also make full use of outstanding SERS activity of Ag nanocrystals as well as uniformity and stability of Au nanocrystals [33–35]. However, the affinity between phenolic molecules such as BPA and noble metals is rather weak when mixing the target analytes with noble metal substrates directly, which is bad for the acquisition of high-quality SERS signals [36]. Because azo dyes have a propensity to bind to noble metal nanocrystals, the coupling reaction of BPA and diazonium ions to produce azo dyes is an ideal solution for the SERS detection of BPA, which can greatly boost the activity of SERS substrates [37]. In addition, the resonance enhancement occurs when the excitation wavelength overlaps or approaches the electronic transition of target molecules, which is called surface-enhanced resonance Raman scattering (SERRS) [38]. Therefore, the combination of azo dyes derived from BPA and SERRS will make it possible for ultra-sensitive detection of BPA.

Herein, we propose magnetic-core–shell–satellite Fe_3O_4-Au@Ag@(Au@Ag) nanocomposites (CSSN NCs), which could serve as ultra-sensitive SERRS substrates for BPA detection. A simple and rapid coupling reaction of Pauly's reagents and BPA was used not only to solve the problem of poor affinity between BPA and noble metals, but also to advance SERRS enhancement performance. BPA azo products were selected as target molecules to discuss the effect of the incremental introduction of noble metals on SERRS activity. The distribution of hot spots was obtained by a finite-difference time-domain (FDTD) theoretical method, and the relevant SERRS enhancement mechanism was discussed. Given that Fe_3O_4 hollow spheres with large specific surface area have a good superparamagnetic property, CSSN NCs could be easily recovered through an external magnet, which provided potential possibilities of recycling and reuse in complex liquid environments [39]. Our work not only enriches research on the tailored design of ultra-sensitive SERRS substrates, but also realizes the rapid and quantitative detection of trace BPA, which may have many possible applications, such as in pollutant detection, environmental monitoring, and food safety.

2. Materials and Methods

2.1. Materials and Characterization

The details of this section can be found in Supplementary Materials.

2.2. Synthesis of PEI-DTC Aqueous Solution, Fe_3O_4 Hollow Spheres, Au Seeds, and Au@Ag Nanocrystals

PEI-DTC aqueous solution, Fe_3O_4 hollow spheres, Au seeds, and Au@Ag nanocrystals were obtained on the basis of our previous work [40].

2.3. Synthesis of Fe_3O_4-Au (FA) NCs

An amount of 20 mg of Fe_3O_4 hollow spheres were added to 20 mL of methanol, followed by dropwise addition of 25 mL of PEI-DTC aqueous solution. After the mixture stood for 60 min, the product was separated and gathered by a magnet. Then 20 mL of Au seeds were added under sonication. Finally, FA NCs were obtained after continuous sonication for 1 h and repeated washing.

2.4. Synthesis of FA@Ag@(Au@Ag) (CSSN) NCs

As illustrated in Scheme 1, the preparation process of CSSN NCs consists of two steps. The first step was the preparation of FA@Ag NCs. First, 10 mg of FA NCs and 5 mL of deionized water were mixed. Subsequently, $AgNO_3$ (0.1 M; 3 mL) and reductant $NH_2OH{\cdot}HCl$ (0.1 M; 12 mL) were added. FA@Ag NCs were obtained after sonication for 2 h and several wash cycles with ethanol.

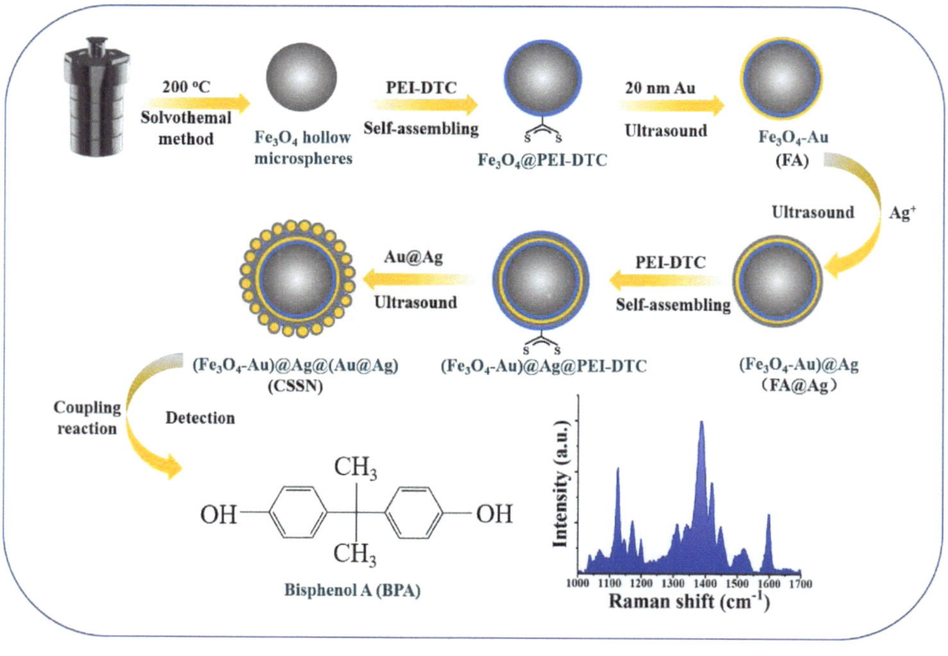

Scheme 1. Scheme of synthetic process of CSSN NCs and SERRS detection protocol for BPA on CSSN SERRS substrates.

The second step was the synthesis of CSSN NCs. The previously obtained FA@Ag NCs were blended with 20 mL of methanol, and 25 mL of PEI-DTC was dropwise-added. The mixture was stored for about 1 h. Then, FA@Ag@PEI-DTC NCs were obtained through washing and drying. After 10 mg of FA@Ag@PEI-DTC NCs was put into 5 mL of deionized water, 40 mL of Au@Ag nanocrystals was poured into the mixture and sonicated for about 2 h. Finally, the product was washed and dried to obtain CSSN NCs.

2.5. Pauly's Reagents and Coupling Reaction

Three kinds of reagents were prepared and kept at 4 °C for further use. Reagent A was a mixture of *p*-aminobenzenesulfonic acid (4.5 g), HCl solution (12 M; 5 mL) and deionized water (500 mL). Reagent B was 5% $NaNO_2$, and reagent C was 10% Na_2CO_3.

Coupling reaction: Reagents A + B + C + BPA ethanol solution in a volume ratio of 1:1:1:2.

2.6. FDTD Algorithm Method

Details of FDTD algorithm method can be found in Supplementary Materials.

2.7. SERRS Measurements

Before SERRS measurements, BPA was changed into azo dyes through a coupling reaction of BPA and diazonium ions to bind noble metals with high affinity and thus achieve highly sensitive detection of BPA. In this work, the phenol group of BPA could be converted into azo dyes with *p*-aminobenzenesulfonic acid through electrophilic aromatic substitution [41,42]. After the coupling reaction, BPA solution (25 μL) with different concentrations of 10^{-10} to 10^{-4} M and CSSN (1 mg) NCs were mixed in the aluminum pan, respectively. SERRS spectroscopy was performed under 514.5 nm laser and acquisition time was about 10 s.

3. Results and Discussion

3.1. Structure and Magnetic Properties of CSSN NCs

XRD technology was used to study the structure and phase purity of Fe_3O_4 hollow spheres, FA, FA@Ag and CSSN NCs. As shown in Figure 1, the diffraction peaks of Fe_3O_4 hollow spheres located at 30.4°, 35.5°, 43.4°, 53.4°, 57.3°, and 62.8° refer to the (112), (211), (220), (024), (303), and (224) planes of Fe_3O_4, respectively, which can be indexed to the cubic inverse spinel structure of Fe_3O_4 (JCPDS 19-0629) [43,44]. In addition, the diffraction peaks are sharp and strong, which indicates that the prepared Fe_3O_4 hollow spheres have high phase purity and good crystallization. However, owing to the fact that Fe_3O_4 as well as γ-Fe_2O_3 have the identical spinel structure, it is not sufficient to identify them only by XRD results [45]. Therefore, the phase structure of Fe_3O_4 hollow spheres was further verified by Mössbauer spectroscopy. As depicted in Figure S1, the Mössbauer spectrum of Fe_3O_4 hollow spheres can be fitted into two sextets, and the magnetic sextets lines illustrate the typical double six peak structure of Fe_3O_4 [46,47]. The corresponding Mössbauer parameters of Fe_3O_4 hollow spheres are presented in Table S1. Hyperfine field is 48.7 and 45.5 Tesla, and the isomer displacement is 0.288 and 0.602 mm/s, which correspond to Fe^{2+} and Fe^{3+} at octahedral interstitial sites and Fe^{3+} at tetrahedral interstitial sites. After Au seeds were loaded on surfaces of Fe_3O_4 hollow spheres, four new XRD diffraction peaks emerged at 38.2°, 44.3°, and 64.5°, which were assigned to (111), (200), and (220) planes of Au (JCPDS 04-0784) [48]. It should be noted that the positions of Au and Ag characteristic peaks are too close to be distinguished [49]. Since the intensities of XRD diffraction peaks are related to the contents of phase in the mixture [50,51], the increase of the intensities of Ag/Au diffraction peaks in the XRD patterns of FA@Ag NCs proves that there is dense Ag adsorbed on the surfaces of FA NCs. The XRD pattern of CSSN NCs shows that the diffraction peaks intensity of Ag/Au further increases significantly when the Au@Ag nanocrystals are adhered to the FA@Ag NCs. By contrast, the XRD diffraction pattern of CSSN NCs exhibits weaker Fe_3O_4 characteristic peaks than that of Fe_3O_4 hollow spheres, FA, and FA@Ag NCs. This may be attributed to the declining proportion of Fe_3O_4 contents caused by the successful modification of the large amount of Au seeds and Au@Ag nanocrystals. Consequently, the information obtained from above XRD and Mössbauer analysis preliminary confirm the successful construction of Fe_3O_4 hollow spheres, FA, FA@Ag, and CSSN NCs.

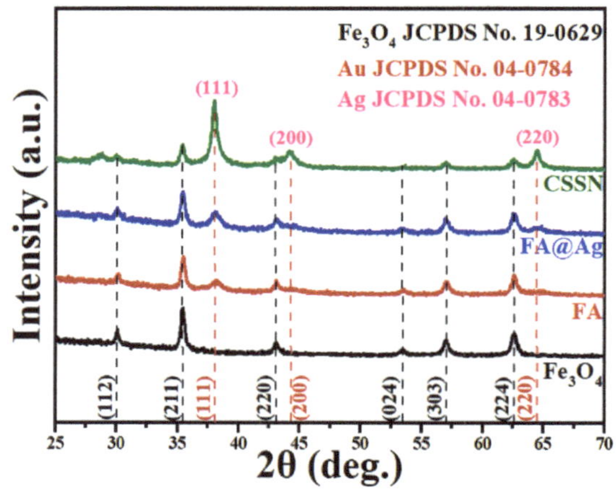

Figure 1. XRD pattern of Fe_3O_4 hollow spheres, FA, FA@Ag, and CSSN NCs.

For researching the magnetic properties of Fe_3O_4 hollow spheres, FA, FA@Ag, and CSSN NCs, their magnetic hysteresis (*M-H*) loops were tested via vibrating sample magnetometer (VSM), as displayed in Figure S2. The *M-H* loops show that Fe_3O_4 hollow spheres, FA, FA@Ag, and CSSN NCs have superparamagnetic property. The saturation magnetization (Ms) values of Fe_3O_4 hollow spheres, FA, FA@Ag, and CSSN NCs were 89.1, 74.5, 63.7, and 53.6 emu·g^{-1}, respectively. It was found that the Ms value gradually reduces with the incremental introduction of noble metal due to the diamagnetism of noble metal nanocrystals [52]. As depicted in the inset of Figure S2, the CSSN NCs can be collected by an external magnet within 50 s even if their Ms value is the lowest among all the materials. The remarkable magnetic response property means that CSSN NCs have great convenience in rapid separation and detection in complex liquid environments.

The morphology, size, and structure of the obtained products were studied by TEM, EDS elemental mapping, and EDS line scanning. As depicted in Figure 2a, the Fe_3O_4 nanocrystals show uniform spherical shapes with size of about 600 nm, and the dark edges and shallow cores of Fe_3O_4 nanocrystals suggest the formation of hollow structures. It can be seen from the TEM image of FA NCs (Figure 2b) that Au seeds (about 20 nm) are densely and uniformly loaded on the surfaces of Fe_3O_4 hollow spheres. Subsequently, the seed-mediated growth technique was used to grow Ag shell on the surfaces of FA NCs. Specifically, Au seeds on Fe_3O_4 hollow spheres were employed as nucleation sites for the formation of Ag shells. As exhibited in the dark-field TEM image of FA@Ag NCs (Figure 2c), with the continuous reduction of Ag^+ on surfaces of FA NCs, Ag shells with subtle roughness formed on the surfaces of FA NCs, and elements of Au and Ag are uniformly distributed on the surfaces of Fe_3O_4 hollow spheres, as shown in the corresponding EDS elemental mapping in Figure 2c. In order to create more hot spots, Au@Ag nanocrystals continued to be assembled on the surfaces of FA@Ag NCs to obtain CSSN NCs by PEI-DTC layers. As presented in Figure 2d, PEI-DTC layers with a thickness of around 15 nm are uniformly coated on the surfaces of FA@Ag NCs. For directly confirming the construction of core–shell–satellite structure in CSSN NCs, we took EDS line scanning profiles (Figure 2f) across CSSN NCs as presented by the orange highlighted line in the dark-field TEM image of Figure 2e. As shown in Figure 2f, it was found that the size of

CSSN NCs increased to around 800 nm due to the presence of Au seeds, Ag shells, and Au@Ag nanocrystals, which is bigger than that of Fe_3O_4 hollow spheres. Moreover, the elements show a symmetrical distribution with the change of detection position. The X-ray intensity of Au and Ag is maximum while Fe and O is minimum in the edge region. By comparison, the relative intensity of Fe and O increases gradually and almost no X-ray intensity of Au and Ag is observed in the direction of the orange arrow. This proves that the Au@Ag nanocrystals are adsorbed firmly on the outermost layer of FA@Ag@PEI-DTC by the affinities between the bidentate ligands with two chelating sulfur groups and Au@Ag nanocrystals [31]. Therefore, the EDS line-scanning results of CSSN NCs are consistent with the conclusions of XRD, TEM, and corresponding EDS mapping, which proves that the formation of CSSN NCs with the core–shell–satellite structure is convincing.

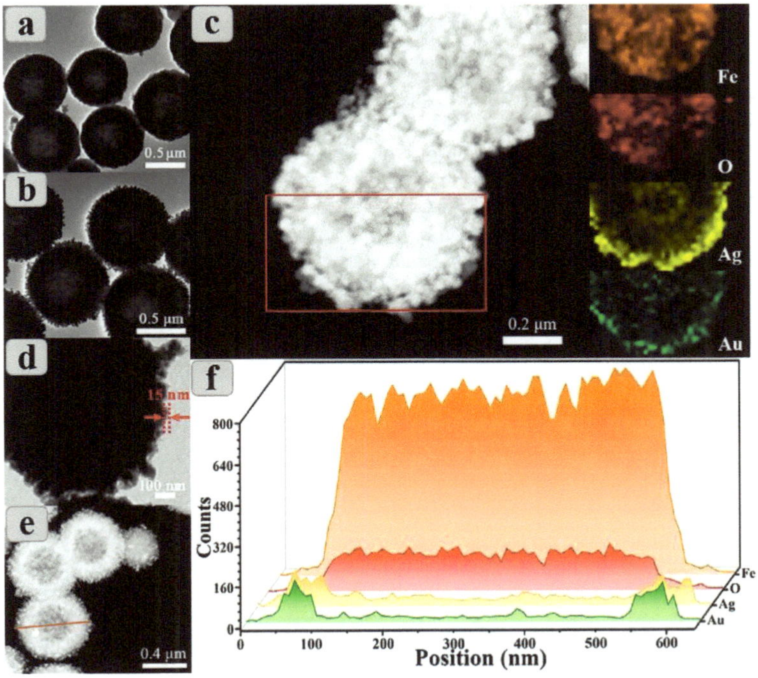

Figure 2. TEM images of Fe_3O_4 hollow spheres (**a**) and FA NCs (**b**). Dark-field TEM images and EDS elemental mapping results (O, Fe, Au and Ag) of FA@Ag NCs (**c**). TEM image of FA@Ag@PEI-DTC NCs (**d**). Dark-field TEM images (**e**) and EDS line scanning spectra of CSSN NCs (**f**).

The valence of elements in CSSN NCs was determined by XPS technology. Full XPS spectra of Fe_3O_4 hollow spheres, FA, FA@Ag, and CSSN NCs are exhibited in Figure S3. Within detection limit of XPS, Fe 2p, O1s, Au 4f, Ag 3d, and C 1s were observed and no impurity was found. High-resolution XPS results of Ag 3d and Au 4f are reflected in Figure 3. Ag 3d spectra in Figure 3a display peaks at 368.2 and 374.2 eV with a spin-orbit splitting of 6 eV for CSSN NCs, which are attributable to characteristics of Ag $3d_{3/2}$ and Ag $3d_{5/2}$ of Ag^0 [53]. As represented in Figure 3b, peaks of CSSN NCs at 84.1 and 87.8 eV with an energy difference of 3.7 eV are attributable to Au $4f_{7/2}$ and Au $4f_{5/2}$ of Au^0 [54,55]. An interesting phenomenon is that the binding energy of Ag 3d as well as Au 4f changes slightly with the incremental introduction of noble metals. The positions of Ag 3d peaks of CSSN NCs are blue-shifted compared with FA@Ag NCs, and the positions of Au 4f peaks are red-shifted compared with FA@Ag and FA NCs. This shift in binding energy may be ascribed to the charge transfer from metallic Au to Ag [56–58].

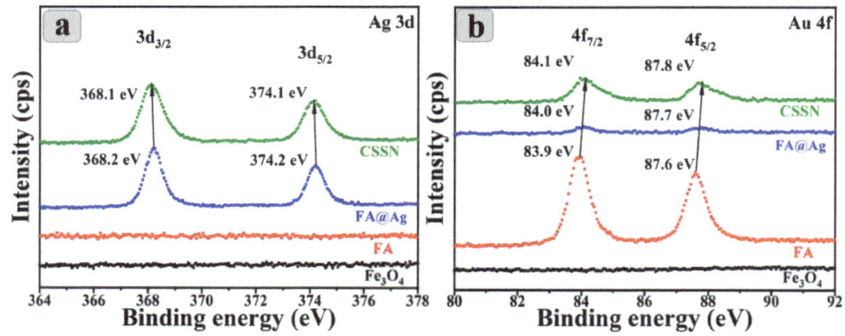

Figure 3. High-resolution XPS spectra of Fe_3O_4 hollow spheres, FA, FA@Ag, and CSSN NCs: Ag 3d (**a**) and Au 4f (**b**).

3.2. Choice of Excitation Source

A widely accepted consensus is that the SERRS method can further improve the sensitivity of Raman scattering spectroscopy, which combines resonance enhancement and SERS [59]. Hence, it is very important to find a suitable excitation source to arouse the SERRS effect. UV–Vis spectra of BPA azo products, FA, FA@Ag, and CSSN NCs were tested to determine a light source with an appropriate wavelength that takes into account the resonance effect of BPA azo products and plasmon resonance effect of CSSN NCs with the laser. As presented in Figure S4, the absorption positions of BPA azo products, FA, FA@Ag, and CSSN NCs are located at 450, 545, 536, and 512 nm, respectively. Because the plasmonic resonance peak of CSSN NCs is closer to the absorption position of BPA azo products compared with FA and FA@Ag NCs, the coupling between BPA azo products and CSSN NCs is considerably easier. Therefore, 514.5 nm laser was selected as the excitation source in this work given that it is more suitable for the coupling absorption.

3.3. SERRS Spectra of BPA Azo Product on FA, FA@Ag, and CSSN NCs

In order to directly evaluate the SERRS performance of various substrates, BPA (10^{-4} M) was chosen as the target molecule and FA, FA@Ag, and CSSN NCs served as SERRS substrates to explore their SERRS-enhancing capabilities, respectively. The detailed band assignments of BPA azo products are exhibited in Table S2 [37,60]. As reflected in Figure 4, with the incremental introduction of noble metals, the SERRS intensity of BPA gradually increases and CSSN NCs exhibit the strongest SERRS sensitivity compared with FA and FA@Ag NCs. The above phenomenon is foreseeable and will be discussed in detail below.

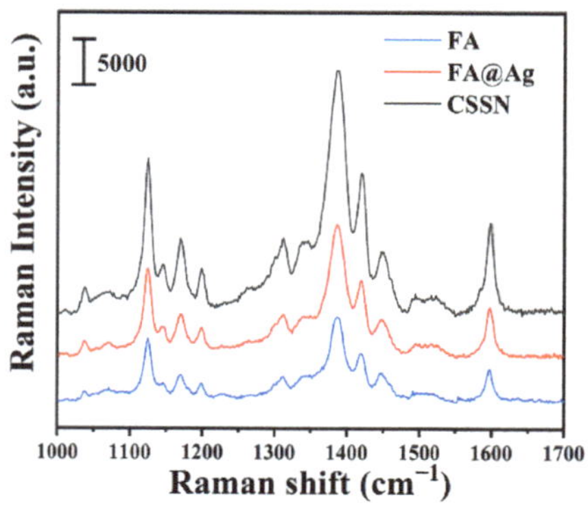

Figure 4. SERRS spectra after azo reaction of BPA azo products (10^{-4} M) on FA, FA@Ag, and CSSN NCs.

3.4. Mechanism of SERRS Enhancement

As mentioned above, the widely accepted theory for SERRS enhancement mechanism is EM mechanism, which derives from LSPR excitation of noble metal nanocrystals [61–63]. Compared with FA NCs, the surfaces of FA@Ag NCs are almost completely covered with Ag shells, which have better LSPR effect than Au, so it is reasonable that SERRS performance of FA@Ag NCs is somewhat better than that of FA NCs. In addition, strong electromagnetic fields will be excited in/between nearby noble metals because of the coupling effect and thus the SERRS signal intensity of target molecule can be significantly enhanced in hot spot regions [64–66]. Consequently, increasing the quantity of hot spots is an effective method to enhance the SERRS activity. To reveal why CSSN NCs have the highest SERRS enhancement, a FDTD theoretical algorithm was employed to visualize the distribution of electromagnetic field. As presented in Figure 5, it can be seen that more hot spots are generated on CSSN NCs. Compared with CSSN NCs, no obvious hot spots are found on the separate FA@Ag NCs (Figure 5a). A reasonable explanation is that the SERRS enhancement of FA@Ag NCs may come from hot spots generated by the aggregation of FA@Ag NCs in an actual detection procedure. As for CSSN NCs, large number of hot spots can also emerge in the region of narrow spacing between two adjacent Au@Ag nanocrystals, as shown in Figure 5b,c. In addition, there are large amounts of hot spots between Ag shells and outermost Au@Ag nanocrystals. It follows that the hot-spots effect is brought into full play through the construction of the core–shell–satellite structure. It also needs to be emphasized here that the introduction of the bimetallic Au@Ag nanocrystals is distinctly important, given that the Au@Ag nanocrystals make full use of excellent SERRS activity of Ag nanocrystals and high stability of Au nanocrystals [67,68]. Therefore, the excellent SERRS performance of CSSN NCs is attributed to a considerable quantity of hot spots generated by the core–shell–satellite structure, as well as excellent SERRS performance of Au@Ag nanocrystals.

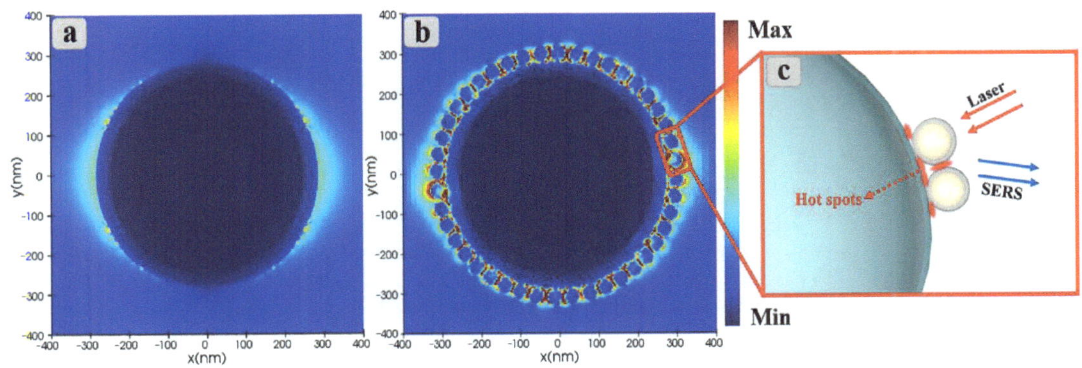

Figure 5. Electric field distribution of FA@Ag (**a**) and CSSN NCs (**b**) obtained by FDTD theoretical algorithm and the diagrammatic sketch of hot spots on CSSN NCs (**c**).

3.5. Quantitative Detection of BPA Azo Products

In order to assess the practicability of CSSN NCs as SERRS substrate for quantitative and sensitive detection of BPA, BPA azo products with different concentration from 10^{-10} to 10^{-4} M were chosen as probe molecules. Sharp and strong characteristic peaks of BPA azo products can be clearly observed from SERRS spectra illustrated in Figure 6a. SERRS intensities of BPA azo products rise monotonously with the increase of concentrations. The limit of detection (LOD) for the detection of BPA is as low as 10^{-10} M (about 0.023 ng/mL), which is well below the safety limit of the European Union (0.6 mg/kg), as well as China (10 ng/mL) [69]. More importantly, compared with the previous reports, our as-prepared CSSN substrate has the highest SERRS enhancement performance (Table 1) [70–75]. Moreover, the relationship between the concentrations of BPA azo products adsorbed on CSSN

NCs and the corresponding SERRS intensities at 1384 cm^{-1} is reflected in Figure 6b. The linear relationship versus the logarithm of the concentrations and correlation coefficient (R^2) is up to 0.96, which further proves that CSSN NCs are high performance SERRS sensors and can realize quantification of BPA down to 10^{-10} M.

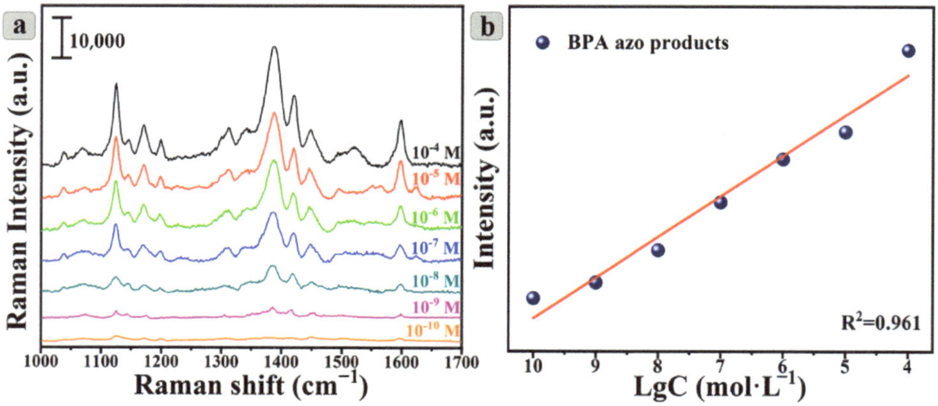

Figure 6. SERRS spectra of BPA azo products (from 10^{-10} to 10^{-4} M) adsorbed on CSSN NCs (**a**). The relationship between the SERRS intensity and concentrations of BPA azo products (**b**).

Table 1. Comparison of the SERS sensitivity of different SERS substrates to BPA.

SERS Substrates	LOD (M)	References
MoS_2/ZnO	10^{-9}	[70]
Au/MBA/PEG/BADGE	3×10^{-9}	[71]
Self-assembled graphitic substrates	10^{-6}	[72]
Ag-thiolated β-cyclodextrin	10^{-7}	[73]
MIPs@Ag NPs	5×10^{-8}	[74]
Ag@MIP	10^{-9}	[75]
CSSN	10^{-10}	this work

4. Conclusions

In conclusion, magnetic core–shell–satellite CSSN NCs for ultra-sensitive SERRS detection of BPA have been successfully developed. The coupling reactions between BPA and Pauly's reagents not only improved the affinity between BPA and substrates, but also amplified the SERRS signals due to the SERRS effect generated by the combination of resonance of BPA azo products and plasma resonance of noble metals. BPA azo products were chosen as target molecules to investigate the effect of incremental introduction of noble metals on SERRS activity. The distribution of electromagnetic field of CSSN NCs was studied through FDTD theoretical algorithm. The results revealed that a considerable number of hot spots were produced on the core–shell–satellite structure. The excellent SERRS activity of CSSN NCs was attributed to abundant hot spots of core–shell–satellite structure as well as outstanding SERRS activity of Au@Ag nanocrystals. BPA azo products were used to evaluate the practicability of CSSN NCs as SERRS substrate. When the concentrations of BPA azo products ranged from 10^{-10} to 10^{-4} M, SERRS intensities followed linear relationship versus the logarithm of the concentrations, and LOD was as low as 10^{-10} M. In addition, the Ms value of superparamagnetic CSSN NCs was 53.6 emu·g^{-1}, which gave CSSN NCs the function of rapid separation and detection in complex liquid environments by an external magnetic field. This study not only provides a novel ultra-sensitive SERRS substrate, but also shows enormous potential for the field of food safety and environmental pollution control.

Supplementary Materials: The following supporting information can be downloaded at: https://www.mdpi.com/article/10.3390/nano12193322/s1, Figure S1: Mössbauer spectrum of Fe_3O_4 hollow spheres; Figure S2: Magnetic hysteresis (*M-H*) loops of Fe_3O_4 hollow spheres, FA, FA@Ag and CSSN NCs (The inset is photograph of CSSN NCs dispersed in deionized water before and after magnet separation); Figure S3: Full XPS spectra of Fe_3O_4 hollow spheres, FA, FA@Ag and CSSN NCs; Figure S4: UV-Vis spectra of BPA azo products, FA, FA@Ag and CSSN NCs; Table S1: Mössbauer spectrum parameters of Fe_3O_4 hollow spheres; Table S2: Band assignments in the SERRS spectra of BPA azo products.

Author Contributions: Investigation, J.H. and T.Z.; Methodology, J.H.; Formal analysis, J.H. and W.Z.; Writing-original draft, J.H.; Conceptualization, M.Z. and Y.L.; Data curation, Z.Z.; Visualization, W.L.; Visualization, N.R.K.; Supervision, H.L. and Y.L.; Funding acquisition, Y.L.; Writing-review & editing; Y.L. All authors have read and agreed to the published version of the manuscript.

Funding: This work was supported by the National Natural Science Foundation of China, China (No. 21676115), Program for the development of Science and Technology of Jilin province, China (No. 20220203021SF, 20200301043RQ and 20200201022JC) and Program for Science and Technology of Education Department of Jilin Province, China (No. JJKH20210611KJ and JJKH20220444KJ).

Data Availability Statement: Not applicable.

Conflicts of Interest: The authors declare no conflict of interest.

References

1. Ismanto, A.; Hadibarata, T.; Kristanti, R.A.; Maslukah, L.; Safinatunnajah, N.; Kusumastuti, W. Endocrine disrupting chemicals (EDCs) in environmental matrices: Occurrence, fate, health impact, physio-chemical and bioremediation technology. *Environ. Pollut.* **2022**, *302*, 119061. [CrossRef] [PubMed]
2. Al Sharabati, M.; Abokwiek, R.; Al-Othman, A.; Tawalbeh, M.; Karaman, C.; Orooji, Y.; Karimi, F. Biodegradable polymers and their nano-composites for the removal of endocrine-disrupting chemicals (EDCs) from wastewater: A review. *Environ. Res.* **2021**, *202*, 111694. [CrossRef] [PubMed]
3. Astrahan, P.; Korzen, L.; Khanin, M.; Sharoni, Y.; Israel, A. Seaweeds fast EDC bioremediation: Supporting evidence of EE2 and BPA degradation by the red seaweed Gracilaria sp., and a proposed model for the remedy of marine-borne phenol pollutants. *Environ. Pollut.* **2021**, *278*, 116853. [CrossRef]
4. Kobroob, A.; Peerapanyasut, W.; Kumfu, S.; Chattipakorn, N.; Wongmekiat, O. Effectiveness of N-Acetylcysteine in the Treatment of Renal Deterioration Caused by Long-Term Exposure to Bisphenol A. *Biomolecules* **2021**, *11*, 655. [CrossRef] [PubMed]
5. Khan, M.R.; Ouladsmane, M.; Alammari, A.M.; Azam, M. Bisphenol A leaches from packaging to fruit juice commercially available in markets. *Food Packag. Shelf* **2021**, *28*, 100678. [CrossRef]
6. Sousa, S.; Maia, M.L.; Delerue-Matos, C.; Calhau, C.; Domingues, V.F. The role of adipose tissue analysis on Environmental Pollutants Biomonitoring in women: The European scenario. *Sci. Total Environ.* **2022**, *806*, 150922. [CrossRef] [PubMed]
7. Din, S.T.U.; Lee, H.; Yang, W. Z-Scheme Heterojunction of 3-Dimensional Hierarchical Bi_3O_4Cl/Bi_5O_7I for a Significant Enhancement in the Photocatalytic Degradation of Organic Pollutants (RhB and BPA). *Nanomaterials* **2022**, *12*, 767. [CrossRef] [PubMed]
8. Wang, X.; Nag, R.; Brunton, N.P.; Siddique, M.A.B.; Harrison, S.M.; Monahan, F.J.; Cummins, E. Human health risk assessment of bisphenol A (BPA) through meat products. *Environ. Res.* **2022**, *213*, 113734. [CrossRef]
9. Makowska, K.; Staniszewska, M.; Bodziach, K.; Calka, J.; Gonkowski, S. Concentrations of bisphenol a (BPA) in fresh pork loin meat under standard stock-farming conditions and after oral exposure—A preliminary study. *Chemosphere* **2022**, *295*, 133816. [CrossRef]
10. Escarda-Castro, E.; Herraez, M.P.; Lombo, M. Effects of bisphenol A exposure during cardiac cell differentiation. *Environ. Pollut.* **2021**, *286*, 117567. [CrossRef]
11. Santhi, V.A.; Hairin, T.; Mustafa, A.M. Simultaneous determination of organochlorine pesticides and bisphenol A in edible marine biota by GC-MS. *Chemosphere* **2012**, *86*, 1066–1071. [CrossRef] [PubMed]
12. Muruganathan, M.; Yoshihara, S.; Rakuma, T.; Shirakashi, T. Mineralization of bisphenol A (BPA) by anodic oxidation with boron-doped diamond (BDD) electrode. *J. Hazard. Mater.* **2008**, *154*, 213–220. [CrossRef] [PubMed]
13. Latif, A.; Maqbool, A.; Sun, K.; Si, Y. Immobilization of Trametes Versicolor laccase on Cu-alginate beads for biocatalytic degradation of bisphenol A in water: Optimized immobilization, degradation and toxicity assessment. *J. Environ.Chem. Eng.* **2022**, *10*, 107089. [CrossRef]
14. Xue, C.S.; Erika, G.; Jiri, H. Surface plasmon resonance biosensor for the ultrasensitive detection of bisphenol A. *Anal. Bioanal. Chem.* **2019**, *411*, 5655–5658. [CrossRef] [PubMed]
15. Jia, M.; Chen, S.; Shi, T.; Li, C.; Wang, Y.; Zhang, H. Competitive plasmonic biomimetic enzyme-linked immunosorbent assay for sensitive detection of bisphenol A. *Food Chem.* **2021**, *344*, 128602. [CrossRef]

16. Dai, J.; Baker, G.L.; Bruening, M.L. Use of Porous Membranes Modified with Polyelectrolyte Multilayers as Substrates for Protein Arrays with Low Nonspecific Adsorption. *Anal. Chem.* **2006**, *78*, 135–140. [CrossRef] [PubMed]
17. Xie, Y.; Ma, L.; Ling, S.; Ouyang, H.; Liang, A.; Jiang, Z. Aptamer-Adjusted Carbon Dot Catalysis-Silver Nanosol SERS Spectrometry for Bisphenol A Detection. *Nanomaterials* **2022**, *12*, 1374. [CrossRef]
18. Simonenko, N.P.; Musaev, A.G.; Simonenko, T.L.; Gorobtsov, P.Y.; Volkov, I.A.; Gulin, A.A.; Simonenko, E.P.; Sevastyanov, V.G. Kuznetsov, N.T. Hydrothermal Synthesis of Ag Thin Films and Their SERS Application. *Nanomaterials* **2021**, *12*, 136. [CrossRef] [PubMed]
19. Zhu, Z.; Bai, B.; You, O.; Li, Q.; Fan, S. Fano resonance boosted cascaded optical field enhancement in a plasmonic nanoparticle in-cavity nanoantenna array and its SERS application. *Light Sci. Appl.* **2015**, *4*, e296. [CrossRef]
20. Wu, R.; Jin, Q.; Storey, C.; Collins, J.; Gomard, G.; Lemmer, U.; Canham, L.; Kling, R.; Kaplan, A. Gold nanoplasmonic particles in tunable porous silicon 3D scaffolds for ultra-low concentration detection by SERS. *Nanoscale Horiz.* **2021**, *6*, 781–790. [CrossRef]
21. Morton, S.M.; Silverstein, D.W.; Jensen, L. Theoretical studies of plasmonics using electronic structure methods. *Chem. Rev.* **2011**, *111*, 3962–3994. [CrossRef] [PubMed]
22. Yang, S.; Yao, J.; Quan, Y.; Hu, M.; Su, R.; Gao, M.; Han, D.; Yang, J. Monitoring the charge-transfer process in a Nd-doped semiconductor based on photoluminescence and SERS technology. *Light Sci. Appl.* **2020**, *9*, 117. [CrossRef] [PubMed]
23. Zong, C.; Xu, M.; Xu, L.; Wei, T.; Ma, X.; Zheng, X.; Hu, R.; Ren, B. Surface-Enhanced Raman Spectroscopy for Bioanalysis: Reliability and Challenges. *Chem. Rev.* **2018**, *118*, 4946–4980. [CrossRef] [PubMed]
24. Pandey, P.; Seo, M.K.; Shin, K.H.; Lee, Y.W.; Sohn, J.I. Hierarchically Assembled Plasmonic Metal-Dielectric-Metal Hybrid Nano-Architectures for High-Sensitivity SERS Detection. *Nanomaterials* **2022**, *12*, 401. [CrossRef] [PubMed]
25. Sun, Y.; Zhang, C.; Yuan, Y.; Xu, M.; Yao, J. The moveable "hot spots" effect in an Au nanoparticles-Au plate coupled system. *Nanoscale* **2020**, *12*, 23789–23798. [CrossRef]
26. Shao, Y.; Li, S.; Niu, Y.; Wang, Z.; Zhang, K.; Mei, L.; Hao, Y. Three-Dimensional Dendritic Au-Ag Substrate for On-Site SERS Detection of Trace Molecules in Liquid Phase. *Nanomaterials* **2022**, *12*, 2002. [CrossRef]
27. Baffou, G.; Bordacchini, I.; Baldi, A.; Quidant, R. Simple experimental procedures to distinguish photothermal from hot-carrier processes in plasmonics. *Light Sci. Appl.* **2020**, *9*, 108. [CrossRef]
28. Weber, M.L.; Willets, K.A. Correlated Super-Resolution Optical and Structural Studies of Surface-Enhanced Raman Scattering Hot Spots in Silver Colloid Aggregates. *J. Phys. Chem. Lett.* **2011**, *2*, 1766–1770. [CrossRef]
29. Zhang, Y.; Wang, Z.; Zhang, Q.; Song, Y.; Zhang, B.; Ren, T.; Yang, H.; Wang, F. Polyethyleneimine mediated interaction for highly sensitive, magnetically assisted detection of tetracycline hydrochloride. *Appl. Surf. Sci.* **2020**, *505*, 144543. [CrossRef]
30. Wang, R.; Yan, X.; Ge, B.; Zhou, J.; Wang, M.; Zhang, L.; Jiao, T. Facile Preparation of Self-Assembled Black Phosphorus-Dye Composite Films for Chemical Gas Sensors and Surface-Enhanced Raman Scattering Performances. *ACS Sustain. Chem. Eng.* **2020**, *8*, 4521–4536. [CrossRef]
31. Liu, Y.; Zhang, Y.; Kou, Q.; Wang, D.; Han, D.; Lu, Z.; Chen, Y.; Chen, L.; Wang, Y.; Zhang, Y.; et al. Fe_3O_4/Au binary nanocrystals: Facile synthesis with diverse structure evolution and highly efficient catalytic reduction with cyclability characteristics in 4-nitrophenol. *Powder Technol.* **2018**, *338*, 26–35. [CrossRef]
32. Ji, Y.; Yang, S.; Guo, S.; Song, X.; Ding, B.; Yang, Z. Bimetallic Ag/Au nanoparticles: A low temperature ripening strategy in aqueous solution. *Colloid. Surface. A* **2010**, *372*, 204–209. [CrossRef]
33. Yang, Y.; Liu, J.; Fu, Z.; Qin, D. Galvanic replacement-free deposition of Au on Ag for core-shell nanocubes with enhanced chemical stability and SERS activity. *J. Am. Chem. Soc.* **2014**, *136*, 8153–8156. [CrossRef]
34. Ke, Y.; Chen, B.; Hu, M.; Zhou, N.; Huang, Z.; Meng, G. In-Situ Monitoring the SERS Spectra of para-Aminothiophenol Adsorbed on Plasmon-Tunable Au@Ag Core-Shell Nanostars. *Nanomaterials* **2022**, *12*, 1156. [CrossRef]
35. Svedendahl, M.; Verre, R.; Käll, M. Refractometric biosensing based on optical phase flips in sparse and short-range-ordered nanoplasmonic layers. *Light Sci. Appl.* **2014**, *3*, e220. [CrossRef]
36. Yin, W.; Wu, L.; Ding, F.; Li, Q.; Wang, P.; Li, J.; Lu, Z.; Han, H. Surface-imprinted SiO_2@Ag nanoparticles for the selective detection of BPA using surface enhanced Raman scattering. *Sensor. Actuat. B Chem.* **2018**, *258*, 566–573. [CrossRef]
37. Han, X.; Pienpinijtham, P.; Zhao, B.; Ozaki, Y. Coupling reaction-based ultrasensitive detection of phenolic estrogens using surface-enhanced resonance Raman scattering. *Anal. Chem.* **2011**, *83*, 8582–8588. [CrossRef] [PubMed]
38. Blanco-Covian, L.; Montes-Garcia, V.; Girard, A.; Fernandez-Abedul, M.T.; Perez-Juste, J.; Pastoriza-Santos, I.; Faulds, K.; Graham, D.; Blanco-Lopez, M.C. Au@Ag SERRS tags coupled to a lateral flow immunoassay for the sensitive detection of pneumolysin. *Nanoscale* **2017**, *9*, 2051–2058. [CrossRef]
39. Ishmukhametov, I.; Batasheva, S.; Rozhina, E.; Akhatova, F.; Mingaleeva, R.; Rozhin, A.; Fakhrullin, R. DNA/Magnetic Nanoparticles Composite to Attenuate Glass Surface Nanotopography for Enhanced Mesenchymal Stem Cell Differentiation. *Polymers* **2022**, *14*, 344. [CrossRef]
40. Kou, Y.; Wu, T.; Zheng, H.; Kadasala, N.R.; Yang, S.; Guo, C.; Chen, L.; Liu, Y.; Yang, J. Recyclable magnetic MIP-based SERS sensors for selective, sensitive, and reliable detection of paclobutrazol residues in complex environments. *ACS Sustain. Chem.* **2020**, *8*, 14549–14556. [CrossRef]
41. Dong, S.; Rene, E.R.; Zhao, L.; Xiaoxiu, L.; Ma, W. Design and preparation of functional azo linked polymers for the adsorptive removal of bisphenol A from water: Performance and analysis of the mechanism. *Environ. Res.* **2022**, *206*, 112601. [CrossRef] [PubMed]

42. Jaguey-Hernandez, Y.; Aguilar-Arteaga, K.; Ojeda-Ramirez, D.; Anorve-Morga, J.; Gonzalez-Olivares, L.G.; Castaneda-Ovando, A. Biogenic amines levels in food processing: Efforts for their control in foodstuffs. *Food Res. Int.* **2021**, *144*, 110341. [CrossRef] [PubMed]
43. Tuo, Y.; Liu, G.; Dong, B.; Zhou, J.; Wang, A.; Wang, J.; Jin, R.; Lv, H.; Dou, Z.; Huang, W. Microbial synthesis of Pd/Fe_3O_4, Au/Fe_3O_4 and PdAu/Fe_3O_4 nanocomposites for catalytic reduction of nitroaromatic compounds. *Sci. Rep.* **2015**, *5*, 13515. [CrossRef]
44. Qu, J.; Liu, G.; Wang, Y.; Hong, R. Preparation of Fe_3O_4–chitosan nanoparticles used for hyperthermia. *Adv. Powder Technol.* **2010**, *21*, 461–467. [CrossRef]
45. Pinto, P.S.; Lanza, G.D.; Souza, M.N.; Ardisson, J.D.; Lago, R.M. Surface restructuring of red mud to produce FeOx(OH)y sites and mesopores for the efficient complexation/adsorption of beta-lactam antibiotics. *Environ. Sci. Pollut. R.* **2018**, *25*, 6762–6771. [CrossRef]
46. Liu, Y.; Zhang, Y.; Kou, Q.; Chen, Y.; Han, D.; Wang, D.; Lu, Z.; Chen, L.; Yang, J.; Xing, G. Eco-friendly seeded Fe_3O_4-Ag nanocrystals: A new type of highly efficient and low cost catalyst for methylene blue reduction. *RSC Adv.* **2018**, *8*, 2209–2218. [CrossRef] [PubMed]
47. Chen, Y.; Zhang, Y.; Kou, Q.; Liu, Y.; Han, D.; Wang, D.; Sun, Y.; Zhang, Y.; Wang, Y.; Lu, Z.; et al. Enhanced Catalytic Reduction of 4-Nitrophenol Driven by Fe_3O_4-Au Magnetic Nanocomposite Interface Engineering: From Facile Preparation to Recyclable Application. *Nanomaterials* **2018**, *8*, 353. [CrossRef] [PubMed]
48. Han, D.; Li, B.; Chen, Y.; Wu, T.; Kou, Y.; Xue, X.; Chen, L.; Liu, Y.; Duan, Q. Facile synthesis of Fe_3O_4@Au core-shell nanocomposite as a recyclable magnetic surface enhanced Raman scattering substrate for thiram detection. *Nanotechnology* **2019**, *30*, 465703. [CrossRef] [PubMed]
49. Shen, J.; Zhu, Y.; Yang, X.; Zong, J.; Li, C. Multifunctional Fe_3O_4@Ag/SiO_2/Au core-shell microspheres as a novel SERS-activity label via long-range plasmon coupling. *Langmuir* **2013**, *29*, 690–695. [CrossRef] [PubMed]
50. Kahle, M.; Kleber, M.; Jahn, R. Review of XRD-based quantitative analyses of clay minerals in soils: The suitability of mineral intensity factors. *Geoderma* **2002**, *109*, 191–205. [CrossRef]
51. Hill, R.J.; Foxworthy, A.M.; White, R.J. PEAKS®: A PC-based method for quantitative X-ray diffraction phase analysis of lead-acid battery materials. *J. Power Sources* **1990**, *32*, 315–328. [CrossRef]
52. Zhao, W.; Zhang, D.; Zhou, T.; Huang, J.; Wang, Y.; Li, B.; Chen, L.; Yang, J.; Liu, Y. Aptamer-conjugated magnetic Fe_3O_4@Au core-shell multifunctional nanoprobe: A three-in-one aptasensor for selective capture, sensitive SERS detection and efficient near-infrared light triggered photothermal therapy of Staphylococcus aureus. *Sensor. Actuat. B Chem.* **2022**, *350*, 130879. [CrossRef]
53. Zhang, F.; Li, Y.; Qi, M.; Tang, Z.; Xu, Y. Boosting the activity and stability of Ag-Cu_2O/ZnO nanorods for photocatalytic CO_2 reduction. *Appl. Catal. B Environ.* **2020**, *268*, 118380. [CrossRef]
54. Li, Z.; Zhang, Y.; Zhang, H.; Yi, J. MOF-derived Au-loaded Co_3O_4 porous hollow nanocages for acetone detection. *Sensor. Actuat. B Chem.* **2021**, *344*, 130182. [CrossRef]
55. Kim, S.-H.; Kim, E.-M.; Lee, C.-M.; Kim, D.W.; Lim, S.T.; Sohn, M.-H.; Jeong, H.-J. Synthesis of PEG-Iodine-Capped Gold Nanoparticles and Their Contrast Enhancement in *In Vitro* and *In Vivo* for X-Ray/CT. *J. Nanomater.* **2012**, *2012*, 504026. [CrossRef]
56. Liu, Y.; Kou, Q.; Wang, D.; Chen, L.; Sun, Y.; Lu, Z.; Zhang, Y.; Wang, Y.; Yang, J.; Xing, S.G. Rational synthesis and tailored optical and magnetic characteristics of Fe_3O_4–Au composite nanoparticles. *J. Mater. Sci.* **2017**, *52*, 10163–10174. [CrossRef]
57. Guo, X.; Sun, X.; Guo, Y.; Guo, Y.; Wang, Y.; Wang, L.; Zhan, W. Enhanced catalytic performance for selective oxidation of propene with O_2 over bimetallic Au–Cu/SiO_2 catalysts. *Rare Met.* **2021**, *40*, 1056–1066. [CrossRef]
58. Kruse, N.; Chenakin, S. XPS characterization of Au/TiO_2 catalysts: Binding energy assessment and irradiation effects. *Appl. Catal. A Gen.* **2011**, *391*, 367–376. [CrossRef]
59. Xu, G.; Guo, N.; Zhang, Q.; Wang, T.; Song, P.; Xia, L. A sensitive surface-enhanced resonance Raman scattering sensor with bifunctional negatively charged gold nanoparticles for the determination of Cr(VI). *Sci. Total Environ.* **2022**, *830*, 154598. [CrossRef]
60. Liu, Y.; Chen, Y.; Zhang, Y.; Kou, Q.; Zhang, Y.; Wang, Y.; Chen, L.; Sun, Y.; Zhang, H.; MeeJung, Y. Detection and Identification of Estrogen Based on Surface-Enhanced Resonance Raman Scattering (SERRS). *Molecules* **2018**, *23*, 1330. [CrossRef]
61. Lin, S.; Guan, H.; Liu, Y.; Huang, S.; Li, J.; Hasi, W.; Xu, Y.; Zou, J.; Dong, B. Binary plasmonic assembly films with hotspot-type-dependent surface-enhanced Raman scattering properties. *ACS Appl. Mater. Inter.* **2021**, *13*, 53289–53299. [CrossRef] [PubMed]
62. Wang, Y.; Zhang, M.; Ma, H.; Su, H.; Li, A.; Ruan, W.; Zhao, B. Surface plasmon resonance from gallium-doped zinc oxide nanoparticles and their electromagnetic enhancement contribution to surface-enhanced Raman scattering. *ACS Appl. Mater. Inter.* **2021**, *13*, 35038–35045. [CrossRef] [PubMed]
63. Dias, E.J.C.; Yu, R.; Garcia de Abajo, F.J. Thermal manipulation of plasmons in atomically thin films. *Light Sci. Appl.* **2020**, *9*, 87. [CrossRef]
64. Li, J.; Zhang, Y.; Ding, S.; Panneerselvam, R.; Tian, Z. Core-shell nanoparticle-enhanced Raman spectroscopy. *Chem. Rev.* **2017**, *117*, 5002–5069. [CrossRef] [PubMed]
65. Yin, B.; Ho, W.K.H.; Zhang, Q.; Li, C.; Huang, Y.; Yan, J.; Yang, H.; Hao, J.; Wong, S.H.D.; Yang, M. Magnetic-Responsive Surface-Enhanced Raman Scattering Platform with Tunable Hot Spot for Ultrasensitive Virus Nucleic Acid Detection. *ACS Appl. Mater. Inter.* **2022**, *14*, 4714–4724. [CrossRef] [PubMed]

66. Zhang, Q.; Zhang, Y.; Chen, H.; Zhang, L.; Li, P.; Xiao, H.; Wu, W. One-dimensional nanohybrids based on cellulose nanocrystal and their SERS performance. *Carbohyd. Polym.* **2022**, *284*, 119140. [CrossRef] [PubMed]
67. Zhang, J.; Wu, C.; Yuan, R.; Huang, J.A.; Yang, X. Gap controlled self-assembly Au@Ag@Au NPs for SERS assay of thiram. *Foo Chem.* **2022**, *390*, 133164. [CrossRef]
68. Li, X.; Lin, X.; Fang, G.; Dong, H.; Li, J.; Cong, S.; Wang, L.; Yang, S. Interfacial layer-by-layer self-assembly of PS nanospheres an Au@Ag nanorods for fabrication of broadband and sensitive SERS substrates. *J. Colloid Interf. Sci.* **2022**, *620*, 388–398. [CrossRef
69. Huang, L.; Wang, X. Rapid and sensitive detection of Bisphenol A in water by LF-NMR based on magnetic relaxation switc sensor. *Microchem. J.* **2021**, *163*, 105911. [CrossRef]
70. Quan, Y.; Yao, J.; Yang, S.; Chen, L.; Li, J.; Liu, Y.; Lang, J.; Shen, H.; Wang, Y.; Wang, Y.; et al. ZnO nanoparticles on MoS microflowers for ultrasensitive SERS detection of bisphenol A. *Microchim. Acta* **2019**, *186*, 593. [CrossRef]
71. Marks, H.L.; Pishko, M.V.; Jackson, G.W.; Coté, G.L. Rational Design of a Bisphenol A Aptamer Selective Surface-Enhance Raman Scattering Nanoprobe. *Anal. Chem.* **2014**, *86*, 11614–11619. [CrossRef] [PubMed]
72. Lin, P.; Hsieh, C.; Hsieh, S. Rapid and Sensitive SERS Detection of Bisphenol A Using Self-assembled Graphitic Substrates. *Sc Rep.* **2017**, *7*, 16698. [CrossRef] [PubMed]
73. Roschi, E.; Gellini, C.; Ricci, M.; Sanchez-Cortes, S.; Focardi, C.; Neri, B.; Otero, J.C.; López-Tocón, I.; Smulevich, G.; Becucci, M Surface-Enhanced Raman Spectroscopy for Bisphenols Detection: Toward a Better Understanding of the Analyte–Nanosysten Interactions. *Nanomaterials* **2021**, *11*, 881. [CrossRef]
74. Wang, Z.; Yan, R.; Liao, S.; Miao, Y.; Zhang, B.; Wang, F.; Yang, H. In situ reduced silver nanoparticles embedded molecularl imprinted reusable sensor for selective and sensitive SERS detection of Bisphenol A. *Appl. Surf. Sci.* **2018**, *457*, 323–331. [CrossRef
75. Ren, X.; Cheshari, E.C.; Qi, J.; Li, X. Silver microspheres coated with a molecularly imprinted polymer as a SERS substrate fo sensitive detection of bisphenol A. *Microchim. Acta* **2018**, *185*, 242. [CrossRef] [PubMed]

nanomaterials

MDPI

Article

Mesenchymal Stem Cell-Mediated Deep Tumor Delivery of Gold Nanorod for Photothermal Therapy

Wan Su Yun [1,2,†], Man Kyu Shim [2,†], Seungho Lim [2], Sukyung Song [2], Jinseong Kim [1,2], Suah Yang [1,2], Hee Sook Hwang [3], Mi Ra Kim [4], Hong Yeol Yoon [2], Dong-Kwon Lim [1], In-Cheol Sun [2] and Kwangmeyung Kim [1,2,5,*]

1 KU-KIST Graduate School of Converging Science and Technology, Korea University, Seoul 02841, Korea
2 Biomedical Research Institute, Korea Institute of Science and Technology (KIST), Seoul 02792, Korea
3 Department of Pharmaceutical Engineering, Dankook University, Cheonan 31116, Korea
4 Department of Otorhinolaryngology-Head and Neck Surgery, Haeundae Paik Hospital, Inje University College of Medicine, Busan 48108, Korea
5 College of Pharmacy, Graduate School of Pharmaceutical Sciences, Ewha Womans University, Seoul 03760, Korea
* Correspondence: kimkm@ewha.ac.kr
† These authors contributed equally to this work.

Abstract: Gold nanoparticles (AuNPs) with various sizes and morphologies have been extensively investigated for effective photothermal therapy (PTT) against multiple cancer types. However, a highly dynamic and complex tumor microenvironment (TME) considerably reduces the efficacy of PTT by limiting deep tumor penetration of AuNPs. Herein, we propose a mesenchymal stem cell (MSC)-mediated deep tumor delivery of gold nanorod (AuNR) for a potent PTT. First, MSCs are treated with tetraacylated N-azidomannosamine ($Ac_4ManNAz$) to introduce modifiable azide (N_3) groups on the cell surface via metabolic glycoengineering. Then, AuNRs modified with bioorthogonal click molecules of bicyclo[6.1.0]nonyne (AuNR@BCN) are chemically conjugated to the N_3 groups on the MSC surface by copper-free click chemistry reaction, resulting in AuNR@MSCs. In cultured MSCs, the appropriate condition to incorporate the AuNR into the MSCs is optimized; in addition, the photothermal efficiency of AuNR-MSCs under light irradiation are assessed, showing efficient heat generation in vitro. In colon tumor-bearing mice, intravenously injected AuNR@MSCs efficiently accumulate within the tumor tissues by allowing deep tissue penetration owing to the tumor homing effect by natural tumor tropism of AuNR@MSCs. Upon localized light irradiation, the AuNR@MSCs significantly inhibit colon tumor growth by the enhanced photothermal effect compared to conventional AuNRs. Collectively, this study shows a promising approach of MSCs-mediated deep tumor delivery of AuNR for effective PTT.

Keywords: mesenchymal stem cell; gold nanorod; drug delivery; deep tumor penetration; photothermal therapy

1. Introduction

Light-mediated therapeutic approaches have emerged as an effective cancer treatment owing to unique advantages, such as minimal invasiveness, high selectivity, and spatiotemporal control [1–3]. Photothermal therapy (PTT), which can convert light to heat via photothermal agents, causes irreversible damage in cancer cells by inducing localized heat depending on the magnitude of the light exposure time and its fluence in a noninvasive manner [4]. Importantly, gold nanoparticles (AuNPs) that have various size and morphology of spheres, rods, shells, clusters, and cages have been extensively investigated for effective PTT in diverse cancer types [5–7]. Importantly, the AuNPs efficiently accumulate within the tumor tissues owing to the enhanced permeability and retention (EPR) effect based on their nano-sized structure; in addition, the tumor targeting efficiency can be further enhanced by modification with active-targeting ligands because AuNPs have great amenability for surface functionalization [8–11]. However, tumor tissues have a highly dynamic and complex microenvironment (TME), which is characterized by (i) regional blood flow to the tumors and increased vessel permeability; (ii) high stiffness, interstitial fluid pressure (IFP), and solid stress; and (iii) physical barriers imposed by the extracellular matrix (ECM) and increased ECM cross-linking [12,13]. These factors considerably reduce the tumor targeting efficiency of passive and active targeting strategies by limiting the deep tumor penetration of nanoparticles [14]. Therefore, there is a desperate need for a promising strategy to deliver AuNR to the deep tumor tissues for improving PTT efficiency.

Mesenchymal stem cells (MSCs) that can be easily isolated from fat liver, muscles, bone marrow, and many other places has become an attractive delivery system as a biovesicles to transport nanoparticles to tumor tissues [15]. The MSCs, that often lack major histocompatibility complex-II (MHC class II), prevent T cell responses and thus evade the immune systems complications in human and animal models [16]. Notably, MSCs exhibit natural tumor tropism premised on the site-specific expression of growth factors, such as epidermal growth factor, platelet-derived growth factor, and stromal-derived factor-1 [17]. Accordingly, they have an intrinsic homing nature to tumors by the activation of tumor-associated chemokine receptors; in addition to their tumor-targeting ability, recent studies have shown that MSCs hold a homing effect against specific tumors, which can be optimized by employing the appropriate types of MSCs to target a specific type of tumor [18]. Based on these distinct advantages, MSCs have been widely studied to enhance the deep tumor penetration of nanoparticles, which is mainly achieved by intracellular loading methods [19]. This is because the MSCs should encapsulate the nanoparticles based on physical and non-specific loading approaches owing to a lack of specific receptors compared to cancer cells [20]. However, this approach can diminish the natural functionality of MSCs and alter their in vivo fate; most importantly, it also shows limited drug loading efficiency, resulting in poor therapeutic efficacy.

Herein, we propose an MSCs-mediated deep tumor delivery of gold nanorod (AuNR) for a potent PTT by the incorporation of AuNR into the MSCs through metabolic glycoengineering and copper-free click chemistry reaction. The metabolic glycoengineering can incorporate modifiable chemical groups into the surface glycans of the cells [21–23]. When the cells are treated with unnatural metabolites such as $Ac_4ManNAz$, $Ac_4GalNAz$, and $Ac_4GlcNAz$, they utilize them to build blocks via an intrinsic glycan mechanism; as a result, azide (N_3) groups in such metabolites are exogenously generated on the cell surface [24]. The successful introduction of N_3 groups on the cell surface using unnatural metabolites was evaluated in stem cells as well as in various cancer (colon, breast, glioma, and prostate cancers) cells and normal (human fibroblast, cardiomyocytes, and human umbilical vein endothelial) cells, which was extensively employed for the biomedical application of tumor-specific imaging and drug delivery [21,23,25]. In particular, nanoparticles containing bio-orthogonal click molecules of bicyclo[6.1.0]nonyne (BCN) or dibenzylcyclooctyne (DBCO) can be chemically conjugated with N_3 group in the unnatural glycans by copper-free click chemistry reaction [25]. The important benefits of this strategy are in allowing high-amount loading of nanoparticles without affecting the intrinsic functions and fates of MSCs. Previous study has demonstrated a successful in vivo tracking of MSCs via magnetic resonance (MR) imaging by labeling the cells with superparamagnetic iron oxide nanoparticles via metabolic glycoengineering [26]. In this study, AuNRs modified with bio-orthogonal click molecules of BCN (AuNR@BCN) are prepared, and modifiable N_3 groups are introduced in MSCs by treating unnatural metabolites ($Ac_4ManNAz$) for metabolic glycoengineering; eventually, AuNR@BCN is incorporated into the MSCs (AuNR@MSCs) via copper-free click chemistry reaction (Scheme 1a). The AuNRs that are incorporated into the MSCs via copper-free click chemistry reaction of BCN and N_3 on the cell surface efficiently internalized into the cytoplasm of the MSCs owing to a membrane turnover mechanism. The biocompatibility of these nanoparticle internalization mechanisms by metabolic glycoengineering and copper-free click chemistry reaction has been evaluated, and it did not affect stem cell function [25]. When the AuNR@MSCs are intravenously injected into the colon tumor models, they efficiently penetrate deep inside the tumor tissues via a stem cell homing effect and induce local hyperthermia upon light irradiation (Scheme 1b). In cell culture systems, the appropriate condition to incorporate the AuNR into the MSCs is optimized. In addition, the photothermal efficiency of AuNR-MSCs are assessed under light irradiation, showing efficient heat generation owing to a high AuNR-loading efficiency compared to conventional intracellular loading methods. The superior antitumor efficacy of AuNR-MSCs by deep tumor penetration is assessed in colon tumor-bearing mice. Collectively, this study shows a promising approach of MSCs-mediated deep tumor delivery of AuNR for an effective PTT.

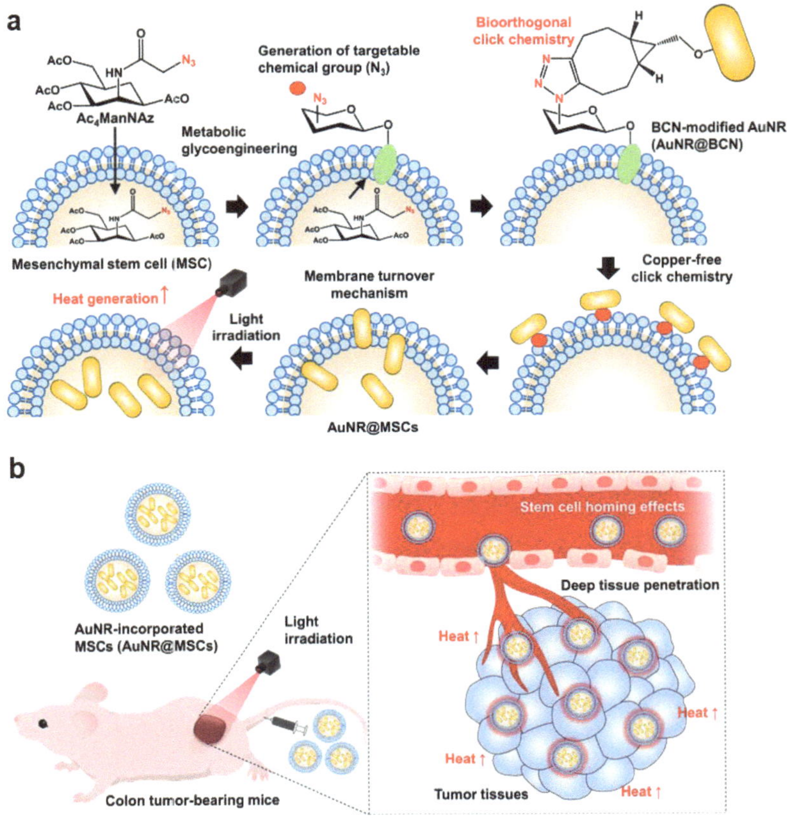

Scheme 1. Mesenchymal stem cells deliver gold nanorod to deep tumor tissues for photothermal therapy. (**a**) AuNRs modified with bio-orthogonal click molecules of BCN (AuNR@BCN) are prepared, and modifiable N_3 groups are introduced into MSCs by treating unnatural metabolites ($Ac_4ManNAz$) for metabolic glycoengineering; eventually, AuNR@BCN is incorporated into the MSCs (AuNR@MSCs) via copper-free click chemistry reaction. The AuNRs that are incorporated into the MSCs via copper-free click chemistry reaction of BCN and N_3 on the cell surface efficiently internalized into the cytoplasm of the MSCs owing to a membrane turnover mechanism. (**b**) When the AuNR@MSCs are intravenously injected into the colon tumor models, they efficiently penetrate deep inside the tumor tissues via a stem cell homing effect and induce local hyperthermia upon light irradiation.

1.1. Reagents

Gold(III) chloride trihydrate, silver nitrate, hydrochloric acid, tetraethyl orthosilicate (TEOS), (3-aminopropyl)trimethoxysilane (APTMS), bicyclo[6.1.0]non-4-yn-9-ylmethyl N-succinimidyl carbonate (BCN-NHS), L-ascorbic acid, and sodium borohydride were purchased from Sigma Aldrich (Oakville, ON, USA). Cetyltrimethylammonium bromide (CTAB) was purchased from Tokyo Chemical Industry (TCI, Tokyo, Japan). Cyanine5.5 NHS ester was purchased from Lumiprobe (Hunt Valley, MD, USA). Tem grid (Carbon Film 200 Mesh copper) was purchased from Electron Microscopy Sciences (Hatfield, PA, USA). Human adipose-derived mesenchymal stem cells (MSCs) and HT29 (human colon adenocarcinoma) were purchased from American Type Culture Collection (ATCC; Manassas, VA, USA). RPMI 1640 medium, fetal bovine serum (FBS), penicillin, and streptomycin were purchased from WELGENE Inc. (Daegu, Korea). Minimum Essential Medium α, fetal bovine serum (FBS), Dulbecco's phosphate-buffered saline (DPBS), EZ-Link phosphine-PEG3-biotin, and streptavidin-conjugated horseradish peroxidase (streptavidin-HRP) were purchased from Thermo Fisher Scientific (Waltham, MA, USA). Cell counting kit-8 (CCK-8) was purchased from Vitascientific (Beltsville, MD, USA). Tetraacetylated N-azidoacetyl mannosamine ($Ac_4ManNAz$) was purchased from Invitrogen (Rockford, IL, USA).

1.2. Preparation and Characterization of AuNR@BCN

To incorporate the AuNRs into the MSCs via copper-free click chemistry reaction, BCN groups were introduced in the AuNRs. First, CTAB (3.6445 g, 0.1 M), $HAuCl_4$ (19.6865 mg, 0.01 M), $AgNO_3$ (1.69 mg), and ascorbic acid (176 mg, 0.1 M) were stirred at 1100 rpm for 3 min, resulting in AuNR@CTAB. Then, AuNR@CTAB was mixed with 0.8 mM CTAB solution, followed by incubation with 0.1 M NaOH and 1 M TEOS (20% *v/v* in MeOH) for 15 min. The resulting $AuNR@SiO_2$ was further mixed with APTMS (0.04 mL) at 4 °C for 24 h to yield $AuNR@NH_2$. Finally, the $AuNR@NH_2$, BCN-NHS ester, and Cy5.5-NHS ester were dissolved in DMSO, and the solutions were stirred for 24 h at 4 °C, resulting in AuNR@BCN.

1.3. Preparation and Characterization of AuNR@MSCs

In order to generate azide (N_3) groups on the MSC surface via metabolic glycoengineering, 5×10^5 cells were seeded in the cell culture dishes, followed by treatment with 20 μM $Ac_4ManNAz$ for 48 h. Then, the cells were further incubated with 10 mM BCN-Cy5.5 at 37 °C for 2 h. For the fluorescence imaging, the cells were washed twice with DPBS, fixed with 4% paraformaldehyde for 20 min, and stained with 4′,6-diamidino-2-phenylindole (DAPI) for 15 min. The N_3 generation in the MSCs was observed via a Leica TCS SP8 laser-scanning confocal microscope (Leica Microsystems GmbH; Wetzlar, Germany) equipped with diode (405 nm), Ar (458, 488, 514 nm), and He-Ne (633 nm) lasers. To directly visualize the N_3 generation, fluorescence dye Cy5.5-conjugated BCN (BCN-Cy5.5, 2 μM), which can be chemically conjugated with N_3 on the cell surface, was further incubated with $Ac_4ManNAz$-treated MSCs. The successful N_3 generation on the MSC surface was also evaluated via Western blot analysis. Briefly, the $Ac_4ManNAz$-treated MSCs were lysed using RIPA buffer (1% SDS, 100 mM Tris-HCl, pH 7.4) with protease inhibitor for 1 h at 4 °C. After protein quantification by bicinchoninic acid protein assay (BCA), each 5 mg/mL cell lysate was mixed with 5 mM phophine-PEG_3-biotin and incubated at room temperature for 12 h. The proteins from each sample were mixed with 1× sodium dodecyl sulfate (SDS) gel-loading dye and boiled for 5 min. Then, 5 μg of proteins was separated by 12% SDS-polyacrylamide gel electrophoresis and subsequently transferred onto polyvinylidene fluoride (PVDF). The membranes were incubated with 1× TBST (10 mol/L Tris, 100 mol/L NaCl, and 0.1% Tween 20, pH 7.4) containing 5% bovine serum albumin (BSA). Finally, the membranes were further incubated with 1× TBST containing streptavidin-HRP for 2 h at room temperature, and the protein band was visualized by an enhanced chemiluminescence kit.

To incorporate the AuNR@BCN in the $Ac_4ManNAz$-treated MSCs via copper-free click chemistry reaction, 5×10^5 $Ac_4ManNAz$-treated MSCs were incubated with AuNR@BCN at a 200 μg/mL concentration for 6 h. Successful AuNR incorporation in the MSCs (AuNR@MSCs) was observed via a Leica TCS SP8 laser-scanning confocal microscope (Leica Microsystems GmbH; Wetzlar, Germany) equipped with diode (405 nm), Ar (458, 488, 514 nm), and He-Ne (633 nm) lasers. The photothermal efficiency of AuNR@MSCs was assessed after light irradiation with a power of 1.0 W/cm^2 (CW laser, Changchun New Industries Optoelectronics Tech. Co. Ltd., Changchun, China). The real-time temperature and thermal images were recorded using a digital thermometer (HH506A, OMEGA, Norwalk, CT 06854, USA) and IR camera (E6, FLIR Systems, Seoul, Korea), respectively.

1.4. Deep Tumor Penetration of AuNR@MSCs in Colon Tumor Models

The deep tumor penetration of AuNR@MSCs was assessed in colon tumor-bearing mice, which were prepared via subcutaneous inoculation of 1×10^7 HT29 cells into the left flank. When the tumor volumes were approximately 200 mm^3, AuNR@MSCs or AuNR@BCN with an equivalent concentration of 5 mg/kg of AuNR were intravenously injected into the mice. To administrate AuNR@MSCs with 5 mg/kg AuNR concentration, the amount of AuNR@MSCs were quantified by comparing with Cy5.5 fluorescence intensities of 5 mg/kg of AuNR@BCN (modified with Cy5.5). The tumor accumulation was observed by noninvasive near-infrared fluorescence (NIRF) imaging via an IVIS Lumina Series III system (PerkinElmer; Waltham, MA, USA). On day 5 after treatment, the major organs and tumor tissues were collected from the mice for ex vivo NIRF imaging, and fluorescence intensities were quantified using Living Image software (PerkinElmer, Waltham, MA, USA). In addition, tumor tissues from mice were cut into 10-μm thick sections for histology. Slide-mounted tumor sections were washed with DPBS three times and stained with GFP fluorescent dye-conjugated CD31 antibody at 4 °C for 12 h. Then, the nuclei of the tumor tissues were stained with DAPI for 15 min at dark condition and analyzed using a Leica TCS SP8 confocal laser scanning microscope (Fasanenstrasse 71, 10719 Berlin, Germany).

1.5. Therapeutic Efficacy and Toxicity Evaluation in Colon Tumor Models

The therapeutic efficacy and toxicity of AuNR@MSCs were assessed in HT29 tumor-bearing mice. Briefly, the mice were randomly divided into four groups: (i) saline, (ii) light irradiation only (Laser only), (iii) AuNR@BCN with light irradiation (AuNR+L), and (iv) AuNR@MSCs with light irradiation (AuNR@MSC+L). When the tumor volumes were approximately 100 mm^3, AuNR@MSCs or AuNR@BCN with an equivalent concentration of 5 mg/kg of AuNR were intravenously injected into the HT29 tumor-bearing mice. In addition, tumor tissues in the Laser only, AuNR+L, and AuNR@MSC+L groups were locally irradiated by light with a power of 1.0 W/cm^2 for 5 min; light irradiation was performed after 3 days of AuNR@MSCs or AuNR@BCN treatment. The therapeutic efficacy was assessed by measuring the tumor volumes, calculated as the largest diameter $\times$ smallest diameter^2 $\times$ 0.53. The tumor volumes and body weights were measured every day, and mice with a tumor size of 2000 mm^3 or higher were counted as dead.

1.6. Statistics

Statistical analyses were performed using GraphPad Prism 9 software (San Diego, CA 92108, USA). The statistical significance between two groups was analyzed using Student's t-test. One-way analysis of variance (ANOVA) was performed for comparisons of more than two groups, and multiple comparisons were analyzed using the Tukey–Kramer *post-hoc* test. Survival data were plotted as Kaplan–Meier curves and analyzed using the log-rank test. In the figures, statistical significance is indicated with asterisks (* $p < 0.05$, ** $p < 0.01$, *** $p < 0.001$).

1.7. Data Availability

All relevant data are available with the article and its Supplementary Information Files, or are available from the corresponding authors upon reasonable request.

2. Results and Discussion

2.1. Preparation and Characterization of AuNR@BCN

To incorporate the AuNR into the human adipose tissue-derived mesenchymal stem cells (MSCs) via metabolic glycoengineering and copper-free click chemistry reaction, AuNRs were modified with BCN (AuNR@BCN; Figure S1). The cetyltrimethylammonium bromide-stabilized gold nanorods (AuNR@CTAB) were used as a platform AuNPs due to their great amenability to modify its size, shape, and surface [5]. First, AuNR@CTAB was coated with silica (AuNR@SiO_2) because the protecting and shielding of the AuNR surface with silica shells can increase their stability and biocompatibility, resulting in the enhanced effectiveness of PTT [27]. The transmission electron microscope (TEM) images of the AuNR@SiO_2 clearly showed the silica-coated areas on the AuNR surface with a thickness of ~20 nm after modification of AuNR@CTAB (Figure 1a). To incorporate the bio-orthogonal click molecules onto the surface of AuNR, the AuNR@SiO_2 was functionalized with 3-aminopropyltriethoxysilane (APTES) using the Stöber method, resulting in AuNR@NH_2 [28]; finally, AuNR@BCN was obtained by the chemical conjugation of AuNR@NH_2 with BCN-succinimidyl carbonate (BCN-NHS ester) via copper-free click chemistry reaction. As shown in Figure 1a, there were no significant morphological changes in the AuNR@NH_2 and AuNR@BCN compared to AuNR@SiO_2, as confirmed by TEM images. The average size of AuNR@CTAB (about 80–90 nm) was slightly increased after silica-coating, wherein the AuNR@SiO_2, AuNR@NH_2, and AuNR@BCN have a similar average size of approximately 130–140 nm (Figure 1b). The zeta potential of AuNR@CTAB with high positive charge was changed to neutral after surface silica-coating, but that was reversed as a positive charge after modification with BCN (Figure 1c). Compared with the AuNR@CTAB (800 nm), the longitudinal SPR peaks of AuNR@SiO_2, AuNR@NH_2, and AuNR@BCN moved to 820 nm owing to the altered local environment around the AuNRs after silica-shell coating (Figure 1d). Next, the biocompatibility of AuNR@BCN was evaluated in the MSCs, showing no significant cytotoxicity after 48 h of treatment, with concentrations from 0 to 200 μg/mL (Figure 1e). Finally, the photothermal efficiency of each AuNR was assessed under light irradiation with a power of 1.0 W/cm^2. As shown in the photothermal images, the local temperature in the tubes was significantly increased up to 56 °C by AuNR@CTAB; in addition, AuNR@SiO_2, AuNR@NH_2, and AuNR@BCN showed comparable photothermal efficiency with AuNR@CTAB in the same experimental condition (Figure 1f,g). These results indicate that surface modification to incorporate the bio-orthogonal click molecules onto the surface of AuNR did not influence their basal photothermal efficiency. Taken together, as a photothermal agent modified with bio-orthogonal click molecules for incorporation into the MSCs, AuNR@BCN was successfully prepared without affecting the intrinsic characteristics of AuNRs, such as morphology, photothermal efficiency, and biocompatibility.

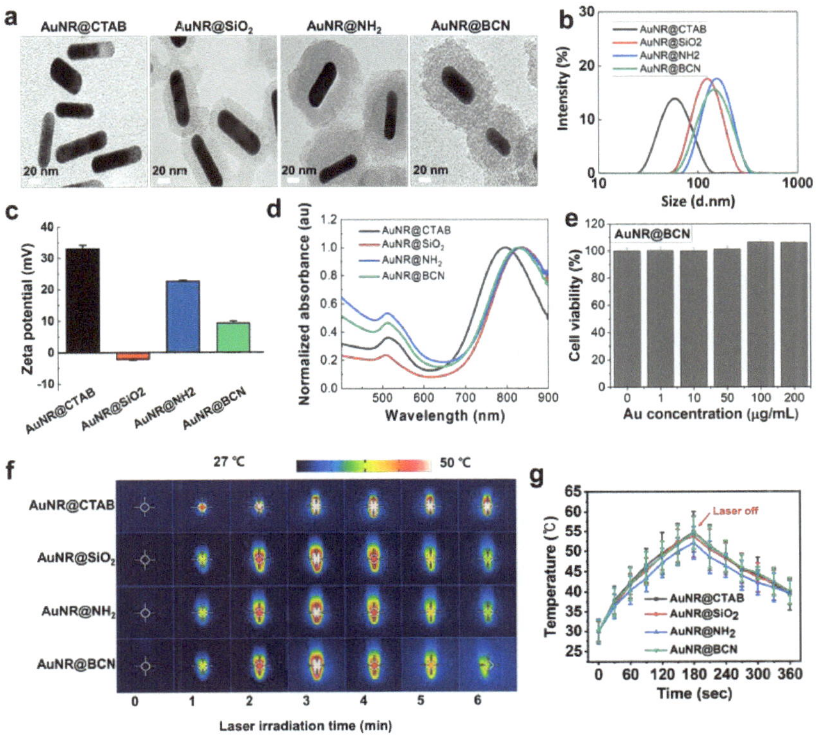

Figure 1. Preparation and characterization of AuNR@BCN. (**a**) TEM images of AuNR@CTAB, AuNR@SiO_2, AuNR@NH_2, and AuNR@BCN. (**b**) The size distribution of AuNR@CTAB, AuNR@SiO_2, AuNR@NH_2, and AuNR@BCN. (**c**) The zeta potential of AuNR@CTAB, AuNR@SiO_2, AuNR@NH_2, and AuNR@BCN. (**d**) UV spectrum of AuNR@CTAB, AuNR@SiO_2, AuNR@NH_2, and AuNR@BCN. (**e**) The viability of MSCs after 48 h of AuNR@BCN treatment with concentrations from 0 to 200 μg/mL. (**f**,**g**) The photothermal efficiency of AuNR@CTAB, AuNR@SiO_2, AuNR@NH_2, and AuNR@BCN under light irradiation with a power of 1.0 W/cm^2.

2.2. Optimization for Generation of Azide Groups on the Stem Cell Surface

In order to generate azide (N_3) groups on the surface of MSCs without affecting their intrinsic functions and fates, appropriate treatment periods and times of unnatural metabolites ($Ac_4ManNAz$) were optimized in vitro. When the MSCs were incubated with different concentrations of $Ac_4ManNAz$ (0–50 μM) for 48 h, the amount of N_3 generated on the cell surface was gradually increased in a dose-dependent manner up to 20 μM, but that was similar in the MSCs treated with 20 or 50 μM (Figure 2a). The N_3 generation on the MSC surface was further visualized via cellular fluorescence imaging, wherein the MSCs were further incubated with BCN modified with fluorescent dye, Cy5.5 (BCN-Cy5.5), for 2 h after treatment and with $Ac_4ManNAz$ (0–50 μM) for 48 h (Figure 2b). As expected, a strong Cy5.5 fluorescence signals on the cell surface were clearly observed in the MSCs treated with 20 μM $Ac_4ManNAz$, and those were similar with 50 μM $Ac_4ManNAz$-treated MSCs. Next, the cytotoxicity was evaluated by the different concentrations of $Ac_4ManNAz$ treatments in MSCs. The result showed that treatment of up to 20 μM $Ac_4ManNAz$ did not induce significant cell death of MSCs, but the cell viability of MSCs was significantly decreased compared to naive cells after 48 h of 50 μM $Ac_4ManNAz$ treatment (Figure 2c). This significant cytotoxicity can potentially cause a negative effect on the intrinsic functions and fates of MSCs. These results are consistent with a previous study that demonstrated the safety of $Ac_4ManNAz$ in stem cells [29]. Over 20 μM $Ac_4ManNAz$ treatment led to the inhibition of the functional properties of stem cells, such as proliferation rate, viability, rate of endocytosis, and genes related to cell adhesion. However, those effects by $Ac_4ManNAz$ treatment were not observed in MSCs when they were treated with 20 μM $Ac_4ManNAz$. Therefore, we further optimized the appropriate treatment time of $Ac_4ManNAz$ via cellular fluorescence imaging of MSCs treated with 20 μM $Ac_4ManNAz$; as described above, BCN-Cy5.5 was subsequently incubated with MSCs after $Ac_4ManNAz$ treatment to visualize the N_3 on the cell surface (Figure 2d). The result indicates that N_3 generation on the MSC surface was gradually increased in an incubation time-dependent manner, but the amount of N_3 groups generated on the cell surface were nearly similar in MSCs after 48 or 72 h of $Ac_4ManNAz$ treatment. In addition, the amounts of BCN-Cy5.5 conjugated with N_3 on the cell surface were significantly larger than the natural uptake of BCN-Cy5.5 that was confirmed in the MSCs without $Ac_4ManNAz$ treatment (Figure S2). From these results, we can expect that drug loading into the MSCs via metabolic glycoengineering and copper-free click chemistry could be considerably higher than conventional intracellular loading methods. Taken together, these results clearly demonstrate that treatment with 20 μM $Ac_4ManNAz$ for 48 h is the optimal condition to generate high amounts of N_3 groups on the MSC surface without affecting their intrinsic functions and fates.

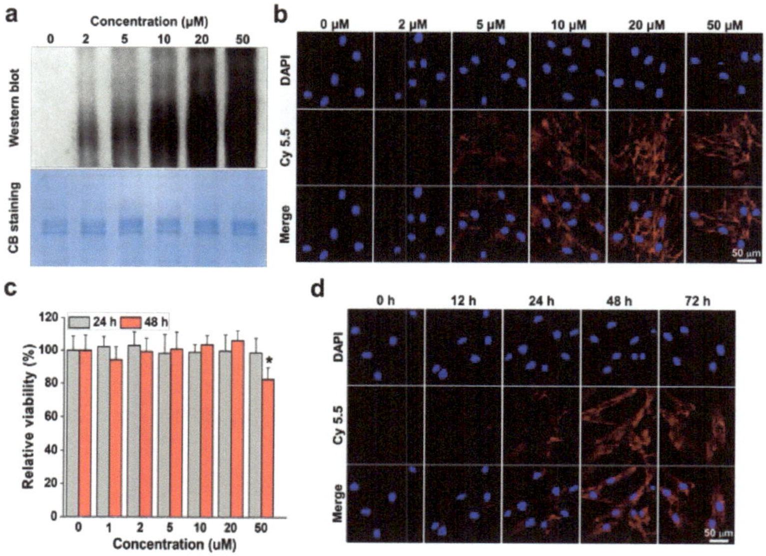

Figure 2. Optimization for the generation of azide groups on the stem cell surface. (**a**,**b**) N_3 generation in the MSCs after 48 h treatment of Ac4ManNAz with concentrations from 0 to 50 μM, confirmed by (**a**) Western blot analysis and (**b**) fluorescence imaging, respectively. (**c**) The viability of MSCs after 48 h of Ac4ManNAz treatment with concentrations from 0 to 50 μM. (**d**) N_3 generation in the MSCs after 20 μM treatment of Ac4ManNAz with different incubation times, * $p < 0.05$.

2.3. Preparation of AuNR-Incorporated MSCs (AuNR@MSCs) in Stem Cell Cultured System

Next, we prepared AuNR-incorporated MSCs (AuNR@MSCs) by the incubation of 20 μM $Ac_4ManNAz$-treated MSCs with AuNR@BCN. To efficiently monitor the AuNR incorporation in the stem cells via fluorescence imaging, AuNR@BCN modified with NHS-Cy5.5 (Cy5.5-AuNR@BCN) was used, wherein the Cy5.5-AuNR@BCN has BCN and Cy5.5 groups with a ratio of 9:1. First, the different concentrations (0–200 μg/mL) of Cy5.5-AuNR@BCN were treated with the $Ac_4ManNAz$-treated MSCs (Man^+). The fluorescence signals (red color) of the AuNRs were clearly observed on the cell surface, and they became gradually stronger in a dose-dependent manner; however, those fluorescence signals in the AuNR@MSCs were similar after 200 or 400 μg/mL of AuNR@BCN treatment (Figure 3a).

Notably, the amount of AuNRs incorporated into the cells was significantly higher in MSCs treated with $Ac_4ManNAz$ compared to naive MSCs (Man^-). These results clearly indicate that metabolic glycoengineering-based nanoparticle incorporation allows a higher loading capacity in the stem cells than conventional intracellular loading methods. The AuNR incorporation efficiency was also evaluated at the different incubation times after the treatment of MSCs with 200 µg/mL AuNR@BCN (Figure 3b). The cellular fluorescence imaging results showed that the amount of AuNRs incorporated into the MSCs was nearly similar after 6 h and 12 h of AuNR@BCN treatment. The successful incorporation of AuNRs into the MSCs was further confirmed via cryogenic electron microscopy, which clearly shows nano-sized rod morphology on the cell surface after 6 h of 200 µg/mL AuNR@BCN treatment (Figure 3c).

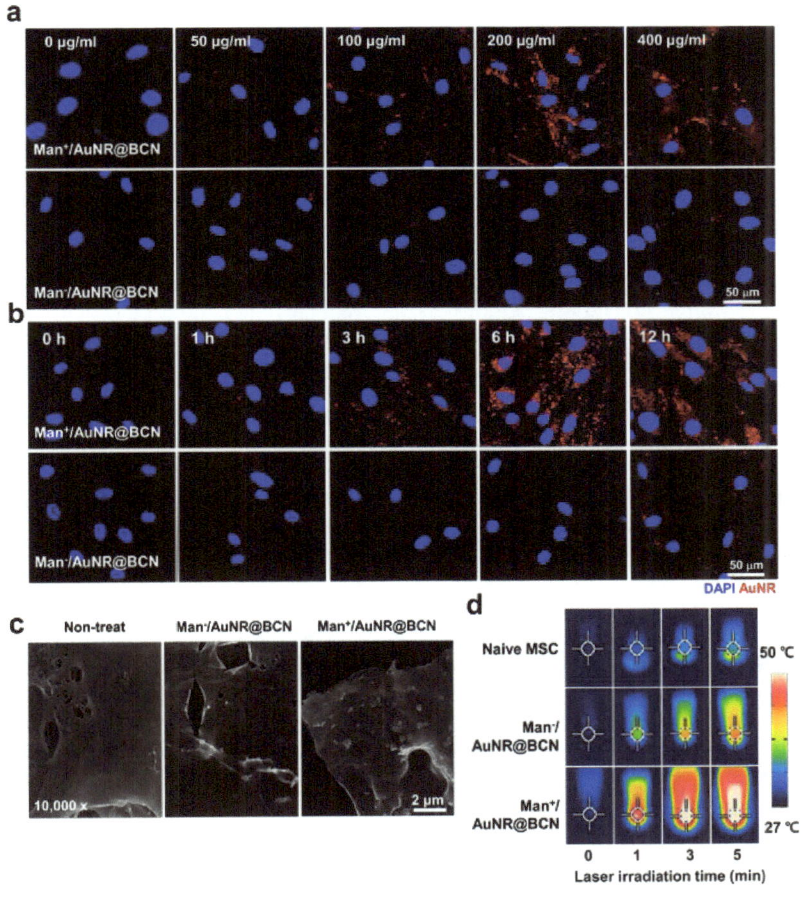

Figure 3. Preparation of AuNR-incorporated MSCs (AuNR@MSCs) in stem cell cultured system. (**a**) Fluorescence images of $Ac_4ManNAz$-treated or naive MSCs after incubation with different concentrations of Cy5.5-AuNR@BCN. (**b**) Fluorescence images of $Ac_4ManNAz$-treated or naive MSCs after incubation with 200 µg/mL of Cy5.5-AuNR@BCN for 0, 1, 3, 6, or 12 h. (**c**) Optical images of $Ac_4ManNAz$-treated or naive MSCs after incubation with 200 µg/mL of Cy5.5-AuNR@BCN for 6 h. (**d**) Photothermal efficiency of Man^+/AuNR@BCN and Man^-/AuNR@BCN under light irradiation with power of 1.0 W/cm^2.

Hence, the photothermal performance of AuNR@MSCs, which were prepared by the treatment of 200 μg/mL AuNRs for 6 h with MSCs pre-treated with 20 μM Ac4ManNAz for 48 h, was evaluated under light irradiation (808 nm, 1.0 W/cm^2). The photothermal images clearly showed a potent heat generation efficiency of AuNR@MSCs (Man$^+$/AuNR@BCN), wherein the local temperature in tubes was significantly increased up to 52 °C, along with the light irradiation time being longer (Figure 3d and Figure S3). Importantly, an increase in local temperature by AuNR@MSCs was significantly higher compared to MSCs incorporating AuNRs via conventional intracellular loading methods (Man$^-$/AuNR@BCN; 39 °C) after 6 min of light irradiation with a power of 1.0 W/cm^2. Taken together, the AuNR@MSCs prepared by the incorporation of AuNRs in the stem cells through metabolic glycoengineering and copper-free click chemistry reaction show considerable heat generation efficiency by allowing a high loading capacity of photothermal agents.

2.4. Deep Tumor Penetration of AuNR@MSCs in Colon Tumor Models

The deep tumor penetration of AuNR@MSCs was assessed in colon tumor models that were prepared by the subcutaneous inoculation of 1 × 10^7 HT29 cells. When the tumor volumes were approximately 200 mm^3, AuNR@MSCs or AuNR@BCN with an equivalent concentration of 5 mg/kg of AuNR were intravenously injected into mice.

Importantly, the tumor accumulation of AuNR@MSCs was significantly higher than that of AuNR@BCN, as confirmed by in vivo NIRF imaging of colon tumor-bearing mice (Figure 4a). The AuNR@MSCs in the tumor tissues was sustainably retained after 5 days of injection, whereas AuNR@BCN passively accumulated within the tumor tissues in the relatively lower levels and was rapidly removed from the tumors from 1 day of injection. These results are attributed to the natural homing effect by tumor tropism of stem cells. Quantitatively, the amount of AuNR@MSCs in the tumor tissues was 1.5–1.71-fold higher than AuNR@BCN on day 5 of injection (Figure 4b). Ex vivo NIRF images further showed 1.47–1.66-fold higher tumor accumulation of AuNR@MSCs than that of AuNR@BCN after 5 days of injection (Figure 4c). Histological analyses were additionally performed to confirm the deep tumor penetration of AuNR@MSCs. Interestingly, a strong Cy5.5 fluorescence (red color) of AuNR@MSCs was observed deep inside the tumor tissues along the CD31-positive blood vessels (green color) on day 5 of treatment (Figure 4d). Notably, AuNR@MSCs showed considerable accumulation in the whole region of the tumors compared with AuNR@BCN, indicating the efficient tumor accumulation of AuNRs by the MSC-mediated delivery strategy. More importantly, the Cy5.5 fluorescence (red color) of AuNR@MSCs was clearly observed in the central core region of the tumor tissues. These results demonstrate an effective deep tumor delivery of AuNRs by MSCs. As a control, only a little AuNR@BCN was observed in the tumor tissues owing to the limited targeting efficiency of nanoparticles by TME. These observations clearly demonstrate that AuNR@MSCs efficiently accumulate deep inside the tumor tissues via the intrinsic homing nature to tumors of stem cells [18]. From these results, we can also expect that AuNR@MSCs would promote a potent PTT under light irradiation owing to their considerable accumulation in whole tumor areas.

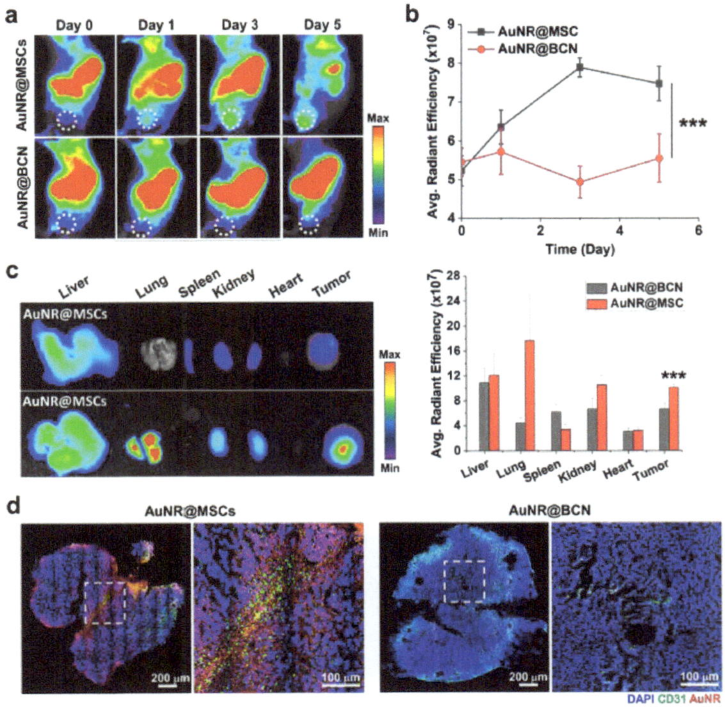

Figure 4. Deep tumor penetration of AuNR@MSCs in colon tumor models. (**a**) NIRF images of HT29 tumor-bearing mice after AuNR@MSCs or AuNR@BCN treatment. (**b**) Fluorescence intensities in the tumor tissues of HT29 tumor-bearing mice after AuNR@MSCs or AuNR@BCN treatment. (**c**) Ex vivo fluorescence images of major organs after 5 days of AuNR@MSCs or AuNR@BCN treatment. (**d**) Tumor tissues stained with fluorescent dye-conjugated anti-CD31 antibody after 5 days of AuNR@MSCs or AuNR@BCN treatment. Statistical significance is indicated with asterisks (*** $p < 0.001$).

2.5. Therapeutic Efficacy of PTT by AuNR@MSCs in Colon Tumor Models

The therapeutic efficacy of PTT by AuNR@MSCs under light irradiation was assessed in HT-29 colon tumor models. First, photothermal efficiency to generate heat under light irradiation was evaluated in mice (Figure 5a). Briefly, the mice were randomly divided into four groups of (i) saline, (ii) light irradiation only (Laser only), (iii) AuNR@BCN with light irradiation (AuNR+L), and (iv) AuNR@MSCs with light irradiation (AuNR@MSC+L). When the tumor volumes were approximately 100 mm^3, AuNR@MSCs or AuNR@BCN with equivalent concentration of 5 mg/kg of AuNR were intravenously injected into the mice. In addition, tumor tissues in the Laser only, AuNR+L, and AuNR@MSC+L groups were locally irradiated by light with a power of 1.0 W/cm^2 for 5 min; light irradiation was performed after 3 days of AuNR@MSCs or AuNR@BCN treatment. The photothermal images of the mice showed significant hyperthermia in the tumor tissues in the AuNR@MSC+L group (47.4 °C) compared to the saline (35.4 °C), Laser only (40.7 °C), and AuNR@BCN+L (43.4 °C) groups. The high photothermal efficiency in localized tumor tissues is attributable to the high loading capacity of AuNR@MSC and their deep tumor penetration, resulting in considerable accumulation. Next, the therapeutic efficacy was assessed by monitoring tumor growth after treatment with the same protocol as described above (Figure 5b). As expected, the mice in the AuNR@MSC+L group (396.06 ± 10.42 mm^3) showed significantly delayed tumor growth compared to those treated with saline (1758.2 ± 380.12 mm^3), Laser only (1190.91 ± 290.75 mm^3), and AuNR@BCN+L (857.08 ± 309.81 mm^3) on day 18 of treatment. The limited therapeutic efficacy of AuNR@BCN+L is due to the low tumor accumulation by highly dynamic and complex ECM that hinder deep tumor penetration of nanoparticles. With the intrinsic biocompatible characteristics of AuNPs, the body weights of mice in the AuNR@MSC+L or AuNR@BCN+L groups showed no significant changes compared to those in the saline group (Figure 5c). As a result, the median survival of mice in the saline, Laser only, and AuNR@BCN+L groups was determined to be 20, 24, and 28 days, respectively; wherein the mice were dead owing to the tumor progression (Figure 5d). In contrast, mice treated with AuNR@MSC+L all survived over 30 days and had significantly inhibited tumor growth. Consequentially, AuNR@MSC can promote intense hyperthermia in the tumor tissues under light irradiation via the MSC-mediated deep tumor delivery of AuNPs, leading to a potent PTT.

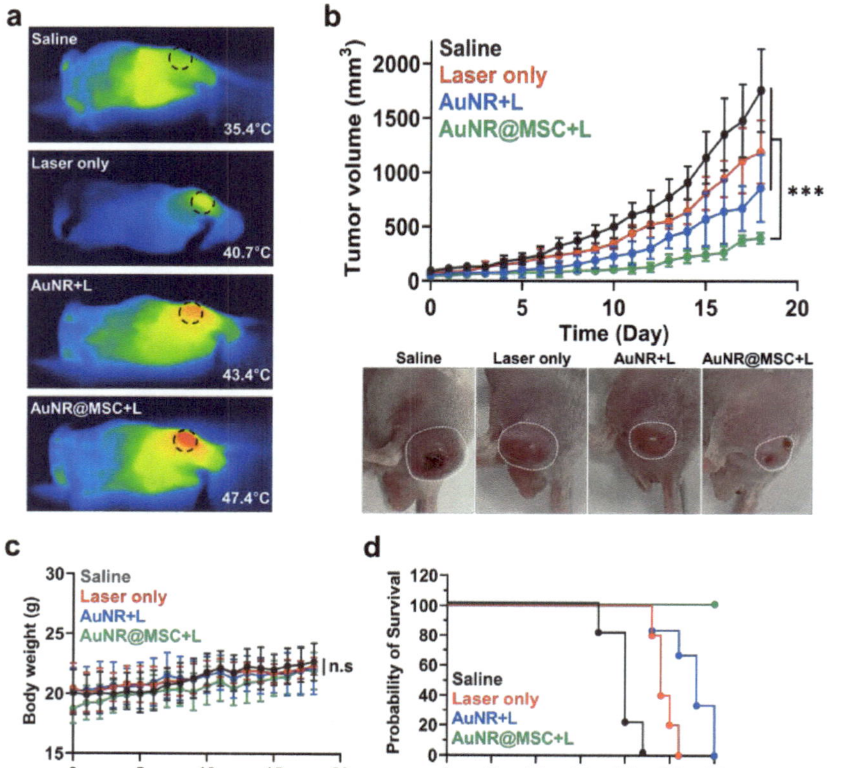

Figure 5. Therapeutic efficacy of PTT by AuNR@MSCs in colon tumor models. (**a**) Photothermal images of HT29 tumor-bearing mice after treatment with saline, Laser only, AuNR+L, or AuNR@MSC+L. (**b**) Tumor growth of HT29 tumor-bearing mice after treatment with saline, Laser only, AuNR+L, or AuNR@MSC+L. (**c**) Body weight changes during treatment. (**d**) Survival of mice after treatment with saline, Laser only, AuNR+L, or AuNR@MSC+L. Statistical significance is indicated with asterisks (*** $p < 0.001$).

3. Conclusions

In this study, we proposed MSC-mediated deep tumor delivery of AuNRs to trigger a potent PTT in colon tumors. First, AuNRs modified with bio-orthogonal click molecules (AuNR@BCN) were prepared, and their size distribution, morphology, photothermal efficiency, and biocompatibility were assessed in vitro. Then, AuNR@BCN was incorporated into the MSC via metabolic glycoengineering and copper-free click chemistry reaction. From these experiments, an appropriate condition to incorporate AuNR in the stem cells was carefully optimized. Importantly, the resulting AuNR@MSC showed high loading capacity compared to the conventional intracellular loading method, resulting in enhanced photothermal efficiency under light irradiation. Notably, the AuNR@MSC efficiently accumulated deep inside the tumor tissues owing to the tumor homing effect by the natural tumor tropism of stem cells when they were intravenously injected into the colon tumor models. As a result, the AuNR@MSC significantly inhibited colon tumor growth under local light irradiation. Overall, these findings suggested that MSC-mediated deep tumor delivery of AuNPs provide a new route for effective PTT. As such, this study introduced a novel technology for efficient delivery of nanoparticles using stem cells, which is potentially applicable for the treatment of a broad spectrum of diseases that require high targeting and deep tissue penetration.

Supplementary Materials: The following supporting information can be downloaded at: https://www.mdpi.com/article/10.3390/nano12193410/s1, Figure S1. Schematic illustration to show the protocol for the preparation of AuNR@MSCs. Figure S2. Fluorescence image of MSCs treated with BCN-Cy5.5 (without $Ac_4ManNAz$ treatment) for 48 h. Figure S3. The photothermal efficiency of Man^+/AuNR@BCN and Man^-/AuNR@BCN under light irradiation with a power of 1.0 W/cm^2.

Author Contributions: Conceptualization, K.K.; methodology, K.K., W.S.Y. and M.K.S.; validation, W.S.Y. and M.K.S.; formal analysis, K.K., W.S.Y. and M.K.S.; investigation, W.S.Y., M.K.S., S.L., S.S., J.K., S.Y., H.S.H., M.R.K., H.Y.Y. and I.-C.S.; resources, K.K. and D.-K.L.; data curation, K.K. and M.K.S.; writing—original draft preparation, K.K. and M.K.S.; visualization, K.K., W.S.Y. and M.K.S.; supervision, K.K.; project administration, K.K.; funding acquisition, K.K. and M.K.S. All authors have read and agreed to the published version of the manuscript.

Funding: This work was supported by grants from the National Research Foundation (NRF) of Korea, funded by the Ministry of Science (NRF-2022M3H4A1A03067401 and NRF-2021R1C1C2005460). This work was supported by a grant from the Research year of Inje University in 2018—0035.

Data Availability Statement: All relevant data are available with the article and its Supplementary Information Files, or are available from the corresponding authors upon reasonable request.

Conflicts of Interest: The authors declare no conflict of interest.

References

1. Choi, J.; Shim, M.K.; Yang, S.; Hwang, H.S.; Cho, H.; Kim, J.; Yun, W.S.; Moon, Y.; Kim, J.; Yoon, H.Y.; et al. Visible-Light-Triggered Prodrug Nanoparticles Combine Chemotherapy and Photodynamic Therapy to Potentiate Checkpoint Blockade Cancer Immunotherapy. *ACS Nano* **2021**, *15*, 12086–12098. [CrossRef]
2. Um, W.; Park, J.; Ko, H.; Lim, S.; Yoon, H.Y.; Shim, M.K.; Lee, S.; Ko, Y.J.; Kim, M.J.; Park, J.H.; et al. Visible light-induced apoptosis activatable nanoparticles of photosensitizer-DEVD-anticancer drug conjugate for targeted cancer therapy. *Biomaterials* **2019**, *224*, 119494. [CrossRef]
3. Cho, I.K.; Shim, M.K.; Um, W.; Kim, J.-H.; Kim, K. Light-Activated Monomethyl Auristatin E Prodrug Nanoparticles for Combinational Photo-Chemotherapy of Pancreatic Cancer. *Molecules* **2022**, *27*, 2529. [CrossRef] [PubMed]
4. Jung, H.S.; Verwilst, P.; Sharma, A.; Shin, J.; Sessler, J.L.; Kim, J.S. Organic molecule-based photothermal agents: An expanding photothermal therapy universe. *Chem. Soc. Rev.* **2018**, *47*, 2280–2297. [CrossRef] [PubMed]
5. Sperling, R.A.; Rivera Gil, P.; Zhang, F.; Zanella, M.; Parak, W.J. Biological applications of gold nanoparticles. *Chem. Soc. Rev.* **2008**, *37*, 1896–1908. [CrossRef] [PubMed]
6. Siddique, S.; Chow, J.C.L. Application of Nanomaterials in Biomedical Imaging and Cancer Therapy. *Nanomaterials* **2020**, *10*, 1700. [CrossRef]
7. Siddique, S.; Chow, J.C.L. Recent Advances in Functionalized Nanoparticles in Cancer Theranostics. *Nanomaterials* **2022**, *12*, 2826. [CrossRef]
8. Torchilin, V. Tumor delivery of macromolecular drugs based on the EPR effect. *Adv. Drug Deliv. Rev.* **2011**, *63*, 131–135. [CrossRef]
9. Shim, M.K.; Park, J.; Yoon, H.Y.; Lee, S.; Um, W.; Kim, J.-H.; Kang, S.-W.; Seo, J.-W.; Hyun, S.-W.; Park, J.H.; et al. Carrier-free nanoparticles of cathepsin B-cleavable peptide-conjugated doxorubicin prodrug for cancer targeting therapy. *J. Control. Release* **2019**, *294*, 376–389. [CrossRef]
10. Shim, M.K.; Na, J.; Cho, I.K.; Jang, E.H.; Park, J.; Lee, S.; Kim, J.-H. Targeting of claudin-4 by Clostridium perfringens enterotoxin-conjugated polysialic acid nanoparticles for pancreatic cancer therapy. *J. Control. Release* **2021**, *331*, 434–442. [CrossRef]

11. Moore, J.A.; Chow, J.C. Recent progress and applications of gold nanotechnology in medical biophysics using artificial intelligence and mathematical modeling. *Nano Express* **2021**, *2*, 022001. [CrossRef]
12. Yhee, J.Y.; Jeon, S.; Yoon, H.Y.; Shim, M.K.; Ko, H.; Min, J.; Na, J.H.; Chang, H.; Han, H.; Kim, J.-H.; et al. Effects of tumor microenvironments on targeted delivery of glycol chitosan nanoparticles. *J. Control. Release* **2017**, *267*, 223–231. [CrossRef] [PubMed]
13. Kyu Shim, M.; Yang, S.; Sun, I.-C.; Kim, K. Tumor-activated carrier-free prodrug nanoparticles for targeted cancer Immunotherapy: Preclinical evidence for safe and effective drug delivery. *Adv. Drug Deliv. Rev.* **2022**, *183*, 114177. [CrossRef] [PubMed]
14. Peng, J.; Yang, Q.; Shi, K.; Xiao, Y.; Wei, X.; Qian, Z. Intratumoral fate of functional nanoparticles in response to microenvironment factor: Implications on cancer diagnosis and therapy. *Adv. Drug Deliv. Rev.* **2019**, *143*, 37–67. [CrossRef]
15. Dalby, M.J.; García, A.J.; Salmeron-Sanchez, M. Receptor control in mesenchymal stem cell engineering. *Nat. Rev. Mater.* **2018**, *3*, 17091. [CrossRef]
16. Le Blanc, K. Immunomodulatory effects of fetal and adult mesenchymal stem cells. *Cytotherapy* **2003**, *5*, 485–489. [CrossRef]
17. Lim, S.; Yoon, H.Y.; Park, S.-J.; Song, S.; Shim, M.K.; Yang, S.; Kang, S.-W.; Lim, D.-K.; Kim, B.-S.; Moon, S.-H.; et al. Predicting in vivo therapeutic efficacy of bioorthogonally labeled endothelial progenitor cells in hind limb ischemia models via non-invasive fluorescence molecular tomography. *Biomaterials* **2021**, *266*, 120472. [CrossRef]
18. Reagan, M.R.; Kaplan, D.L. Concise Review: Mesenchymal Stem Cell Tumor-Homing: Detection Methods in Disease Model Systems. *Stem Cells* **2011**, *29*, 920–927. [CrossRef]
19. Zhang, X.; Yao, S.; Liu, C.; Jiang, Y. Tumor tropic delivery of doxorubicin-polymer conjugates using mesenchymal stem cells for glioma therapy. *Biomaterials* **2015**, *39*, 269–281. [CrossRef]
20. Lv, F.-J.; Tuan, R.S.; Cheung, K.M.C.; Leung, V.Y.L. Concise Review: The Surface Markers and Identity of Human Mesenchymal Stem Cells. *Stem Cells* **2014**, *32*, 1408–1419. [CrossRef]
21. Shim, M.K.; Yoon, H.Y.; Ryu, J.H.; Koo, H.; Lee, S.; Park, J.H.; Kim, J.-H.; Lee, S.; Pomper, M.G.; Kwon, I.C.; et al. Cathepsin B-Specific Metabolic Precursor for In Vivo Tumor-Specific Fluorescence Imaging. *Angew. Chem. Int. Ed.* **2016**, *55*, 14698–14703. [CrossRef]
22. Shim, M.K.; Yoon, H.Y.; Lee, S.; Jo, M.K.; Park, J.; Kim, J.-H.; Jeong, S.Y.; Kwon, I.C.; Kim, K. Caspase-3/-7-Specific Metabolic Precursor for Bioorthogonal Tracking of Tumor Apoptosis. *Sci. Rep.* **2017**, *7*, 16635. [CrossRef] [PubMed]
23. Lee, S.; Jung, S.; Koo, H.; Na, J.H.; Yoon, H.Y.; Shim, M.K.; Park, J.; Kim, J.-H.; Lee, S.; Pomper, M.G.; et al. Nano-sized metabolic precursors for heterogeneous tumor-targeting strategy using bioorthogonal click chemistry in vivo. *Biomaterials* **2017**, *148*, 1–15. [CrossRef]
24. Yoon, H.Y.; Shin, M.L.; Shim, M.K.; Lee, S.; Na, J.H.; Koo, H.; Lee, H.; Kim, J.-H.; Lee, K.Y.; Kim, K.; et al. Artificial Chemical Reporter Targeting Strategy Using Bioorthogonal Click Reaction for Improving Active-Targeting Efficiency of Tumor. *Mol. Pharm.* **2017**, *14*, 1558–1570. [CrossRef]
25. Lim, S.; Kim, W.; Song, S.; Shim, M.K.; Yoon, H.Y.; Kim, B.-S.; Kwon, I.C.; Kim, K. Intracellular Uptake Mechanism of Bioorthogonally Conjugated Nanoparticles on Metabolically Engineered Mesenchymal Stem Cells. *Bioconjugate Chem.* **2021**, *32*, 199–214. [CrossRef] [PubMed]
26. Lim, S.; Yoon, H.Y.; Jang, H.J.; Song, S.; Kim, W.; Park, J.; Lee, K.E.; Jeon, S.; Lee, S.; Lim, D.-K.; et al. Dual-Modal Imaging-Guided Precise Tracking of Bioorthogonally Labeled Mesenchymal Stem Cells in Mouse Brain Stroke. *ACS Nano* **2019**, *13*, 10991–11007. [CrossRef] [PubMed]
27. Mine, E.; Yamada, A.; Kobayashi, Y.; Konno, M.; Liz-Marzán, L.M. Direct coating of gold nanoparticles with silica by a seeded polymerization technique. *J. Colloid Interface Sci.* **2003**, *264*, 385–390. [CrossRef]
28. Wong, Y.J.; Zhu, L.; Teo, W.S.; Tan, Y.W.; Yang, Y.; Wang, C.; Chen, H. Revisiting the Stöber Method: Inhomogeneity in Silica Shells. *J. Am. Chem. Soc.* **2011**, *133*, 11422–11425. [CrossRef]
29. Han, S.-S.; Shim, H.-E.; Park, S.-J.; Kim, B.-C.; Lee, D.-E.; Chung, H.-M.; Moon, S.-H.; Kang, S.-W. Safety and Optimization of Metabolic Labeling of Endothelial Progenitor Cells for Tracking. *Sci. Rep.* **2018**, *8*, 13212. [CrossRef]

nanomaterials

MDPI

Article

Boosted Radiation Bystander Effect of PSMA-Targeted Gold Nanoparticles in Prostate Cancer Radiosensitization

Daiki Hara [1,2], Wensi Tao [1,3], Ryder M. Schmidt [1,2], Yu-Ping Yang [3,4], Sylvia Daunert [3,4], Nesrin Dogan [1,2], John Chetley Ford [1,2,*], Alan Pollack [1,3] and Junwei Shi [1,3,*]

1 Department of Radiation Oncology, Miller School of Medicine, University of Miami, Miami, FL 33136, USA
2 Department of Biomedical Engineering, College of Engineering, University of Miami, Miami, FL 33146, USA
3 Sylvester Comprehensive Cancer Center, Miller School of Medicine, University of Miami, Miami, FL 33136, USA
4 Department of Biochemistry and Molecular Biology, Miller School of Medicine, University of Miami, Miami, FL 33136, USA
* Correspondence: jcf137@med.miami.edu (J.C.F.); jxs1725@med.miami.edu (J.S.)

Abstract: Metal nanoparticles are effective radiosensitizers that locally enhance radiation doses in targeted cancer cells. Compared with other metal nanoparticles, gold nanoparticles (GNPs) exhibit high biocompatibility, low toxicity, and they increase secondary electron scatter. Herein, we investigated the effects of active-targeting GNPs on the radiation-induced bystander effect (RIBE) in prostate cancer cells. The impact of GNPs on the RIBE presents implications for secondary cancers or spatially fractionated radiotherapy treatments. Anti-prostate-specific membrane antigen (PSMA) antibodies were conjugated with PEGylated GNPs through EDC–NHS chemistry. The media transfer technique was performed to induce the RIBE on the non-irradiated bystander cells. This study focused on the LNCaP cell line, because it can model a wide range of stages relating to prostate cancer progression, including the transition from androgen dependence to castration resistance and bone metastasis. First, LNCaP cells were pretreated with phosphate buffered saline (PBS) or PSMA-targeted GNPs (PGNPs) for 24 h and irradiated with 160 kVp X-rays (0–8 Gy). Following that, the collected culture media were filtered (sterile 0.45 μm polyethersulfone) in order to acquire PBS- and PGNP- conditioned media (CM). Then, PBS- and PGNP-CM were transferred to the bystander cells that were loaded with/without PGNPs. MTT, γ-H2AX, clonogenic assays and reactive oxygen species assessments were performed to compare RIBE responses under different treatments. Compared with 2 Gy-PBS-CM, 8 Gy-PBS-CM demonstrated a much higher RIBE response, thus validating the dose dependence of RIBE in LNCaP cells. Compared with PBS-CM, PGNP-CM exhibited lower cell viability, higher DNA damage, and a smaller survival fraction. In the presence of PBS-CM, bystander cells loaded with PGNPs showed increased cell death compared with cells that did not have PGNPs. These results demonstrate the PGNP-boosted expression and sensitivity of RIBE in prostate cancer cells.

Keywords: gold nanoparticle; radiosensitization; radiation-induced bystander effect; prostate LNCaP cancer cells; prostate-specific membrane antigen; active targeting

1. Introduction

Prostate cancers contributed to about 34,500 deaths in the United States in 2022, according to American Cancer Society's estimates [1]. As a mainstay of cancer treatment, radiotherapy (RT) is commonly used to offer both definitive and palliative strategies for prostate cancer management [2]. The efficacy of RT in cancer therapy stems from the fact that ionizing radiation can directly and indirectly damage DNA and disrupt the atomic structure of biomolecules in the cellular environment [3]. In recent decades, we have witnessed a development boom in terms of high-precision RT techniques, such as intensity modulated radiotherapy (IMRT) and real-time adaptive MRI-guided radiation;

these techniques allow improved dose conformity to the tumor target as well as a decreased dose to adjacent healthy tissues [4]. Nevertheless, due to the similar mass energy absorption properties of both cancer and healthy tissues, physical radiation dose escalation and beam conformality has approached an upper limit with regard to prostate cancer external beam RT. Radiosensitizers, including small molecules, macromolecules, and nanomaterials, are promising agents that offer the means for further tumor dose escalation with improved normal tissue sparing [5].

Due to their high mass energy absorption coefficient relative to soft tissue, nanoparticles of high atomic number (Z) materials (such as: iodine, gadolinium, hafnium, tantalum, tungsten, bismuth) have been implemented to improve the contrast between tumors and healthy tissues, thus enabling tumor-specific radiosensitization with reduced side effects [6]. As a promising high-Z nano-radiosensitizer, gold nanoparticles (GNPs) have lately garnered attention due to their special properties; these include high biocompatibility with low toxicity, and the facile attachment of a variety of biological ligands [7]. The efficiency of GNP radiosensitization has been extensively validated in both in vitro and in vivo scenarios using numerous types of ionizing radiation, including kilovoltage (kV) and megavoltage (MV) photons as well as charged particles [8]. Compared with other nanoparticles, GNPs have been well studied for their efficient radiosensitizing effects, their multitude of mechanisms which allows radiosensitization to be carried out, and their comparatively limited toxicity [9]. Although dose enhancement factors vary based on radiation source, Jones et al. reported Monte Carlo simulated microscopic dose enhancement factors of 80× in 50 kVp photon beams and 9.8× in 6 MV photon beams, up to 100 μm, from the GNP surfaces [10]. Similarly, Lin et al. measured a dose enhancement factor of up to 14× in 150 MeV protons at a distance of 10 μm from the nanoparticle surface [11]. Aside from the incident beam, further optimization studies on GNP shape and size suggest that spherical nanoparticles between 10 nm to 20 nm provide the most optimal increase in secondary electron scatter and they minimize the level of toxicity in normal tissue [12].

Blood circulation pathway/time and the extent of tumor accumulation dictates the level of biodistribution, toxicity, and radiosensitization from GNPs. Polyethylene glycol (PEG) is ubiquitously used to coat GNPs, which significantly reduces nonspecific binding with cells and serum proteins, improves GNPs' stability and biocompatibility under physiological conditions, and it greatly lengthens the circulation half-life of GNPs in vivo [13]. GNPs passively leak into the tumor interstitium from blood vessels feeding the tumor via the enhanced permeation and retention (EPR) effect; this occurs due to the tumors' leaky vasculature and poor lymphatic drainage [14]. The EPR effect is the rationale behind the passive targeting approach; however, the efficiency of passive targeting is low, and it is still controversial as to whether the EPR effect is relevant in humans [15]. Therefore, the improved accumulation of GNPs with regard to tumors, beyond reliance on the EPR effect, is critical for GNP-based therapies. GNPs are ideal candidates for conjugating tumor-targeting agents because GNP surface chemistry enables a multitude of chemicals to bind at high densities [16]. Among potential targeting agents for prostate cancer, prostate specific membrane antigen (PSMA) ligands have been employed with great success in preclinical and clinical studies for PET imaging and radionuclide therapies; this is because prostate cancers consistently overexpress PSMA [17,18]. Our previous studies have shown the effectiveness of PSMA-targeted GNPs and their ability to accumulate at higher concentrations and to be retained for longer in prostate tumors [19].

Multiple studies have confirmed the radiosensitization of tumor-targeted GNPs in a multitude of tumor tissues, including prostate cancers [8]. It is worth mentioning that the radiosensitization observed in both in vitro and in vivo studies is often significantly greater than the dose enhancement predicted by Monte Carlo computational models [20]. These suggest that complex chemical and biological components, including enhanced ROS production, change during cell cycle distribution. Moreover, these components also affect the overall toxicity levels in cancer cells, and they are involved in GNP-induced radiosensitization [20]. Of these biochemical effects, the GNPs' effects on the radiation-induced bystander effect (RIBE) have recently received attention. RIBE is the phenomenon by which non-irradiated cells exhibit similar ionizing radiation damage as a result of signals received from nearby irradiated cells [21]. The bystander signals involved in this process may cause altered gene expression, DNA and chromosomal damage, cell proliferation alteration, or cell death in non-irradiated cells [22]. RIBEs have been demonstrated using a range of experimental systems with multiple biological endpoints; this has been the case since they were first identified by Nagasawa and Little in 1992 [23]. The underlying mechanisms mediating RIBE responses have been extensively studied and it has been shown that reactive oxygen and nitrogen species (ROS/NOS), DNA repair proteins, cytokines, ligands, extracellular DNA (ecDNA), microRNA (miRNA), and membrane molecules are the main materials that are released from targeted cells. Moreover, they are transferred to non-targeted cells via gap junction intercellular communication (GJIC) and the media/circulatory system [24].

In standard radiotherapy, where uniform fields are delivered and all cells are directly exposed to radiation, RIBE phenomena can be neglected; however, the role of RIBEs may become more influential when heterogeneous or non-uniform fields are considered. Although most clinical radiotherapy focuses on uniform exposures, there are some examples of non-uniform plans being utilized in the clinic, including spatially fractionated radiotherapy (or GRID), mini-beam radiotherapy (MBRT), microbeam radiotherapy (MRT), and dose-painting radiotherapy [25]. Healthy tissue has been shown to be more tolerant of spatially fractionated dose fields than tumor tissue, thus allowing for a high dose to be delivered in a single fraction. Spatially fractionated GRID purposely irradiates the tumor with highly non-uniform dose fields containing steep dose gradients [26,27]. Similarly to GRID therapy, MBRT and MRT are also characterized by alternating distributions of high and low doses, but on a much smaller scale (a hundred micrometers) [28]. Dose-painting radiotherapy allows for the heterogeneous delivery of high radiation doses within the tumor by targeting one or more regions of interest that are defined by functional imaging [29]. Furthermore, Monte Carlo simulation studies have shown that GNP-induced dose enhancement can be increased by a factor of 3000 compared with doses originating from a hypothetical water nanoparticle, but only at microscopic distances of 10 μm [10,11]. The heterogenous distribution of GNPs, combined with the highly localized energy depositions made by GNPs, resemble the dose pattern of spatially fractionated radiotherapy [30]; thus, RIBE plays an important role in GNP-enhanced radiation therapy.

RIBE responses have been observed in many cell types such as lymphocytes, fibroblasts, endothelial, and cancer cells [31]. Rostami et al. first investigated the effects of glucose-coated GNPs (Glu-GNPs) on the RIBE in MCF-7 (human breast cancer) and QUDB (human lung cancer) cell lines [32]. Their results demonstrated Glu-GNPs' enhanced RIBEs in QUDB cells, but there was no RIBE enhancement in MCF-7 cells. This observation suggests that the impact of GNPs on the RIBE is cell type specific (i.e., some cell types are unable to produce bystander signals whereas others are unable to respond to bystander signals). To the best of our knowledge, there are no studies focusing upon the impact of GNP-induced radiosensitization on the RIBE in prostate cancer cells. The present study was carried out to investigate how actively targeting GNPs impacts the RIBE response in prostate cancer LNCaP cells. LNCaP cells were selected as the focus in this study because they can model a wide range of prostate cancer stages, including the transition from androgen dependence to castration resistance and bone metastasis. Moreover, the high expression of PSMA means that the LNCaP cancer model is a good choice with which to develop active-targeted nanotherapeutic strategies for prostate cancer. Anti-PSMA antibodies were conjugated with PEGylated GNPs to develop PSMA-targeted GNPs (PGNPs). A conditioned medium transfer technique was employed to evaluate the RIBE responses. The yield of the RIBE signal from irradiated cells, and the sensitivity of non-irradiated cells with regard to the RIBE signal, was used to investigate the impact of PGNPs on the RIBE in prostate cancer cells. As control groups, RIBE responses were also implemented for use on other prostate cancer cell lines (including PC3, and 22Rv1) as well on normal prostate cell lines (RWPE-1).

2. Materials and Methods

2.1. Synthesis of PSMA-Targeted GNPs (PGNPs)

Commercially available PEGylated GNPs (Creative Diagnostics, NY, USA) were used as substrates for functionalization in this study. The GNP concentration was quantified by optical absorption spectra which were determined using UV-vis spectroscopy (240 nm–780 nm, Nanodrop, Thermo Fisher Scientific, Waltham, MA, USA). To realize active and passive prostate cancer targeting, anti-PSMA and mouse IgG antibodies (Creative Diagnostics, NY, USA) were coupled with PEGylated GNPs using EDC/NHS chemistry. First, 0.2 M 1-ethyl-3-(-3-dimethylaminopropyl) carbodiimide hydrochloride (EDC) (#22980; Thermo Fisher Scientific, Waltham, MA, USA) and 0.2 M N-hydroxysuccinimide (NHS) (#24500; Thermo Fisher Scientific, Waltham, MA, USA) were simultaneously added to a solution that included OD = 1 GNPs, 0.1 mg/mL antibody, and 0.1 M sodium borate buffer. This mixture was incubated at room temperature for 24 h. Then, the conjugation solution was washed out by three centrifugations at 15,000× *g* for 30 min, and the final GNP pellet was stored in Milli-Q water (Millipore, Bedford, MA, USA) to obtain the desired concentration.

2.2. Characterization of PGNPs

Transmission electron microscopy was used to determine the shape and size of the developed PGNPs. PGNPs were cast onto a carbon Formvar-coated copper grid for 30 min. Excess liquid was absorbed using filter paper, and the grid was allowed to air dry overnight. The grids were viewed at 80 kV in a JEOL JEM-1400 transmission electron microscope (JEOL USA, Peabody, MA, USA) and images were captured with an AMT BioSprint digital camera (Advanced Microscopy Techniques, Woburn, MA, USA). Additional characterizations of nanoparticles have been detailed in our previously published works that focus on PGNPs [19].

2.3. Cell Culture

Human prostate cancer cells (LNCaP, PC3, and 22Rv1) were obtained from ATCC. Cells were cultured in RPMI1640 media (ThermoFisher Scientific, Waltham, MA, USA) with 10% fetal bovine serum (FBS, GeminiBio, Sacramento, CA, USA) and 1% penicillin-streptomycin (ThermoFisher Scientific, Waltham, MA, USA). The cells were grown in a CO_2 incubator at 37 °C and subcultured when they reached 80–90% confluency. Normal prostate cells (RWPE-1) were cultured in the same conditions, although keratinocyte serum-free media (ThermoFisher Scientific, Waltham, MA, USA) was used instead of RPMI1640.

2.4. PGNP Cytotoxicity

A MTT viability assay was used to assess the cytotoxicity of treatments on LNCaP cells. Moreover, 24 h after the treatments, LNCaP cells in 96-well plates were stained with 1 M yellow tetrazolium substrate (3-(4,5-dimethylthiazol-2-yl)-2, 5-diphenyltetrazolium bromide) (MTT, Thermo Fisher Scientific, Waltham, MA, USA) in serum-free RPMI1640, for 4 h, at 37 °C. The supernatant was removed from the wells and formazan crystals were dissolved with the addition of 100 µL DMSO. This was added to each well and left for 30 min at 37 °C. An additional column of empty wells received 100µL of DMSO, which were to be read as blanks. Microplates were read on a Clariostar microplate (BMG Labtech, Ortenberg, Germany) reader for absorbance at 570 nm. MTT assay results were normalized to ensure no treatment absorptions.

2.5. Cellular PGNP Uptake Assay

LNCaP cells were treated with 50, 100, 150, 250, 500, and 750 µg/mL PGNPs in serum-free RPMI1640 media for 24 h. Serum-free media were used to avoid changes to the nanoparticle surface as a result of FBS proteins. After 24 h, the cells were washed three times with PBS, and they were replenished with fresh media supplemented with FBS and pen-strep prior to running assays. PGNP concentrations were measured using

UV-Vis spectrophotometry on a Nanodrop Spectrophotometer before and after treatment. Spectrophotometry was conducted using a range between 280 nm to 700 nm, where the peak intensity at 520 nm was used to quantify PGNP concentration; this was achieved by benchmarking against the manufacturer's recorded GNP optical density. The total cellular uptake of PGNP was calculated as the PGNP's concentration difference in media before and after 24 h of treatment. Cells plated on coverslips were used to identify the intracellular distribution of PGNPs with TEM. PGNP-treated cells were fixed overnight in 2% glutaraldehyde, in 0.1 M phosphate buffer; then, they were post-fixed for 1 h in 2% osmium tetroxide in 0.1 M phosphate buffer, dehydrated through a series of graded ethanols, and embedded in an EM-bed (Electron Microscopy Sciences, Fort Washington, PA, USA). The glass coverslip was dissolved in hydrofluoric acid. In addition, 100 nm sections were cut on a Leica Ultracut EM UC7 ultramicrotome and stained with uranyl acetate and lead citrate. Images were captured on the same setup as the PGNPs on grids.

2.6. RIBE on LINCaP Cells

Conditioned media was collected from LNCaP cells to induce RIBE in LNCaP bystander cells. Prior to irradiation, cells were washed three times with PBS, and serum-free RPMI1640 was added to the culture plates. Cells were irradiated between 0–8 Gy at 2.0 Gy/min on a 160 kV (0.0775 Å) Radsource RS-2000 Biological Irradiator (Rad Source Technologies, Buford, GA, USA) with cells placed at 28 cm SSD (25 cm diameter circular field size). Following irradiation, cells were incubated at 37 °C for 1 h. Media from sham and irradiated cells were harvested and filtered through a 0.45 μm polyethersulfone filter (VWR, Radnor, PA, USA). All conditioned media were prepared immediately before use.

2.7. Effects of PGNPs on RIBE

To assess the effect of PGNPs on the RIBE, LNCaP cells were divided into three groups; bystander cells treated with conditioned media from irradiated cells (PBS-CM), bystander cells treated with conditioned media from PGNP treated irradiated cells (PGNP-CM), and PGNP treated bystander cells treated with conditioned media (PBS-CM+PGNP).

2.8. Clonogenic Assay

LNCaP cells were grown in 6-well dishes, and they were treated as mentioned above. Irradiated or PBS-CM treated cells were trypsinized, then replated on 6 cm dishes with varying cell numbers. Following 3 weeks of growth, plates were washed three times with cold PBS, fixed in 4% paraformaldehyde (Thermo Fisher Scientific, Waltham, MA, USA), and stained with 0.5% crystal violet (Thermo Fisher Scientific, Waltham, MA, USA) for 30 min. Plates were washed with tap water, imaged on a Chemidoc touch (Bio-rad, Hercules, CA, USA), and counted using particle counting on ImageJ.

2.9. γH2AX Assay

LNCaP cells were grown in chamber slides (Thermo Fisher Scientific, Waltham, MA, USA), treated as mentioned above, and fixed in 4% neutral buffered formaldehyde, permeabilized with 0.1% Triton X-100, and blocked with 1% bovine serum albumin in PBS-tween containing 5% goat serum. Slides were incubated with an antibody in order to hosphor-H2AX (1:500, CST). This was followed by incubation with goat-anti-rabbit Alexa488 (1:2000, Invitrogen, Waltham, MA, USA) and the slides were then mounted with a Prolong gold antifade reagent with DAPI (Invitrogen, Waltham, MA, USA). Cells were analyzed on a Leica confocal microscope (Leica Microsystems, Deerfield, IL, USA) with ×63 magnification. All γH2AX foci were counted using ImageJ particle counting.

2.10. ROS Quantification

LNCaP cells in 96-well plates were preloaded with 20 μM of DCFH-DA (Sigma-Aldrich, St. Louis, MO, USA) in serum-free RPMI1640 for 45 min. Afterwards, cells were washed three times with PBS and fresh serum-free RPMI1640 media; alternatively, conditioned media was added. For direct irradiation ROS studies, 0–8 Gy of radiation was administered to the cells. DCFH-DA fluorescence at 480 nm excitation/535 nm emission was scanned on a Clariostar microplate reader 1 h post irradiation or post conditioned media treatment.

2.11. Statistical Analysis

GraphPad Prism version 9.0 software was used for all statistical analyses. Differences between experimental groups were assessed using an unpaired *t*-test during a comparison of the two groups.

3. Results

3.1. Characterization, Cell Uptake, and Cytotoxicity of PGNPs

The performance of PGNPs with respect to internalization and biocompatibility was assessed for effective usage as radiosensitizers. A TEM image shows the spherical monodispersed nanoparticles following conjugation with the PSMA-antibody, with a mean diameter of 17.1 nm ± 0.28 nm (Figure 1A,B).

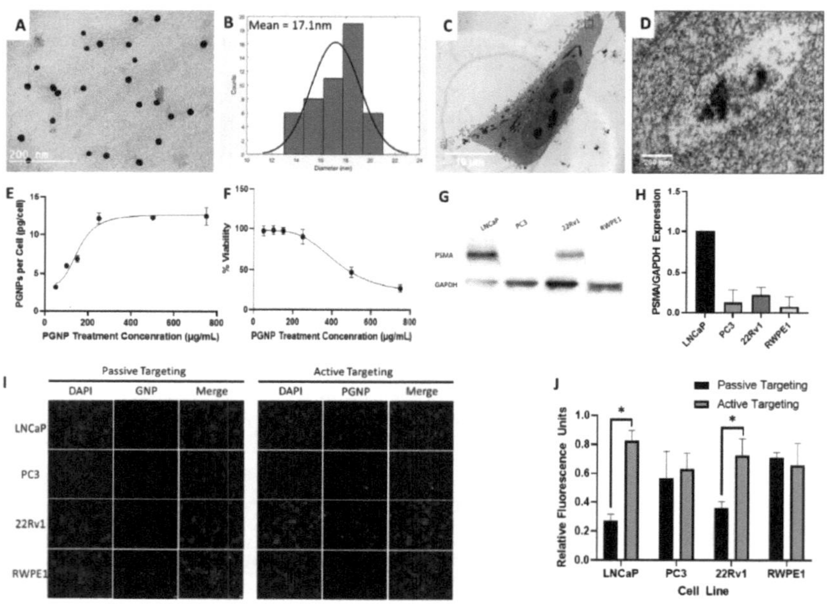

Figure 1. (**A**) TEM image of PSMA-antibody conjugated gold nanoparticles (PGNPs). (**B**) Histogram of PGNP diameters (mean = 17.1 nm ± 0.28 nm) from TEM images. (**C**) Representative TEM image of a LNCaP cell treated with 250 µg/mL of PGNPs. (**D**) Magnified region with arrow heads indicating internalized PGNPs. (**E**) PGNP uptake in LNCaP cells after 24 h of treatment at various PGNP doses (50 µg/mL, 100 µg/mL, 150 µg/mL, 250 µg/mL, 500 µg/mL, 750 µg/mL). (**F**) Percentage viability of the LNCaP cells incubated with different concentrations of PGNP (50 µg/mL, 100 µg/mL, 150 µg/mL, 250 µg/mL, 500 µg/mL, 750 µg/mL). (**G**) Western blot for PSMA expression in whole cell lysate of LNCaP, PC3, 22Rv1 (prostate cancer), and RWPE1 (normal prostate) cells. (**H**) Quantification of PSMA expression in the western blot normalized to a GAPDH signal. (**I**) Fluorescence microscopy image of LNCaP, PC3, 22Rv1, and RWPE1 cells treated with 250 µg/mL of passive targeting (mouse IgG conjugated) GNPs and active targeting (PSMA-antibody conjugated) PGNPs. (**J**) Quantification of fluorescence microscopy normalized to the highest signal in each cell line. * Denotes $p < 0.05$ from Welch's *t*-test.

Incubating PSMA-expressing LNCaP cells with the developed PGNPs revealed the internalization of PGNPs in small clusters throughout the cell (Figure 1C,D). Cellular uptake analysis (Figure 1E) demonstrated the enhanced PGNP levels at increasing concentrations, with the maximum saturation point starting from a concentration of 250 μg/mL. On average, a single LNCaP cell accumulated 12 picograms PGNP under 250 μg/mL of PGNP treatment (Figure 1E). The percentage cell viability (Figure 1F) demonstrated that there exists no statistically significant cytotoxicity for PGNPs with concentrations up to 250 μg/mL; therefore, the biocompatible concentration of the 250 μg/mL PGNPs was applied to the following RIBE experiments, which falls under the calculated IC50 value of 420 μg/mL.

To validate the specificity of active-targeting PGNPs, GNPs conjugated with mouse IgG (passive-targeting) and anti-PSMA IgG (active-targeting) were used for treatment on prostate cancer cell lines (including LNCaP, PC-3, and 22Rv1) and normal prostate endothelial cell lines (RWPE-1). First, we compared the expression of PSMA with different prostate cancer cell lines, as well as normal prostate cell lines. Western blot analysis demonstrated that LNCaP and 22Rv1 express significant levels of PSMA, whereas PC-3 and RWPE-1 are PSMA negative (Figure 1G,H). Then, we compared the targeting efficiency of PSMA-passive and -active GNPs on different cell lines by labelling GNPs with fluorophores and treating each cell line with 250 μg/mL of GNPs for 24 hrs. Compared with PSMA-passive GNPs, PSMA-active GNPs exhibited significantly increased levels of fluorescence signals in LNCaP and 22Rv1 cell lines, but not in PC3 and RWPE-1 cells; therefore, moving forward, LNCaP cells were identified as the optimal cell line with which to study the RIBE because of their high PSMA expression (Figure 1I,J).

3.2. *Effect of Radiation Dose or PGNP Concentration on Radiosensitization*

PGNP-induced radiosensitization in LNCaP cells was assessed through a clonogenic cell survival assay and γ-H2AX assays. To examine the effects of radiation doses and PGNP concentrations on in vitro radiosensitization, the LNCaP cells were incubated with PGNPs at different concentrations (0, 50, 100, 150, 250 μg/mL) and irradiated with different doses (0, 2, 4, 6, 8 Gy). Clonogenic cell survival assay results (Figure 2A) demonstrated enhanced radiosensitization levels in LNCaP cells when subjected to increased PGNP concentrations. Moreover, for a specific PGNP concentration, a higher radiosensitization level was observed at a higher radiation dose. Figure 2B compares the radiosensitization of PGNPs, using various concentrations, under 2-Gy irradiation. When treated with 150 μg/mL PGNPs or more, there was a statistically significant difference regarding the radiosensitization levels between the groups that were treated without and with PGNPs, thus suggesting that 150 μg/mL was the minimum concentration required for PGNP-induced radiosensitization under 2 Gy radiation.

γ-H2AX foci were assessed as biomarkers for double-strand DNA damages. As shown by the fluorescence images in Figure 2C, γ-H2AX foci could be clearly distinguished after the irradiation (2 Gy) of LNCaP cells incubated with/without PGNPs (250 μg/mL). Additionally, the non-irradiated LNCaP cells that were incubated with 250 μg/mL PGNPs showed no signs of foci formation, thus showing that these nanoparticles have no DNA damage effects without radiation. The average number of γ-H2AX foci per cell was counted during 2-Gy irradiation at different PGNP concentrations (0, 50, 100, 150, 250 μg/mL), and the results are presented in Figure 2D. PGNP treatment showed a dose dependent increase in foci count with an increase in GNP concentration (Figure 2D). A statistically significant increase in foci count was seen at 250 μg/mL.

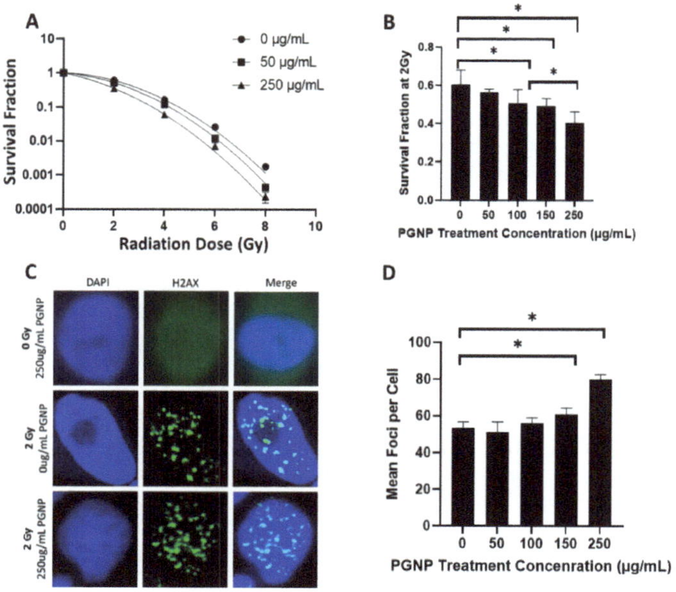

Figure 2. (**A**) Representative survival curve for LNCaP cells irradiated with various doses (0, 2, 4, 6, 8 Gy) following a 24 h pretreatment of PGNPs at different concentrations. (**B**) Survival fraction of 2 Gy-irradiated LNCaP cells, 24 h after the treatment with PGNPs at different concentrations. (**C**) γ-H2AX staining of LNCaP cells pretreated with 0 or 250 µg/mL PGNPs and irradiated with 0 Gy or 2 Gy. (**D**) Comparison of γ-H2AX foci counts for groups with different PGNP concentrations after 2Gy irradiation. * Denotes $p < 0.05$ from Welch's t-test.

3.3. Effect of PGNP on RIBE Signaling Intensity

The effect of PGNPs on RIBE signaling was assessed through the conditioned medium (CM) transfer procedure (Figure 3A), where the bystander LNCaP cells were treated with PBS- or PGNP-CM that was extracted from irradiated cells that were pretreated with PBS or 250 µg/mL PGNPs. Figure 3B shows the radiation response of PBS-CM treatments on LNCaP/PC3/22Rv1/RWPE-1 bystander cells. The RIBE response of the bystander cells treated with the PBS-CM from the irradiation groups (2, 4, 6, 8 Gy) were statistically different from the non-irradiation group ($p < 0.05$); this indicated the radiation dose dependence of the RIBE in LNCaP cells. This dose-dependence effect was also observed in PC-3 and 22Rv1 cells, but not in RWPE-1 cells. Moreover, 22Rv1 cells showed a weaker RIBE response than LNCaP/PC3 prostate cancer cell lines. MTT assays (Figure 3C) showed the statistically different cell viability between LNCaP bystander cells treated with 2 Gy-PGNP-CM and 2 Gy-PBS-CM. This aligns with the PGNP-enhanced radiosensitization (as shown in Figure 2) and the radiation dose dependence of the RIBE (as shown in Figure 3B). γ-H2AX assays (Figure 3D) demonstrated the significantly different number of foci, or the DNA damage between bystander cells, treated with 2 Gy-PGNP-CM and 2 Gy-PBS-CM. Furthermore, the clonogenic cell survival assay showed significantly decreased colony formations of bystander cells treated with 2 Gy-PGNP-CM compared with the 2 Gy-PBS-CM group. (Figure 3E); therefore, PGNP-induced radiosensitization further enhanced the RIBE signaling in LNCaP cells. In Figure 3C–E, no significant differences were observed between the LNCaP bystander cells treated with 0 Gy-PGNP-CM or 0 Gy-PBS-CM, which further illustrates the non-cytotoxicity of the 250 µg/mL PGNPs.

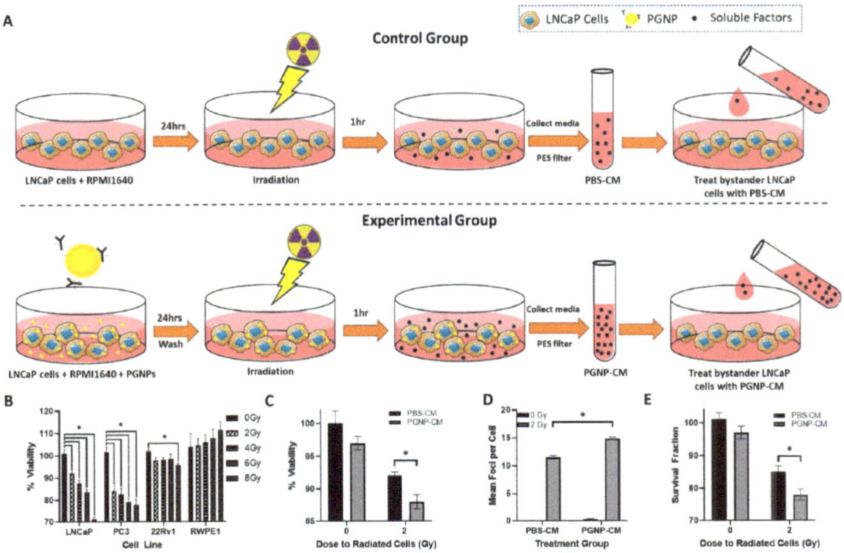

Figure 3. (**A**) Workflow of the conditioned medium transfer procedure for a RIBE response investigation. (**B**) MTT cell viability assays of LNCaP/PC3/22Rv1 (prostate cancer) and RWPE1 (normal prostate) bystander cells treated with PBS-CM extracted from their respective irradiated cell line cells under different radiation doses. (**C–E**) MTT cell viability, γH2AX, and clonogenic assays of bystander cells treated with 0 Gy-PBS-CM, 2 Gy-PBS-CM, 0 Gy-PGNP-CM, and 2 Gy-PGNP-CM. * Denotes $p < 0.05$ from Welch's t-test.

3.4. Effect of PGNP on the Sensitivity of Bystander Cells to the RIBE

The influence of PGNPs on the sensitivity of bystander cells to the RIBE was assessed on the PBS-CM treated LNCaP cells that were pretreated with or without 250μg/mL PGNPs. The detailed workflow is outlined in Figure 4A. Moreover, 2 Gy irradiation was adopted here because 2 Gy was the minimum RT dose that enabled the generation of a significant RIBE response in LNCaP bystander cells, as shown in Figure 3B. The RIBE response was evaluated in the following four groups: 0 Gy-PBS-CM on no-PGNPs-loaded cells (0 Gy-PBS-CM), 0 Gy-PBS-CM treatment on PGNPs-loaded bystander cells (0 Gy-PBS-CM+PGNPs), 2 Gy-PBS-CM on no-PGNPs-loaded cells (2 Gy-PBS-CM+PGNPs), and 2 Gy-PBS-CM on PGNPs-loaded cells (2 Gy-PBS-CM+PGNPs). MTT assays in Figure 4B demonstrated that there was significantly lower cell viability in the 2 Gy-PBS-CM+PGNPs group (85%) compared with the 2 Gy-PBS-CM group (92%). Additionally, compared with the 2 Gy-PBS-CM treatment group, the 2 Gy-PBS-CM+PGNPs group showed both an increased foci count in the γH2AX assay and decreased colony formation in the clonogenic assay (Figure 4C,D), thus indicating the PGNP-boosted susceptibility of LNCaP bystander cells to RIBE damage.

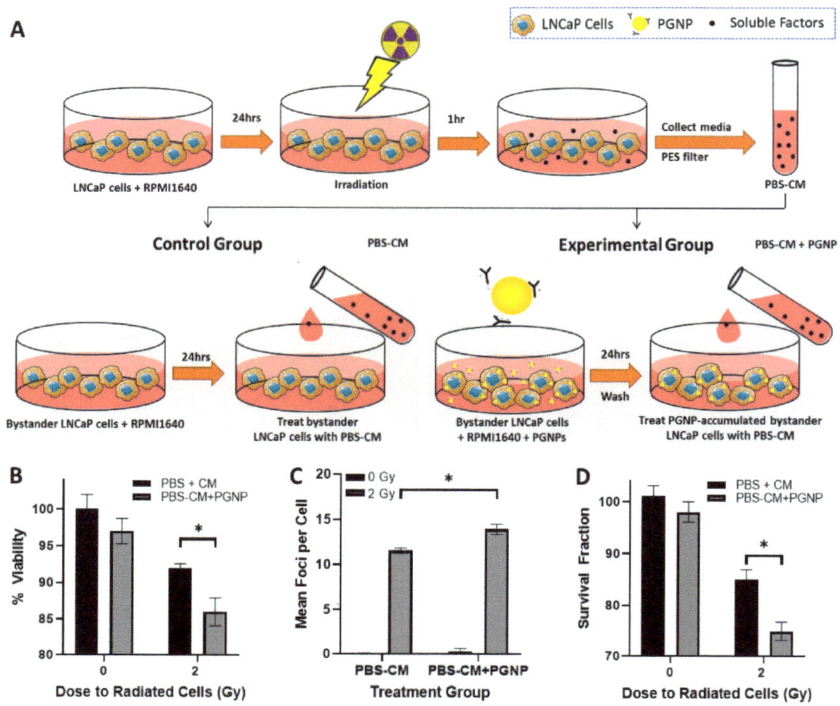

Figure 4. (**A**) Workflow of the conditioned medium transfer procedure in order to investigate the effect of PGNPs on the susceptibility of LNCaP bystander cells to RIBE damage. (**B–D**) MTT cell viability, γH2AX, and clonogenic assays of bystander cells under different treatments: 0 Gy-PBS-CM: 0 Gy-PBS-CM on bystander cells pretreated without PGNPs; 0 Gy-PBS-CM+PGNPs: 0 Gy-PBS-CM treatment on bystander cells pretreated with 250 μg/mL PGNPs; 2 Gy-PBS-CM+PGNPs: 2 Gy-PBS-CM on bystander cells pretreated without PGNPs; 2 Gy-PBS-CM+PGNPs: 2 Gy-PBS-CM on bystander cells pretreated with 250 μg/mL PGNPs. * Denotes $p < 0.05$ from Welch's t-test.

3.5. Effect of PGNP on ROS Production

As a common and potent type of RIBE mediator, ROS was measured with DCFDA in order to study the impact of PGNPs on ROS production in relation to the RIBE response. Figure 5A shows the percentage of ROS change between LNCaP cells treated with PGNPs at various nontoxic concentrations (10, 50, 100, 150, 200, 250 μg/mL). There were no significant differences between the different treatment groups, thus indicating that no clear relationship between PGNP concentration and ROS production exists. Figure 5B compares the intracellular ROS levels of irradiated LNCaP cells that were pretreated with and without PGNPs (250 μg/mL). As shown, irradiation-induced ROS production that was approximately 1.2× higher at every radiation dose (Figure 5B) in LNCaP cells pretreated with PGNPs, compared with LNCaP cells pretreated without PGNPs; this indicates that PGNPs boosted ROS production during radiation treatment. Figure 5C exhibits the effect of PGNPs on ROS production in LNCaP bystander cells in the following groups: treated with PBS-CM, treated with PGNP-CM, incubated with PGNPs for 24 h, and treated with PBS-CM (PBS-CM+PGNPs). For each treatment group, the bystander cells treated with conditioned media harvested from irradiated cells that were subjected to higher doses produced more ROS, which highlights the radiation dose dependence of ROS production with regard to the RIBE. Due to the GNP-induced radiosensitization, the bystander cells treated with PGNP-CM showed a larger ROS fold change, compared with the cells treated with PBS-CM. Compared with the PBS- and PGNP-CM treatment groups, the PBS-CM+PGNP group demonstrated a much higher level of ROS production, thus indicating the effect of internalized PGNPs on the sensitivity of bystander cells to the RIBE response.

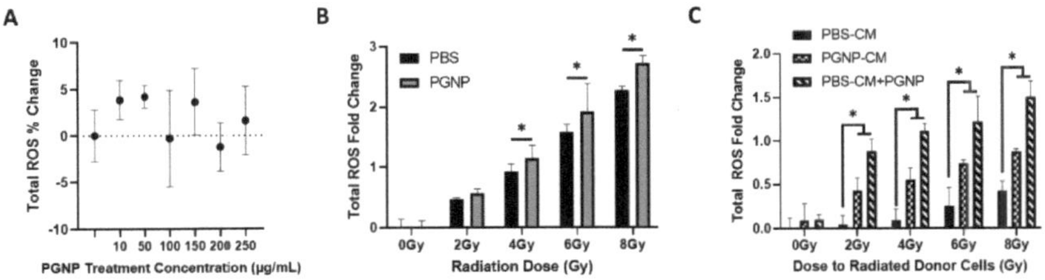

Figure 5. DCFDA total ROS production measurements normalized in accordance with the measurements of the untreated group. (**A**) Percentage of ROS change in LNCaP cells incubated with PGNPs under various concentrations. (**B**) Total ROS change induced by irradiation in LNCaP cells pretreated with and without 250 µg/mL PGNPs ($p < 0.05$ at 4, 6, and 8 Gy). (**C**) Total ROS change in bystander LNCaP cells in the following groups: treated with PBS-CM, treated with PGNP-CM, and bystander cells incubated with PGNPs for 24 h and treated with PBS-CM (PBS-CM+PGNPs) ($p < 0.05$ at 2, 4, 6, and 8 Gy). * Denotes $p < 0.05$ from Welch's *t*-test.

4. Discussion

Actively targeting tumors is crucial for GNP-aided radiosensitization; indeed, it enhances tumor specific accumulation and cell killing while sparing surrounding healthy tissue. In this study, internalization analysis on TEM images demonstrated the intracellular biodistribution of the developed PGNPs in cell cytoplasm. Although nuclear localization would theoretically induce a greater number of DNA double strand breaks through increased secondary electron scatter around GNPs, recent studies have suggested mitochondrial damage poses an additional threat to long-term cancer cell proliferation [33]; therefore, the cytoplasmic effects of radiosensitization are equally valuable in terms of direct DNA damage. Furthermore, uptake experiments indicated that cellular GNP saturation occurred at approximately 250 µg/mL, with minimal toxicity occurring within the LNCaP cells, and accumulations reaching 12 pg/cell. This value indicates lower toxicity levels than other studies performed with GNPs and LNCaP cells, which reported IC50 values between 100–200 µg/mL after 24 h [34–36]. This difference is mainly because of the media used during incubation with GNPs. In other studies, complete growth media were used, which allows the control group to continue growing; however, in our study, serum free media was used, which significantly reduces cell proliferation in the control cells. GNPs are known to induce cell cycle arrest; therefore, viability studies using complete media generate an IC50 based on a control cell line with cell proliferation, whereas GNP treated cells can become senescent [37]. However, this study employs serum-free media, which ensures that the MTT assay specifically measures cell death. This toxicity and uptake data agree with previous targeted nanoparticle studies [38]. Furthermore, this study confirmed that the anti-PSMA antibody functionalized PSMA-targeting GNPs are effective radiosensitizers for prostate cancer cells through a clonogenic assay and γ-H2AX assay. Both the clonogenic assay and γ-H2AX results showed a GNP dose dependent response to LNCaP cell radiosensitivity. The focus of the LNCaP prostate cancer cell line in the present study was motivated by the high expression of PSMA in LNCaP cells, and studies showing the high sensitivity of LNCaP cells to radiotherapy. Western blot confirmed the highest expression of PSMA in LNCaP cells compared with other prostate cancer cell lines (including PC3 and 22Rv1), as well as a normal prostate cell line (i.e., RWPE-1). Both LNCaP and 22Rv1 cell lines demonstrated an elevated uptake of PGNPs. Interestingly, the normal prostate cells (i.e., RWPE-1) demonstrated a high uptake of the GNPs used in this study, under the same in vitro incubation conditions as the prostate cancer cell groups. During GNP-assisted radiotherapy, tumor specificity occurs because 5–150 nm-sized GNPs accumulate in the tumor tissues as a result of the enhanced permeation and retention (EPR) effect; however, in contrast to the in vitro scenario, normal prostate tissues accumulate significantly fewer GNPs than prostate tumors in vivo due to their low levels of EPR. Therefore, in the in vivo scenario, normal prostate cells only accumulate a small number of GNPs, even though they do show a high uptake of GNPs in vitro.

Moreover, GNPs affected the RIBE in LNCaP cells. LNCaP bystander cell viability decreased with increased radiation doses up to 8 Gy when using the media transfer technique. The dose dependence of the RIBE response was also observed in other bystander prostate cancer cells (PC3, 22Rv1), but not in normal prostate cells (RWPE-1). As expected, radioresistant 22Rv1 cells showed a minimal response to the RIBE compared with LNCaP and PC3 cells. Additionally, PGNP-CM treatment induced increased cell death in bystander cells compared with bystander cells treated with PBS-CM. The increased cell death in bystander cells after PGNP-CM treatment suggests that the GNPs increased the radiation damage in irradiated cells, and thereby increased RIBE signaling because the RIBE in LNCaP cells was dose dependent for a wide range of cells. Previous studies on breast cancer cells showed that the dose dependence of cell lines in relation to the RIBE is critical to a GNP-enhanced RIBE [32]. γ-H2AX and clonogenic cell survival assays further confirmed the radiosensitization of LNCaP cells through the RIBE. This study only investigated the effects of the cytotoxicity of active-targeting PGNPs on LNCaP cells. Future relevant studies are required to investigate the effects of the cytotoxicity of PGNPs on PC3, 22Rv1, and RWPE-1 cell lines, which would enable an exploration of the impact of PGNPs on different prostate cancer cells as well as normal prostate cells.

PGNPs also increased the LNCaP cells' sensitivity to RIBE signals in a cell viability assay, γ-H2AX, and a clonogenic assay. The increased total cellular ROS following conditioned media exposure in PGNP-treated bystander cells demonstrated that imbalanced cellular redox potential plays a role. Previous studies indicate that metal nanoparticle treatment augments antioxidant and ROS-generating enzyme levels and leads to increased cellular ROS production [39]. Culcasi et al. and Wilhelmi et al. identified an upregulation in ROS-generating NADPH oxidase (NOX) enzymes that played a key role in metal nanoparticle cytotoxicity through increased cellular ROS, which thus led to DNA fragmentation [40]. Enzymes such as NOX enzymes, which can induce oxidative stress, play a key role in radiation therapy and the RIBE because studies show that the inhibition of NOX and other ROS-generating enzymes can significantly reduce DNA damage [41,42]. Furthermore, ROS clearance through enzymes such as superoxide dismutase (SOD), glutathione peroxidase (GPx), and catalases maintain cellular redox potential. A variety of metal nanoparticle studies in various tissues demonstrate a decrease in the expression of these enzymes, including a decrease in GPx in TiO_2, a decrease in SOD and catalases in CuO, and a decrease in SOD in iron oxide [39]. Although there is a dearth of experiments with regard to GNPs and antioxidant enzyme changes, it is a potential parameter requiring further investigation; therefore, the dose dependent increase in total cellular ROS upon the irradiation of PGNP-treated cells likely occurs as a result of a two-fold effect. First, the physical dose enhancement of PGNPs increased local dose deposition, leading to physical radiation dose dependent ROS production, and PGNP-impaired cellular redox maintenance with biologically boosted ROS levels. The latter was observed in bystander cells with conditioned media transfer studies. Ultimately, the elevated ROS levels occurred as a result of significantly increased enzymatic ROS production, which produced a surplus, and the conditioned media treatment in the bystander cells, which is likely to be the reason for the increased DNA damage seen in the γ-H2AX assay; this is also likely to be the case for the increased radiosensitization seen in the clonogenic survival assays.

A key element of this study was the use of conditioned media as a tool for inducing RIBE. Conditioned media utilizes soluble cytokines such as IL-1, IL-2, IL-6, IL-8, TNF-alpha, and TGF-beta [43]. Many of these cytokines are highly expressed in immune cells, such as macrophages, but in this study, we only studied the soluble factors in conditioned media produced by LNCaP cells, and we treated them to LNCaP cells; situations such as these can only be observed locally within the tumor. Therefore, the impact of the RIBE in distal regions or RIBE signals from immune cells could augment the observed effect. Furthermore, using the conditioned media technique caused the loss of two major RIBE signals: free radicals and gap junction signals [24]. Most free radicals are eliminated quickly before the conditioned media is transferred. Additionally, the strongest known signaling of the RIBE occurs through gap junctions. Previous RIBE studies have utilized partial culture irradiation, whereby a few cells in a single culture dish are radiated [44]; however, this RIBE is very localized, with an unclear, limited range [45]. Therefore, RIBE enhancement with PGNPs may have a different effect if studied with the partial irradiation method. Furthermore, recent studies focus on the impact of exosomes on the radiation-induced bystander effect [46,47]. A major impact of filtering the conditioned media with 0.45 μm sterile filter is the preservation of exosomes in the conditioned media. Lastly, the use of conditioned media provided in vivo translation opportunities for this study. The injection of conditioned media into mice could potentially show tumor treatment effects, and vice versa, the serum from irradiated mice can be used in vitro. These studies reveal the degree and range of RIBE signaling changes that GNPs can affect. The results presented here suggest that the bystander prostate cancer cells would experience greater cell death rates when these cells are enriched with GNPs and treated with RIBE signals from irradiated cancer cells with internalized GNPs.

A potential caveat of this study is the fact that the RIBE enhancement of cancer cells treated with kV photons is limited to LNCaP prostate cancer cells. Previous studies have demonstrated that various cells have different responses to the RIBE [46]. For instance, some cells (such as the breast cancer cell line, MCF7) are known to show a radiation-induced bystander effect but not in a dose dependent manner. On the other hand, QUDB cells have a dose dependent response [32]. Another study showed that there is a radiation-induced cell growth inhibitory bystander response, but only at high radiation doses, not low radiation does [48]. This is particularly interesting as LNCaP demonstrated a consistent RIBE, including low and high radiation doses. To consider the full effect of the radiation-induced signaling of the RIBE in prostate cancer cells, studies into other prostate cancer cells, as well as normal tissue prostate fibroblast cells, can provide an understanding of how the RIBE could be used in a clinical setting. Additionally, they could also demonstrate the effect of GNPs on these bystander effects. Moreover, RIBEs have been known to differ based on their radiation quality and LET. The RIBE has been studied using various radiation sources, including photons, particles, and heavy ions at varying degrees of intensity [49–51]. Although many of these have not been compared against each other, several studies indicate that the magnitude of bystander effect is dependent on LET [52]. Changes between radiation sources have been shown to change dose dependence as well as relieve the RIBE burden entirely [53]. This is particularly important when studying GNP-induced radiosensitization because the mechanism of GNP-mediated radiosensitization is different depending on the radiation source. For instance, interactions between kV energy beams and GNPs leads to the emission of short-range secondary electrons and radiosensitization. The use of kV energies in this study maximizes the effect of secondary electron scatter compared with other modalities [8]. Moreover, MV irradiated GNPs induce radiosensitization through increased mitochondrial toxicity, and Monte Carlo and in vitro studies suggest protons or heavy ions generate a massive increase in ROS production [54]. All these modalities are currently used clinically for prostate cancer treatments in the form of brachytherapy or external beam radiotherapy, and the 2 Gy increments match the conventional fractionation schemes of standard treatment schedules. As our study suggests, the RIBE contributes to GNP-mediated radiosensitivity through ROS production, and the increase in sensitivity may be further enhanced when protons and heavy ion-irradiated cells are the donors for the conditioned media. Lastly, this study focuses on GNPs, but a multitude of metal nanoparticles have demonstrated radiosensitization capabilities. A diverse group of these metal nanoparticles induce cellular redox changes, which suggest that the effects observed in this study are not limited to GNPs. In addition, several studies indicate that metal nanoparticles introduce cellular toxicity through ion shedding heavy metal poisoning with TiO2, carcinogenesis from ZnO, and inflammation from impurities in Ag [55–57]. GNPs are free of these problems, but these toxicities relate largely to ROS generation from the nanoparticles, which could further impact RIBE sensitivity in a similar manner to GNPs.

The clinical benefit of the bystander effect can be observed through a series of in vivo experiments, including the use of spatially fractionated radiotherapy [25]. Spatially fractionated radiotherapy is the delivery of a radiation dose to smaller fields without delivering radiation to the entire tumor to reduce toxicity. Preclinical studies in mouse models demonstrated that a 10 Gy dose to a small region of the tumor would cause radiation damage to the surrounding tumor tissue [58]. Furthermore, this bystander effect showed changes in gene expression for DNA repair, cell cycle arrest, and apoptosis, as is the case with the changes seen in irradiated tumor cells. Although studies on spatially fractionated radiotherapy have been conducted since the 1950s, inventions of multi-leaf collimators have recently shown promising results regarding prostate tumor responses [59]. The implication of enhanced bystander effects from GNPs suggests that they can be valuable in combination with spatially fractionated radiotherapy to enhance the bystander effect. The results presented here suggest that spatially fractioned radiotherapy would benefit from GNPs, even if GNPs are not homogenously distributed throughout the tumor. Factors such as hypoxia, nanoparticle targeting agents, and tumor microenvironments can affect the distribution of GNPs, but a spatially fractionated study would enhance tumor treatment efficacy, regardless of the GNP content in each beamlet. Furthermore, it is possible to use imaging studies to target regions with high GNP concentrations in order to selectively utilize the effects that GNPs have on the RIBE for dose painting. In vivo experiments concerning GNPs, combined with spatially fractionated radiotherapy, could also reveal the extent of the GNP-enhanced RIBE to the surrounding tissue. An additional weakness in this study, for clinical reference, is the use of a single radiation dose to induce a bystander effect. For clinical use, the bystander effect would require fractionated radiation. It is important to note that fractionated radiation doses have had mixed results in RIBE studies [60]; however, previous studies on the pharmacodynamics of gold nanoparticles indicate that fractionated radiation therapy is completely feasible for GNP-enhanced radiation therapies [19].

Finally, this study suggested that the radiosensitization effects of off-targeted GNPs could be a potential double-edged sword. The increased effect of RIBE sensitivity as a result of nanoparticles could induce a greater number of secondary tumorigeneses, as nanoparticles are known to accumulate in various organs such as the liver, spleen, and kidneys [33,61]. These effects can be particularly worrisome since secondary cancers in the lung are prevalent with prostate cancer treatments [62]. Previous studies have demonstrated low accumulations of GNPs in the lung, which may therefore mitigate this risk, but the high accumulations in the liver, kidney, and spleen may increase the risk of secondary cancers in these organs when treated with GNPs. This form of distant organ damage may be described more specifically as an 'abscopal effect', and it is not particular to the bystander effect; however, the proper control of nanoparticle biodistribution with targeted nanoparticles could be critical for GNP radiotherapy safety.

Studies on the RIBE present further biological explanations on the variances observed between in silico GNP studies and in vivo/in vitro experiments. Several Monte Carlo studies have demonstrated that GNPs can increase local dose deposition by up to 200× per nanoparticle with 250 kvP photon beams [63]; however, the increased secondary electron scatter fails to account for the total radiosensitization observed in vitro, especially for megavoltage energies [20]. Studies show that GNP radiosensitization can occur due to a variety of reasons, including physical, chemical, and biological [20]. Physically, GNPs increase the secondary electron scatter, and chemically, the GNPs form free radicals to make DNA more susceptible to radiation damage. A myriad of biological effects occur, including cell cycle changes, mitochondrial damage, and DNA repair inhibition. Studies in glucose-capped GNPs increased the cell population in the G2/M phase, which is the most radiosensitive phase. Secondly, several studies on the GNPs' impact on mitochondria show radiosensitization through the loss of mitochondrial potential, thus leading to necrosis, caspase-related mitochondrial dysfunction, and mitochondrial membrane polarization; this leads to apoptosis when combined with radiation. Thirdly, H2AX studies have shown complications with DNA repair when GNPs are present, based on increased double stranded DNA damage to the foci after 24 h. RIBE changes from GNPs add to the list of biological changes which lead to increased radiosensitization, and the biological factors affecting radiosensitization from direct radiation further enhance the RIBE. Nanoparticles have recently garnered attention in radiotherapy as valuable agents for imaging and radiosensitization. As such, the direct effects of metallic nanoparticle radiosensitization and toxicity have been heavily studied [38,64]; however, the nanoparticles' impact on the RIBE has not been well documented. The radiation-induced bystander effect can have a critical role in improving the treatment efficacy of treatment modalities, such as spatially fractionated radiotherapy, and it can also have detrimental effects in the form of off-target secondary cancers. This study introduced the potential for combining GNPs to augment the effects of the RIBE in prostate cancer treatment.

5. Conclusions

Prostate cancer-targeting GNPs were used to radiosensitize LNCaP prostate cancer cells. This study elucidated the impact of the bystander effect on the biological radiosensitization of prostate cancer cells with targeted GNPs. Additionally, this study highlights the increased signaling intensity of the bystander effect from GNPs in the irradiated cells, as well as the greater sensitivity to the bystander effect in non-irradiated cells as a result of GNP treatment. A plethora of studies have demonstrated that radiosensitization from metal nanoparticles occurs as a result of a multitude of biological and chemical factors; these factors impact a greater area than the originally proposed site for local dose deposition. The findings from this study contribute to knowledge concerning another biological component of GNP radiosensitization. Ultimately, these findings suggest further research should be undertaken to investigate the GNP-enhanced bystander effects, abscopal effects, distant tissue toxicity risks, and combinations of GNPs and heterogenous dose distribution treatments.

Author Contributions: Conceptualization, J.S. and W.T.; methodology, J.S., W.T., D.H. and Y.-P.Y.; software, D.H. and R.M.S.; validation, D.H., W.T. and J.S.; formal analysis, D.H.; investigation, D.H., J.S., W.T., A.P. and J.C.F.; resources, J.S., A.P., J.C.F., S.D. and N.D.; data curation, D.H. and J.S.; writing—original draft preparation, D.H. and J.S.; writing—review and editing, J.S., W.T. and J.C.F.; supervision, J.C.F. and J.S.; funding acquisition, A.P., J.C.F., S.D. and J.S. All authors have read and agreed to the published version of the manuscript.

Funding: Financial support was received from Department of Radiation Oncology at University of Miami. This work was supported by Sylvester Comprehensive Cancer Center and the American Cancer Society. This work was also partially supported by Sylvester CCSG transdisciplinary pilot grant, and cancer center core grant SP30 240139-02.

Data Availability Statement: Not applicable.

Conflicts of Interest: The authors declare no conflict of interest.

References

1. Siegel, R.L.; Miller, K.D.; Fuchs, H.E.; Jemal, A. Cancer statistics, 2022. *CA Cancer J. Clin.* **2022**, *72*, 7–33. [CrossRef] [PubMed]
2. Attard, G.; Parker, C.; Eeles, R.A.; Schroder, F.; Tomlins, S.A.; Tannock, I.; Drake, C.G.; de Bono, J.S. Prostate cancer. *Lancet* **2016**, *387*, 70–82. [CrossRef] [PubMed]
3. Hall, E.J.; Giaccia, A.J. *Radiobiology for the Radiologist*; Wolters Kluwer: Philadelphia, PA, USA, 2019; 597p.
4. Baskar, R.; Lee, K.A.; Yeo, R.; Yeoh, K.W. Cancer and radiation therapy: Current advances and future directions. *Int. J. Med. Sci.* **2012**, *9*, 193–199. [CrossRef] [PubMed]
5. Song, G.; Cheng, L.; Chao, Y.; Yang, K.; Liu, Z. Emerging Nanotechnology and Advanced Materials for Cancer Radiation Therapy. *Adv. Mater.* **2017**, *29*, 1700996. [CrossRef] [PubMed]
6. Sun, B.; Hagan, C.T.t.; Caster, J.; Wang, A.Z. Nanotechnology in Radiation Oncology. *Hematol. Oncol. Clin. N. Am.* **2019**, *33*, 1071–1093. [CrossRef] [PubMed]

7. Dreaden, E.C.; Alkilany, A.M.; Huang, X.; Murphy, C.J.; El-Sayed, M.A. The golden age: Gold nanoparticles for biomedicine. *Chem. Soc. Rev.* **2012**, *41*, 2740–2779. [CrossRef] [PubMed]
8. Her, S.; Jaffray, D.A.; Allen, C. Gold nanoparticles for applications in cancer radiotherapy: Mechanisms and recent advancements. *Adv. Drug Deliv. Rev.* **2017**, *109*, 84–101. [CrossRef]
9. Choi, J.; Kim, G.; Cho, S.B.; Im, H.J. Radiosensitizing high-Z metal nanoparticles for enhanced radiotherapy of glioblastoma multiforme. *J. Nanobiotechnol.* **2020**, *18*, 122. [CrossRef]
10. Jones, B.L.; Krishnan, S.; Cho, S.H. Estimation of microscopic dose enhancement factor around gold nanoparticles by Monte Carlo calculations. *Med. Phys.* **2010**, *37*, 3809–3816. [CrossRef]
11. Lin, Y.; McMahon, S.J.; Scarpelli, M.; Paganetti, H.; Schuemann, J. Comparing gold nano-particle enhanced radiotherapy with protons, megavoltage photons and kilovoltage photons: A Monte Carlo simulation. *Phys. Med. Biol.* **2014**, *59*, 7675–7689. [CrossRef]
12. Ma, N.; Wu, F.G.; Zhang, X.; Jiang, Y.W.; Jia, H.R.; Wang, H.Y.; Li, Y.H.; Liu, P.; Gu, N.; Chen, Z. Shape-Dependent Radiosensitization Effect of Gold Nanostructures in Cancer Radiotherapy: Comparison of Gold Nanoparticles, Nanospikes, and Nanorods. *ACS Appl. Mater. Interfaces* **2017**, *9*, 13037–13048. [CrossRef] [PubMed]
13. Jokerst, J.V.; Lobovkina, T.; Zare, R.N.; Gambhir, S.S. Nanoparticle PEGylation for imaging and therapy. *Nanomedicine* **2011**, 715–728. [CrossRef] [PubMed]
14. Iyer, A.K.; Khaled, G.; Fang, J.; Maeda, H. Exploiting the enhanced permeability and retention effect for tumor targeting. *Drug Discov. Today* **2006**, *11*, 812–818. [CrossRef] [PubMed]
15. Danhier, F. To exploit the tumor microenvironment: Since the EPR effect fails in the clinic, what is the future of nanomedicine? *Control. Release* **2016**, *244*, 108–121. [CrossRef]
16. Boisselier, E.; Astruc, D. Gold nanoparticles in nanomedicine: Preparations, imaging, diagnostics, therapies and toxicity. *Chem Soc. Rev.* **2009**, *38*, 1759–1782. [CrossRef]
17. Hofman, M.S.; Violet, J.; Hicks, R.J.; Ferdinandus, J.; Thang, S.P.; Akhurst, T.; Iravani, A.; Kong, G.; Ravi Kumar, A.; Murphy, D.G et al. [(177)Lu]-PSMA-617 radionuclide treatment in patients with metastatic castration-resistant prostate cancer (LuPSMA trial A single-centre, single-arm, phase 2 study. *Lancet Oncol.* **2018**, *19*, 825–833. [CrossRef]
18. Hofman, M.S.; Lawrentschuk, N.; Francis, R.J.; Tang, C.; Vela, I.; Thomas, P.; Rutherford, N.; Martin, J.M.; Frydenberg, M Shakher, R.; et al. Prostate-specific membrane antigen PET-CT in patients with high-risk prostate cancer before curative-intent surgery or radiotherapy (proPSMA): A prospective, randomised, multicentre study. *Lancet* **2020**, *395*, 1208–1216. [CrossRef]
19. Hara, D.; Tao, W.; Totiger, T.M.; Pourmand, A.; Dogan, N.; Ford, J.C.; Shi, J.; Pollack, A. Prostate Cancer Targeted X-ray Fluorescence Imaging via Gold Nanoparticles Functionalized With Prostate-Specific Membrane Antigen (PSMA). *Int. J. Radiat. Oncol. Biol. Phys.* **2021**, *111*, 220–232. [CrossRef]
20. Rosa, S.; Connolly, C.; Schettino, G.; Butterworth, K.T.; Prise, K.M. Biological mechanisms of gold nanoparticle radiosensitization *Cancer Nanotechnol.* **2017**, *8*, 2. [CrossRef]
21. Prise, K.M.; O'Sullivan, J.M. Radiation-induced bystander signalling in cancer therapy. *Nat. Rev. Cancer* **2009**, *9*, 351–360 [CrossRef]
22. Hei, T.K.; Zhou, H.; Ivanov, V.N.; Hong, M.; Lieberman, H.B.; Brenner, D.J.; Amundson, S.A.; Geard, C.R. Mechanism of radiation-induced bystander effects: A unifying model. *J. Pharm. Pharmacol.* **2008**, *60*, 943–950. [CrossRef] [PubMed]
23. Nagasawa, H.; Little, J.B. Induction of sister chromatid exchanges by extremely low doses of alpha-particles. *Cancer Res.* **1992**, *52* 6394–6396. [PubMed]
24. Najafi, M.; Fardid, R.; Hadadi, G.; Fardid, M. The mechanisms of radiation-induced bystander effect. *J. Biomed. Phys. Eng.* **2014**, *4* 163–172. [PubMed]
25. Asur, R.; Butterworth, K.T.; Penagaricano, J.A.; Prise, K.M.; Griffin, R.J. High dose bystander effects in spatially fractionated radiation therapy. *Cancer Lett.* **2015**, *356*, 52–57. [CrossRef]
26. Schultke, E.; Balosso, J.; Breslin, T.; Cavaletti, G.; Djonov, V.; Esteve, F.; Grotzer, M.; Hildebrandt, G.; Valdman, A.; Laissue, J Microbeam radiation therapy—Grid therapy and beyond: A clinical perspective. *Br. J. Radiol.* **2017**, *90*, 20170073. [CrossRef]
27. Penagaricano, J.A.; Moros, E.G.; Ratanatharathorn, V.; Yan, Y.; Corry, P. Evaluation of spatially fractionated radiotherapy (GRID and definitive chemoradiotherapy with curative intent for locally advanced squamous cell carcinoma of the head and neck Initial response rates and toxicity. *Int. J. Radiat. Oncol. Biol. Phys.* **2010**, *76*, 1369–1375. [CrossRef]
28. Griffin, R.J.; Ahmed, M.M.; Amendola, B.; Belyakov, O.; Bentzen, S.M.; Butterworth, K.T.; Chang, S.; Coleman, C.N.; Djonov, V Formenti, S.C.; et al. Understanding High-Dose, Ultra-High Dose Rate, and Spatially Fractionated Radiation Therapy. *Int. J. Radiat. Oncol. Biol. Phys.* **2020**, *107*, 766–778. [CrossRef]
29. Webb, S. The physical basis of IMRT and inverse planning. *Br. J. Radiol.* **2003**, *76*, 678–689. [CrossRef]
30. Schuemann, J.; Berbeco, R.; Chithrani, D.B.; Cho, S.H.; Kumar, R.; McMahon, S.J.; Sridhar, S.; Krishnan, S. Roadmap to Clinical Use of Gold Nanoparticles for Radiation Sensitization. *Int. J. Radiat. Oncol. Biol. Phys.* **2016**, *94*, 189–205. [CrossRef]
31. Havaki, S.; Kotsinas, A.; Chronopoulos, E.; Kletsas, D.; Georgakilas, A.; Gorgoulis, V.G. The role of oxidative DNA damage in radiation induced bystander effect. *Cancer Lett.* **2015**, *356*, 43–51. [CrossRef]
32. Rostami, A.; Toossi, M.T.; Sazgarnia, A.; Soleymanifard, S. The effect of glucose-coated gold nanoparticles on radiation bystander effect induced in MCF-7 and QUDB cell lines. *Radiat. Environ. Biophys.* **2016**, *55*, 461–466. [CrossRef] [PubMed]

33. Hatoyama, K.; Kitamura, N.; Takano-Kasuya, M.; Tokunaga, M.; Oikawa, T.; Ohta, M.; Hamada, Y.; Tada, H.; Kobayashi, Y.; Kamei, T.; et al. Quantitative analyses of amount and localization of radiosensitizer gold nanoparticles interacting with cancer cells to optimize radiation therapy. *Biochem. Biophys. Res. Commun.* **2019**, *508*, 1093–1100. [CrossRef] [PubMed]
34. Guo, J.; O'Driscoll, C.M.; Holmes, J.D.; Rahme, K. Bioconjugated gold nanoparticles enhance cellular uptake: A proof of concept study for siRNA delivery in prostate cancer cells. *Int. J. Pharm.* **2016**, *509*, 16–27. [CrossRef] [PubMed]
35. Tangthong, T.; Piroonpan, T.; Thipe, V.C.; Khoobchandani, M.; Katti, K.; Katti, K.V.; Pasanphan, W. Water-Soluble Chitosan Conjugated DOTA-Bombesin Peptide Capped Gold Nanoparticles as a Targeted Therapeutic Agent for Prostate Cancer. *Nanotechnol. Sci. Appl.* **2021**, *14*, 69–89. [CrossRef]
36. Changizi, O.; Khoei, S.; Mahdavian, A.; Shirvalilou, S.; Mahdavi, S.R.; Keyvan Rad, J. Enhanced radiosensitivity of LNCaP prostate cancer cell line by gold-photoactive nanoparticles modified with folic acid. *Photodiagn. Photodyn. Ther.* **2020**, *29*, 101602. [CrossRef]
37. Li, Q.; Huang, C.; Liu, L.; Hu, R.; Qu, J. Effect of Surface Coating of Gold Nanoparticles on Cytotoxicity and Cell Cycle Progression. *Nanomaterials* **2018**, *8*, 1063. [CrossRef]
38. Khlebtsov, N.; Dykman, L. Biodistribution and toxicity of engineered gold nanoparticles: A review of in vitro and in vivo studies. *Chem. Soc. Rev.* **2011**, *40*, 1647–1671. [CrossRef]
39. Feng, X.; Chen, A.; Zhang, Y.; Wang, J.; Shao, L.; Wei, L. Central nervous system toxicity of metallic nanoparticles. *Int. J. Nanomed.* **2015**, *10*, 4321–4340. [CrossRef]
40. Culcasi, M.; Benameur, L.; Mercier, A.; Lucchesi, C.; Rahmouni, H.; Asteian, A.; Casano, G.; Botta, A.; Kovacic, H.; Pietri, S. EPR spin trapping evaluation of ROS production in human fibroblasts exposed to cerium oxide nanoparticles: Evidence for NADPH oxidase and mitochondrial stimulation. *Chem. Biol. Interact.* **2012**, *199*, 161–176. [CrossRef]
41. Weyemi, U.; Redon, C.E.; Aziz, T.; Choudhuri, R.; Maeda, D.; Parekh, P.R.; Bonner, M.Y.; Arbiser, J.L.; Bonner, W.M. Inactivation of NADPH oxidases NOX4 and NOX5 protects human primary fibroblasts from ionizing radiation-induced DNA damage. *Radiat. Res.* **2015**, *183*, 262–270. [CrossRef]
42. Brieger, K.; Schiavone, S.; Miller, F.J., Jr.; Krause, K.H. Reactive oxygen species: From health to disease. *Swiss Med. Wkly.* **2012**, *142*, w13659. [CrossRef] [PubMed]
43. Blyth, B.J.; Sykes, P.J. Radiation-induced bystander effects: What are they, and how relevant are they to human radiation exposures? *Radiat. Res.* **2011**, *176*, 139–157. [CrossRef] [PubMed]
44. Gerashchenko, B.I.; Howell, R.W. Bystander cell proliferation is modulated by the number of adjacent cells that were exposed to ionizing radiation. *Cytom. A* **2005**, *66*, 62–70. [CrossRef] [PubMed]
45. Hu, B.; Wu, L.; Han, W.; Zhang, L.; Chen, S.; Xu, A.; Hei, T.K.; Yu, Z. The time and spatial effects of bystander response in mammalian cells induced by low dose radiation. *Carcinogenesis* **2006**, *27*, 245–251. [CrossRef] [PubMed]
46. Smolarz, M.; Skoczylas, L.; Gawin, M.; Krzyzowska, M.; Pietrowska, M.; Widlak, P. Radiation-Induced Bystander Effect Mediated by Exosomes Involves the Replication Stress in Recipient Cells. *Int. J. Mol. Sci.* **2022**, *23*, 4149. [CrossRef]
47. Nakaoka, A.; Nakahana, M.; Inubushi, S.; Akasaka, H.; Salah, M.; Fujita, Y.; Kubota, H.; Hassan, M.; Nishikawa, R.; Mukumoto, N.; et al. Exosome-mediated radiosensitizing effect on neighboring cancer cells via increase in intracellular levels of reactive oxygen species. *Oncol. Rep.* **2021**, *45*, 13. [CrossRef] [PubMed]
48. Liu, S.Z.; Jin, S.Z.; Liu, X.D. Radiation-induced bystander effect in immune response. *Biomed. Environ. Sci.* **2004**, *17*, 40–46. [PubMed]
49. Shao, C.; Furusawa, Y.; Kobayashi, Y.; Funayama, T.; Wada, S. Bystander effect induced by counted high-LET particles in confluent human fibroblasts: A mechanistic study. *FASEB J.* **2003**, *17*, 1422–1427. [CrossRef]
50. Gow, M.D.; Seymour, C.B.; Ryan, L.A.; Mothersill, C.E. Induction of bystander response in human glioma cells using high-energy electrons: A role for TGF-beta1. *Radiat. Res.* **2010**, *173*, 769–778. [CrossRef]
51. Frankenberg, D.; Greif, K.D.; Giesen, U. Radiation response of primary human skin fibroblasts and their bystander cells after exposure to counted particles at low and high LET. *Int. J. Radiat. Biol.* **2006**, *82*, 59–67. [CrossRef]
52. Shao, C.; Aoki, M.; Furusawa, Y. Bystander effect on cell growth stimulation in neoplastic HSGc cells induced by heavy-ion irradiation. *Radiat. Environ. Biophys.* **2003**, *42*, 183–187. [CrossRef] [PubMed]
53. Anzenberg, V.; Chandiramani, S.; Coderre, J.A. LET-dependent bystander effects caused by irradiation of human prostate carcinoma cells with X rays or alpha particles. *Radiat. Res.* **2008**, *170*, 467–476. [CrossRef] [PubMed]
54. Rudek, B.; McNamara, A.; Ramos-Mendez, J.; Byrne, H.; Kuncic, Z.; Schuemann, J. Radio-enhancement by gold nanoparticles and their impact on water radiolysis for X-ray, proton and carbon-ion beams. *Phys. Med. Biol.* **2019**, *64*, 175005. [CrossRef] [PubMed]
55. Awasthi, K.K.; John, P.J.; Awasthi, A.; Awasthi, K. Multi walled carbon nano tubes induced hepatotoxicity in Swiss albino mice. *Micron* **2013**, *44*, 359–364. [CrossRef] [PubMed]
56. Hardas, S.S.; Butterfield, D.A.; Sultana, R.; Tseng, M.T.; Dan, M.; Florence, R.L.; Unrine, J.M.; Graham, U.M.; Wu, P.; Grulke, E.A.; et al. Brain distribution and toxicological evaluation of a systemically delivered engineered nanoscale ceria. *Toxicol. Sci.* **2010**, *116*, 562–576. [CrossRef] [PubMed]
57. Freyre-Fonseca, V.; Delgado-Buenrostro, N.L.; Gutierrez-Cirlos, E.B.; Calderon-Torres, C.M.; Cabellos-Avelar, T.; Sanchez-Perez, Y.; Pinzon, E.; Torres, I.; Molina-Jijon, E.; Zazueta, C.; et al. Titanium dioxide nanoparticles impair lung mitochondrial function. *Toxicol. Lett.* **2011**, *202*, 111–119. [CrossRef]

58. Asur, R.S.; Sharma, S.; Chang, C.W.; Penagaricano, J.; Kommuru, I.M.; Moros, E.G.; Corry, P.M.; Griffin, R.J. Spatially fractionated radiation induces cytotoxicity and changes in gene expression in bystander and radiation adjacent murine carcinoma cells. *Radiat. Res.* **2012**, *177*, 751–765. [CrossRef]
59. Pollack, A.; Chinea, F.M.; Bossart, E.; Kwon, D.; Abramowitz, M.C.; Lynne, C.; Jorda, M.; Marples, B.; Patel, V.N.; Wu, X.; et a Phase I Trial of MRI-Guided Prostate Cancer Lattice Extreme Ablative Dose (LEAD) Boost Radiation Therapy. *Int. J. Radiat. Onco Biol. Phys.* **2020**, *107*, 305–315. [CrossRef]
60. Daguenet, E.; Louati, S.; Wozny, A.S.; Vial, N.; Gras, M.; Guy, J.B.; Vallard, A.; Rodriguez-Lafrasse, C.; Magne, N. Radiation induced bystander and abscopal effects: Important lessons from preclinical models. *Br. J. Cancer* **2020**, *123*, 339–348. [CrossRef]
61. Albanese, A.; Tang, P.S.; Chan, W.C. The effect of nanoparticle size, shape, and surface chemistry on biological systems. *Annu Rev. Biomed. Eng.* **2012**, *14*, 1–16. [CrossRef]
62. Brenner, D.J.; Curtis, R.E.; Hall, E.J.; Ron, E. Second malignancies in prostate carcinoma patients after radiotherapy compared with surgery. *Cancer* **2000**, *88*, 398–406. [CrossRef]
63. Pagacova, E.; Stefancikova, L.; Schmidt-Kaler, F.; Hildenbrand, G.; Vicar, T.; Depes, D.; Lee, J.H.; Bestvater, F.; Lacombe, S Porcel, E.; et al. Challenges and Contradictions of Metal Nano-Particle Applications for Radio-Sensitivity Enhancement in Cance Therapy. *Int. J. Mol. Sci.* **2019**, *20*, 588. [CrossRef] [PubMed]
64. Butterworth, K.T.; Coulter, J.A.; Jain, S.; Forker, J.; McMahon, S.J.; Schettino, G.; Prise, K.M.; Currell, F.J.; Hirst, D.G. Evaluation of cytotoxicity and radiation enhancement using 1.9 nm gold particles: Potential application for cancer therapy. *Nanotechnology* **2010**, *21*, 295101. [CrossRef] [PubMed]

nanomaterials

MDPI

Article

Nano-Conversion of Ineffective Cephalosporins into Potent One against Resistant Clinical Uro-Pathogens via Gold Nanoparticles

Syed Mohd Danish Rizvi [1,2], Talib Hussain [2,3], Farhan Alshammari [1], Hana Sonbol [4], Nabeel Ahmad [5], Syed Shah Mohammed Faiyaz [6], Mohammad Amjad Kamal [7,8,9,10], El-Sayed Khafagy [11,12], Afrasim Moin [1,2,*] and Amr Selim Abu Lila [1,2,*]

1 Department of Pharmaceutics, College of Pharmacy, University of Ha'il, Ha'il 81442, Saudi Arabia
2 Molecular Diagnostic & Personalized Therapeutic Unit, University of Ha'il, Ha'il 81442, Saudi Arabia
3 Department of Pharmacology and Toxicology, College of Pharmacy, University of Ha'il, Ha'il 81442, Saudi Arabia
4 Department of Biology, College of Science, Princess Nourah bint Abdulrahman University, Riyadh 11671, Saudi Arabia
5 Department of Biotechnology, School of Allied Sciences, Dev Bhoomi Uttarakhand University, Dehradun 248007, India
6 Department of Physiology, College of Medicine, University of Ha'il, Ha'il 81442, Saudi Arabia
7 Institutes for Systems Genetics, Frontiers Science Center for Disease-Related Molecular Network, West China Hospital, Sichuan University, Chengdu 610065, China
8 King Fahd Medical Research Center, King Abdulaziz University, Jeddah 21589, Saudi Arabia
9 Department of Pharmacy, Faculty of Allied Health Sciences, Daffodil International University, Dhaka 1207, Bangladesh
10 Novel Global Community Educational Foundation, Hebersham, NSW 2770, Australia
11 Department of Pharmaceutics, College of Pharmacy, Prince Sattam Bin Abdulaziz University, Al-Kharj 11942, Saudi Arabia
12 Department of Pharmaceutics and Industrial Pharmacy, Faculty of Pharmacy, Suez Canal University, Ismailia 41522, Egypt
* Correspondence: a.moinuddin@uoh.edu.sa (A.M.); a.abulila@uoh.edu.sa (A.S.A.L.)

Abstract: Infections caused by resistant bacterial pathogens have increased the complications of clinicians worldwide. The quest for effective antibacterial agents against resistant pathogens has prompted researchers to develop new classes of antibiotics. Unfortunately, pathogens have acted more smartly by developing resistance to even the newest class of antibiotics with time. The culture sensitivity analysis of the clinical samples revealed that pathogens are gaining resistance toward the new generations of cephalosporins at a very fast rate globally. The current study developed gold nanoparticles (AuNPs) that could efficiently deliver the 2nd (cefotetan-CT) and 3rd (cefixime-CX) generation cephalosporins to resistant clinical pathogens. In fact, both CT and CX were used to reduce and stabilize AuNPs by applying a one-pot synthesis approach, and their characterization was performed via spectrophotometry, dynamic light scattering and electron microscopy. Moreover, the synthesized AuNPs were tested against uro-pathogenic resistant clinical strains of *Escherichia coli* and *Klebsiella pneumoniae*. CT-AuNPs characteristic SPR peak was observed at 542 nm, and CX-AuNPs showed the same at 522 nm. The stability measurement showed ζ potential as -24.9 mV and -25.2 mV for CT-AuNPs and CX-AuNPs, respectively. Scanning electron microscopy revealed the spherical shape of both the AuNPs, whereas, the size by transmission electron microscopy for CT-AuNPs and CX-AuNPs were estimated to be 45 ± 19 nm and 35 ± 17 nm, respectively. Importantly, once loaded onto AuNPs, both the cephalosporin antibiotics become extremely potent against the resistant strains of *E. coli* and *K. pneumoniae* with MIC_{50} in the range of 0.5 to 0.8 µg/mL. The findings propose that old-generation unresponsive antibiotics could be revived into potent nano-antibiotics via AuNPs. Thus, investing efforts, intellect, time and funds for a nano-antibiotic strategy might be a better approach to overcome resistance than investing the same in the development of newer antibiotic molecule(s).

Keywords: antibiotic resistance; cefotetan; cefixime; clinical pathogens; gold nanoparticles

1. Introduction

The global prevalence of infections caused by antibiotic-resistant bacterial pathogens has markedly increased in the past few decades [1]. There is a plethora of reports that suggested "antibiotic resistance" as an impending threat to the human population [2–4]. In addition, WHO has listed resistance in microorganisms as one of the ten most serious health threat to the global population [5]. Despite continuous warnings, the inappropriate use of antibiotics and self-medication has not stopped. On the other hand, bacterial pathogens have acted smartly by developing different types of resistance mechanisms to evade the action of conventional antibiotics [6–10]. Thus, the current grave situation demands better alternative therapeutic options. Interestingly, nanoparticles could deliver antibiotics effectively to the resistant pathogen, and aid in converting unresponsive antibiotics into potent ones [11–19]. In fact, a large surface area: volume ratio of nanoparticles enables a number of antibiotic molecules to attach to it and prepare a multivalent nano-antibiotic against the resistant pathogen [20]. Further, nanoparticles themselves possess antibacterial potential via the generation of ROS and interaction with bacterial biomolecules (DNA, RNA, protein, enzymes), cell walls, and cell membranes. Although the mechanism of antibacterial action of nanoparticles is not fixed, this can be considered as a blessing in disguise to tackle resistance aspects of bacterial pathogens.

Nanoparticle that has gained the limelight during the past decade due to their exceptional features is gold nanoparticles (AuNPs). AuNPs are not only receiving attention from researchers all over the world, but big companies such as Sigma-Aldrich, NanoHybrids, Cytodiagnostics, Goldsol, etc. are also investing in them [21]. The worldwide market for AuNPs was approximately four and a half billion US dollars in 2021 which is expected to be eight billion US dollars by 2027 [22,23]. Importantly, AuNPs have been widely explored for their potential against antibiotic-resistant strains of different pathogens. In fact, AuNPs have helped to form successful nano-antibiotics by effectively grafting antibiotic molecules on their surface without compromising or damaging the active moiety of the attached antibiotic [19,20,24,25]. Moreover, the synthesis of AuNPs-based nano-antibiotics does not require a complex technique, a simple one-pot synthesis approach where the antibiotic itself acts as a reducing and capping agent could be adequate [11–15,24]. In the present study, AuNPs were explored as a delivery tool for two different generations of cephalosporin antibiotics i.e., 2nd generation- cefotetan (CT) and 3rd generation-cefixime (CX). The successive generations (1st to 5th) of cephalosporins have been developed to increase their potency against different bacterial pathogens (with a prime focus on gram-negative bacteria). However, bacterial pathogens have gained resistance to almost all the generations of cephalosporins with time. Thus, converting unresponsive old-generation cephalosporin antibiotics into effective nano-antibiotics via AuNPs has its due clinical relevance.

The present study successfully developed nano-antibiotics of CT and CX and compared their antibacterial potential against resistant gram-negative uro-pathogenic strains. It is noteworthy to mention that *Escherichia coli* and *Klebsiella pneumoniae* clinical strains used in the present study were not only resistant to cephalosporins but showed resistance towards β-lactamase inhibitor combination and other classes of antibiotics. However, prior to the exploration of antibacterial potential, both the nano-antibiotics (CT-AuNPs and CX-AuNPs) were duly characterized by spectrophotometry, dynamic light scattering and electron microscopy. In addition, loading efficiency was also calculated for both the nano-formulations and taken into consideration while calculating their antibacterial strength or MIC_{50}.

2. Materials and Methods

2.1. Materials

Cefotetan (CT), cefixime (CX), gold chloride salt, culture media, chemicals and solvents were procured from Sigma Aldrich (St. Louis, MO, USA).

2.2. AuNPs Synthesis

Second-generation (CT) and third-generation (CX) cephalosporin antibiotics at 250 μg concentration were added to a 3 mL reaction mixture containing 1 mM gold chloride salt (prepared in 7.2 pH phosphate buffer), separately [11]. Further, both the reaction mixtures were incubated at 40 °C for two days. The synthesis of AuNPs was visually confirmed by the color change i.e., from pale yellow to ruby red. Centrifugation at 30,000× g was performed for 30 min to collect synthesized AuNPs. Centrifuged AuNPs were further washed with milli-Q water and ethanol.

2.3. AuNPs Characterization

Characterization of CT and CX synthesized AuNPs were performed by spectrophotometry, zeta sizing and potential, and electron microscopy.

2.3.1. Spectrophotometry

UV-Visible double-beam spectrophotometer (UV-1601, Shimadzu, Tokyo, Japan) was used to scan (from 200 nm to 800 nm) the transformation of gold salt into AuNPs at 1 nm resolution.

2.3.2. Zeta Sizing and Zeta Potential

The synthesized CT-AuNPs and CX-AuNPs were sonicated (1 min) and filtered (0.45 μm membrane filters) before analyzing them on Malvern Nano Zetasizer (ZEN3600, Malvern Instrument Ltd., Malvern, UK). The mean hydrodynamic diameter of both the AuNPs was calculated by using DTS0112 cuvette, whereas the surface zeta potential for each AuNPs was estimated by using DTS1070 cuvette [16].

2.3.3. Electron Microscopy

Scanning electron microscopy (SEM) and Transmission electron microscopy (TEM) analysis was performed for both CT-AuNPs and CX-AuNPs.

FEI quanta 250 SEM (FEI Company, Hillsboro, OR, USA) at 30 kV acceleration voltage was used to collect SEM images of the synthesized AuNPs. Prior to SEM analysis, samples of CT-AuNPs and CX-AuNPs were deposited on the surface of a silicon substrate (conductive) and dried by using a hotplate at 60 °C.

TEM (Tecnai G2 Spirit) with a fitted BioTwin lens (Hillsboro, OR, USA) at 80 kV accelerating voltage was used to collect TEM images. However, the CT-AuNPs and CX-AuNPs samples were fixed on a carbon-coated copper grid before TEM analysis.

2.4. Estimation of Loading Efficiency

The loading efficiency of CT and CX onto AuNPs was calculated with the help of UV-Vis spectrophotometer as described by Alshammari et al. [15]. After AuNPs synthesis by CT and CX, the 3 mL reaction mixtures were centrifuged for 30 min at 30,000× g and the supernatants were cautiously taken into a separate falcon tube. The concentration of unbound CT and CX were estimated in the supernatant at 254 nm [26] and 290 nm [27], respectively by using their pre-determined calibration curve. Further, the loading efficiency % for each AuNPs sample was estimated by applying the following formula [28]:

$$\% \text{ of loading efficiency} = [(Y - Z)/Y] \times 100 \quad (1)$$

where, Y is the initial concentration of CT or CX used for the AuNPs synthesis, and Z is the remaining (unbound) concentration of CT or CX in the supernatant.

2.5. Assessment of Antibacterial Potential of CT-AuNPs and CX-AuNPs

2.5.1. Resistant Uropathogenic Strains

Uro-pathogenic resistant strains of *Klebsiella pneumoniae* (Seq# 427811998026) and *Escherichia coli* (Seq# 427812372404) were obtained from Hail General Hospital, Hail, Saudi Arabia. Both the pathogenic strains were inoculated in a fresh nutrient broth medium and incubated at 37 °C for 18 h. Prior to antibacterial evaluation, the turbidity for each strain was maintained up to 1.5 × 108 CFU/mL (0.5 McFarland standard).

2.5.2. Antibacterial Assessment by Agar Well Diffusion

Preliminary antibacterial assessment of CT-AuNPs and CX-AuNPs was performed by agar well diffusion method [29]. Mueller–Hinton agar plates were swabbed with a fresh inoculum of each uro-pathogenic strain. After that well cutter was used to aseptically punch four holes (two holes of 4 mm and two holes of 8 mm) on the swabbed agar plates. Pure CT (without AuNPs) and CT-AuNPs were added to the wells at 3.25 µg/well and 6.5 µg/well. Similar concentrations were applied for pure CX and CX-AuNPs. All the Petri plates were incubated for 18 h at 37 °C, and the zone of inhibition was measured. The final zone of inhibition in mm is a mean of triplicate experiments.

2.5.3. Antibacterial Assessment by Calculating Minimal Inhibitory Concentration

The MIC_{50} of pure antibiotic (CT and CX) and its gold nanoformulation (CT-AuNPs and CX-AuNPs) against uropathogenic strains of *K. pneumoniae* and *E. coli* were estimated by using microbroth dilution technique [30]. The concentration of both pure antibiotic and antibiotic after loading to AuNPs was kept in a range from 0.126 to 65 µg/mL in 96-well microtiter plates. After concentration adjustment, 10 µL of test strain (at 0.5 McFarland standard) was added to each well. All the microtiter plates were incubated for 18 h at 37 °C, and MIC_{50} was calculated. The minimum concentration at which the growth was inhibited was noted as MIC, however, triplicate experiments were performed to calculate the mean ± standard deviation of MIC values.

3. Results and Discussion

New-generation cephalosporins have been developed from time to time with an aim to provide better antibiotic therapy for the human population [31]. However, bacterial pathogens have gained resistance against almost all generations of cephalosporin [32–34]. Pathogens have developed different resistance mechanisms such as antibiotic efflux, decrease in antibiotic uptake, disarming antibiotics and manipulating antibiotic targets to escape the cidal/static effect of antibiotics [10]. All these mechanisms could work for developing cephalosporin resistance but particularly disarming by β-lactamase enzyme is considered the major resistance mechanism. Hence, to combat β-lactamase enzyme, β-lactamase inhibitors were added as an adjuvant with the antibiotics. It showed a remarkable effect and provided a strong hope against resistant bacterial pathogens [35]. Unfortunately, pathogens were smart enough to develop resistance towards even the combination of antibiotic and β-lactamase inhibitors [36,37]. Thus, scientists are working hard to explore alternative therapeutic options against resistant bacterial pathogens. Here, in the present study, gold nanoformulations of old-generation (2nd and 3rd) cephalosporins were developed and tested against resistant strains of *E. coli* and *K. pneumoniae*.

3.1. Synthesis of CT-AuNPs and CX-AuNPs

AuNPs of various sizes and features are synthesized *via* chemical reducing agents or natural extracts/enzymes (green synthesis) [38–42]. In chemical synthesis approaches, chemicals such as sodium borohydrate, trisodium citrate, and hydrazine are used as reducing agents; however, AuNPs synthesized after chemical reduction sometimes need additional capping agents for stabilization [43]. On the other hand, plant extracts, microorganisms and natural enzymes are applied as alternatives to chemicals for green synthesis of AuNPs [44,45]. However, the exact reducing and capping agent is difficult to decipher from a plethora of compounds from the natural extract. Moreover, loading/attachment of desired drug onto AuNPs is itself a tedious job that might include the use of harsh chemicals such as 1-ethyl-3-(3-dimethylaminopropyl) carbodiimide [19,46]. Hence, researchers developed one pot-AuNPs synthesis approach where the drug itself acts as a reducing and stabilizing agent to lessen the use of chemicals and ease the process of drug loading [11–15,24]. A similar approach was applied in the current study to synthesize AuNPs by two different generations of cephalosporins [2nd generation-Cefotetan (CT) and 3rd generation-Cefixime (CX)]. Both the cephalosporin antibiotics reduced and stabilized AuNPs in a single step (Figure 1). It is noteworthy to mention that the amine group of β-lactam antibiotics (cephalosporins) usually contributes to gold/silver salts reduction, and the β-lactam ring remains intact for its activity on the surface of nanoparticles even after stabilization or capping [20,24,25,47].

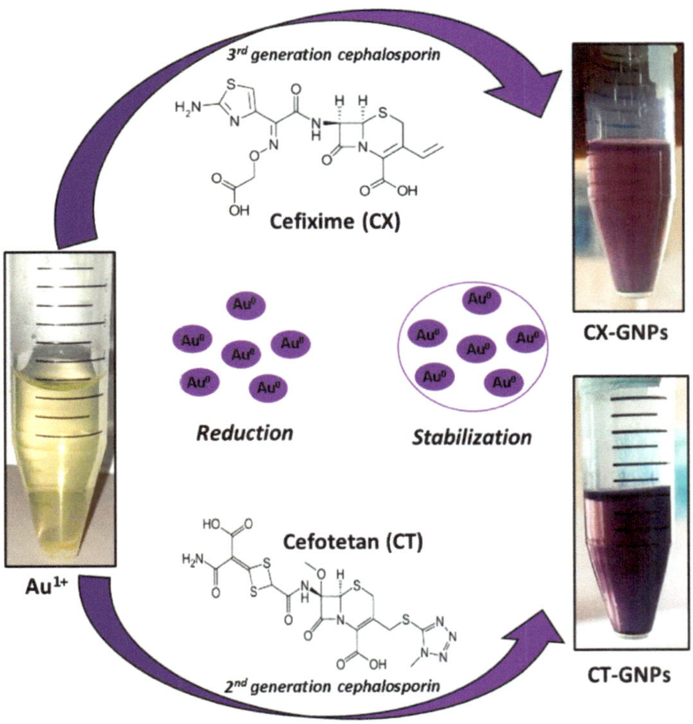

Figure 1. Schematic representation of AuNPs synthesis by cefotetan and cefixime.

To synthesize AuNPs, CT and CX were added at 250 μg concentration to the reaction mixture containing gold salt in pH 7.2 buffer. It has been reported that pH could influence AuNPs zeta potential and stability [48]. In fact, a recent study [49] has shown the influence of pH change on the size and zeta potential of AuNPs. The authors showed that a pH value from 6 to 10 was most appropriate for the synthesis of stable AuNPs. They also observed that acidic pH was unfavorable for AuNPs synthesis with almost neutral zeta potential, large size and aggregation. However, in the current study, the pH was kept at 7.2 (physiological pH) on the basis of earlier reports [11–16,42]. The color change from pale yellow to ruby red (Figure 1) visually confirmed the synthesis of CT- and CX-AuNPs. Both CT-AuNPs and CX-AuNPs were further characterized via spectrophotometry, zeta sizer and electron microscopy.

3.2. Characterization of CT-AuNPs and CX-AuNPs

The first characterization of synthesized AuNPs was performed by UV-Vis spectrophotometer on the basis of characteristic surface plasma resonance (SPR) of AuNPs (i.e., in between the visible range of 500 nm to 600 nm). SPR band peaks for CT-AuNPs and CX-AuNPS were observed at 542 nm and 522 nm, respectively (Figure 2). Additional peaks at 254 nm [26] and 290 nm [27] were also observed during spectrophotometric scanning that corresponds to CT and CX, respectively, which suggested the successful loading of CT and CX onto AuNPs (Figure 2). In accordance, similar dual peaks have been reported by several recent investigations on antibiotic-mediated AuNPs synthesis [11–15]. The AuNPs synthesized by cefoxitin, delafloxacin, cefotaxime, vancomycin and ceftriaxone showed a second peak at 235, 290, 260, 278 and 241 nm, respectively, along with the characteristic AuNPs peak between 500 to 600 nm. Even the conjugation of antibiotics onto AuNPs with chemical agent EDC (instead of a one-step synthesis approach) also showed dual peaks that confirm the successful loading of the antibiotics to AuNPs [19,46].

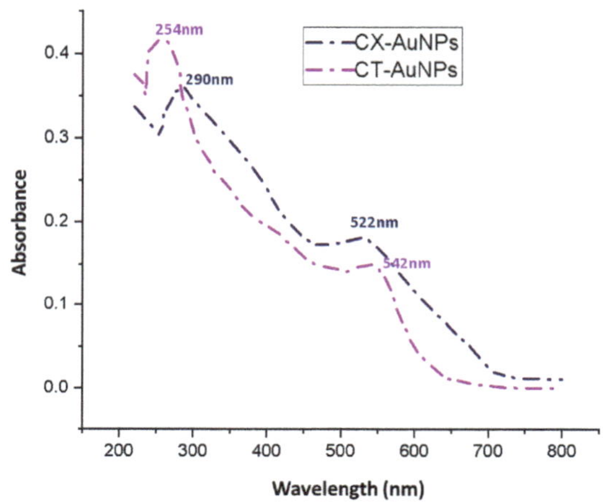

Figure 2. UV-Visible spectrophotometric scan of CX-AuNPs and CT-AuNPs.

The zeta size distribution of CX-AuNPs and CT-AuNPs in the dispersion was estimated by a dynamic light scattering approach. The size was estimated by measuring the arbitrary variations of the scattered intensity of light due to AuNPs dispersion [50]. The Z-average mean size for CX-AuNPs and CT-AuNPs was 117 nm and 119.6 nm, respectively (Figure 3). In addition, the colloidal stability of CX-AuNPs and CT-AuNPs was measured by zeta potential that estimates the surface charge on AuNPs. In fact, it depends on the electrostatic repulsion strength of the similarly charged particles in the dispersion [51]. The Zeta potential of CX-AuNPs and CT-AuNPs was estimated as −25.2 mV and −24.9 mV, respectively. The higher positive or negative zeta potential i.e., more than ±20 mV, repels the particles and minimizes the probability of aggregation [52]. Thus, CX-AuNPs and CT-AuNPs synthesized in the current study showed long-term stability even at room temperature for months. However, the functional group present on the antibiotics (CX and CT) might have attributed to the negative charge on the surface of AuNPs.

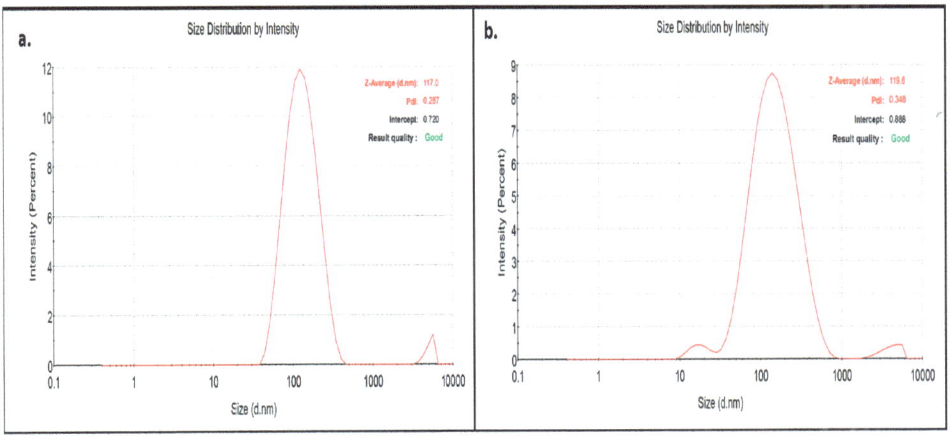

Figure 3. Size distribution vs intensity plot of (**a**) CX-AuNPs (**b**) CT-AuNPs.

Scanning Electron Microscopy (SEM) scans the nanomaterial surface with a focused beam of high-energy electrons. The electron beam interacts with the nanomaterial to create signals that could be further translated into information pertinent to the surface morphology of the nanomaterial [53]. Here, in the present investigation, SEM was used to observe the surface morphology of the synthesized CX-AuNPs and CT-AuNPs, and both the AuNPs appeared spherical in shape (Figure 4). These findings were in accordance with the previous investigations, where antibiotics were loaded/conjugated to AuNPs and a similar spherical shape pattern was observed [15,19,46].

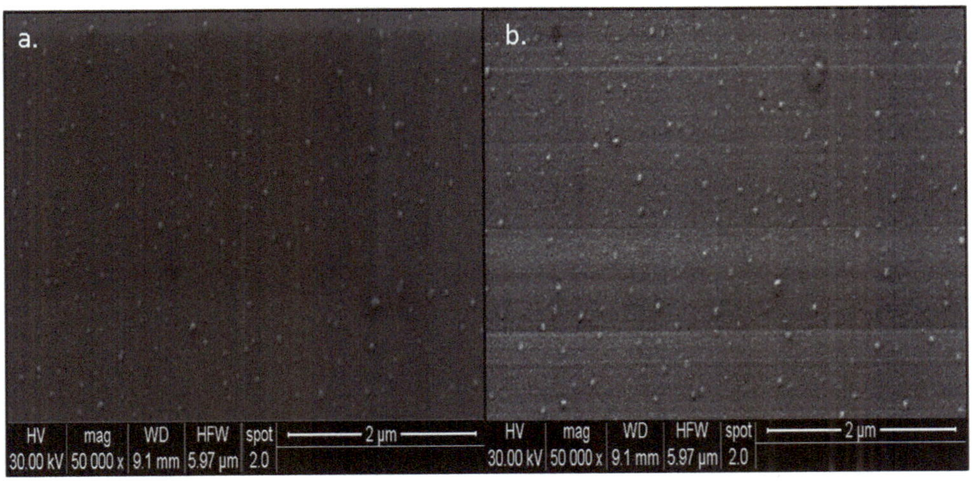

Figure 4. SEM images of (**a**) CX-AuNPs and (**b**) CT-AuNPs.

To estimate the size of inorganic core of CX-AuNPs and CT-AuNPs, transmission electron microscopy (TEM) was applied. TEM images were taken at 1,000,000× magnification for both the AuNPs (Figure 5). TEM analysis revealed that both AuNPs samples were poly-dispersed, with mean size of 35 ± 17 nm and 45 ± 19 nm for CX-AuNPs and CT-AuNPs, respectively. There was no observed aggregation in the TEM and SEM images that corresponds to the successful capping of AuNPs by both the antibiotics (CX and CT). It is a fact that different measurement techniques will show variation in size of nanoparticles. Dynamic light scattering approach calculate the size including the adhered solvent layer, while TEM estimates the size of inorganic core [54,55]. Hence, size determination by TEM is comparatively lower than size by DLS. Similar aspects of size differences have been observed in various previous reports on antibiotic loaded AuNPs [11–15,19].

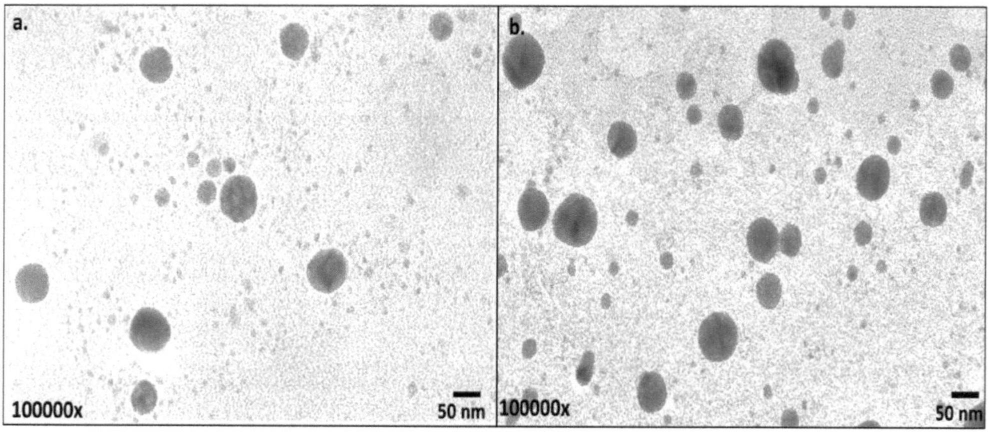

Figure 5. TEM images of (**a**) CX-AuNPs and (**b**) CT-AuNPs.

Furthermore, the loading efficiency of CX-AuNPs and CT-AuNPs was estimated by using the formula described by Gomes et al. [28]. In fact, the calculation of loading efficiency is a crucial factor for the nanoformulation characterization and their biological application. The loading efficiency of CX and CT on AuNPs was calculated as 77.48% and 79.84%, respectively. Initially, 250 µg of both antibiotics (CX and CT) were added to the reaction mixture. Out of 250 µg, 193.7 µg of CX was loaded onto CX-AuNPs and 199.6 µg of CT was loaded onto CT-AuNPs. There was no significant loss of antibiotics observed during the synthesis of AuNPs, and an appropriate amount has been successfully loaded onto AuNPs.

3.3. Comparative Antibacterial Assessment of CT-AuNPs and CX-AuNPs

In the past, several prevalence studies have confirmed the rapid expansion of cephalosporin resistance among gram-negative bacterial pathogens, particularly in clinical strains of *K. pneumoniae* and *E. coli* [56–60]. The present study designed nanoformulations of old-generation (second and third) cephalosporins against cephalosporin-resistant clinical strains of *K. pneumoniae* and *E. coli*. Comparative antibacterial assessment of AuNPs loaded with second generation-cefotetan (CT) and third-generation-cefixime (CX) was performed initially by well-diffusion technique (Figure 6 and Table 1) followed by MIC_{50} calculation (Figure 7). At 3.25 mg and 6.5 mg concentrations of CT-AuNPs, the zone of inhibitions against *E. coli* were observed as 18mm and 23mm, respectively (Figure 6a). Whereas, inhibition zones against *K. pneumoniae* were 19 mm and 24 mm at the same concentrations of CT-AuNPs (Figure 6b). On the other hand, CX-AuNPs at 3.25 mg and 6.5 mg concentrations showed inhibition zones against *E. coli* as 17 mm and 22 mm, respectively (Figure 6c). However, the zone of inhibitions against *K. pneumoniae* of CX-AuNPs at the same concentrations was estimated as 18 mm and 23 mm, respectively (Figure 6d). It is noteworthy to mention that at 3.25 mg and 6.5 mg concentrations neither pure CT nor pure CX showed any activity against the tested strains (Figure 6). Thus, it could be inferred from the results that CT and CX become potent against the tested strain after loading onto AuNPs at 3.25 mg and 6.5 mg concentrations. Both the tested bacterial pathogens belong to *Enterobactericiae* family, and according to Clinical and Laboratory Standards Institute (CLSI 2020) guidelines, the breakpoint of CT (30 mg) for sensitivity is ≥16 mm zone of inhibition and for CX (5 mg) sensitivity breakpoint is ≥19mm against *Enterobacterales*. In the present study, cefotetan (after loading onto AuNPs) at very low concentration (3.25 mg) showed significant activity i.e., more than 16 mm inhibition zone (CLSI Breakpoint) against both *K. pneumoniae* and *E. coli*. Hence, it can be safely stated that a second-generation cephalosporin (CT) has become relatively more potent than third-generation cephalosporin (CX) after loading onto AuNPs.

Agar well diffusion technique provided qualitative antibacterial assessment, and quantitative assessment was further performed by calculating the MIC_{50} concentration of antibiotic nanoformulations and pure antibiotics against the tested strains. A comparative analysis of MIC values of pure CT and CX, and their gold nanoformulations were performed on *E. coli* (Figure 7a) and *K. pneumoniae* (Figure 7b). MIC_{50} of CX-AuNPs and CT-AuNPs against *E. coli* were estimated as 0.8 mg/mL and 0.65 mg/mL, respectively; whereas, pure CX and CT showed MIC_{50} as 19mg/mL and 32.5 mg/mL, respectively. On the other hand, CX-AuNPs and CT-AuNPs showed MIC_{50} as 0.75 mg/mL and 0.5 mg/mL against *K. pneumoniae*; while, MIC_{50} of pure antibiotics, CX and CT were estimated as 17 mg/mL and 27 mg/mL, respectively. MIC sensitivity breakpoint for CT against *Enterobactericiae* is $\leq$16 mg/mL, however, for CX it is $\leq$1 mg/mL (CLSI 2020). MIC of both AuNPs comes well under the limit of sensitivity set by the CLSI guidelines. It is to be noted that there is a huge difference in MIC values of the pure antibiotics and their gold nanoformulations. MIC_{50} values for CX were decreased by 23.75 and 22.66 times when loaded onto AuNPs against *E. coli* and *K. pneumoniae*, respectively. In contrast, MIC_{50} values for CT were decreased by 50 and 54 times when loaded onto AuNPs against *E. coli* and *K. pneumoniae*, respectively. MIC findings were in accordance with the well-diffusion assay results that suggested CT showed more potency than CX when loaded with AuNPs against the tested strains.

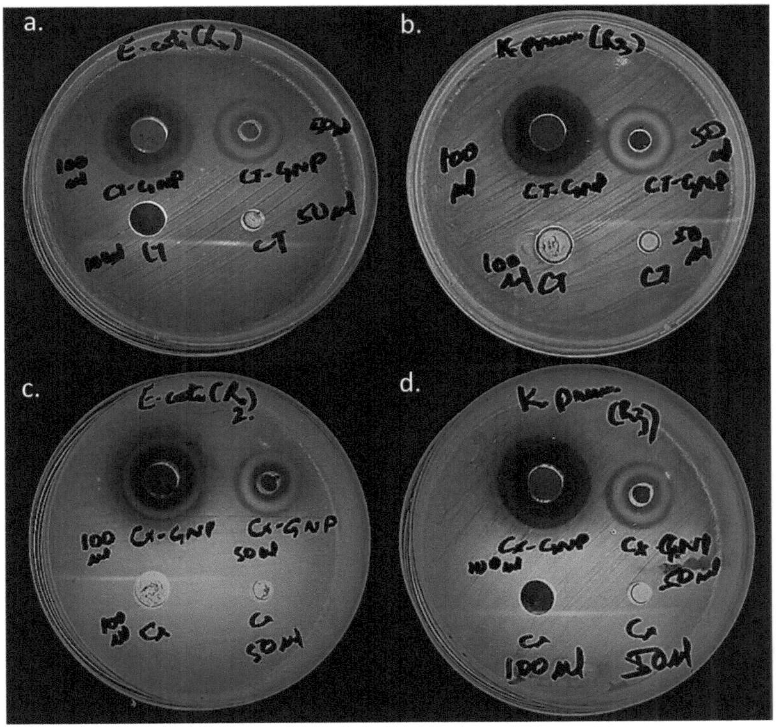

Figure 6. Antibacterial activity of (**a**) pure CT and CT-AuNPs against *E. coli*, (**b**) pure CT and CT-AuNPs against *K.pneumoniae*, (**c**) pure CX and CX-AuNPs against *E. coli*, and (**d**) pure CX and CX-AuNPs against *K.pneumoniae*.

Table 1. : Inhibition zones of CT-AuNPs and CX-AuNPs against *E. coli* and *K. pneumoniae*.

	E. coli	*K. pneumoniae*
CX-GNPs (3.25 mg)	17 mm	18 mm
CX-GNPs (6.5 mg)	22 mm	23 mm
CT-GNPs (3.25 mg)	18 mm	19 mm
CT-GNPs (6.5 mg)	23 mm	24 mm

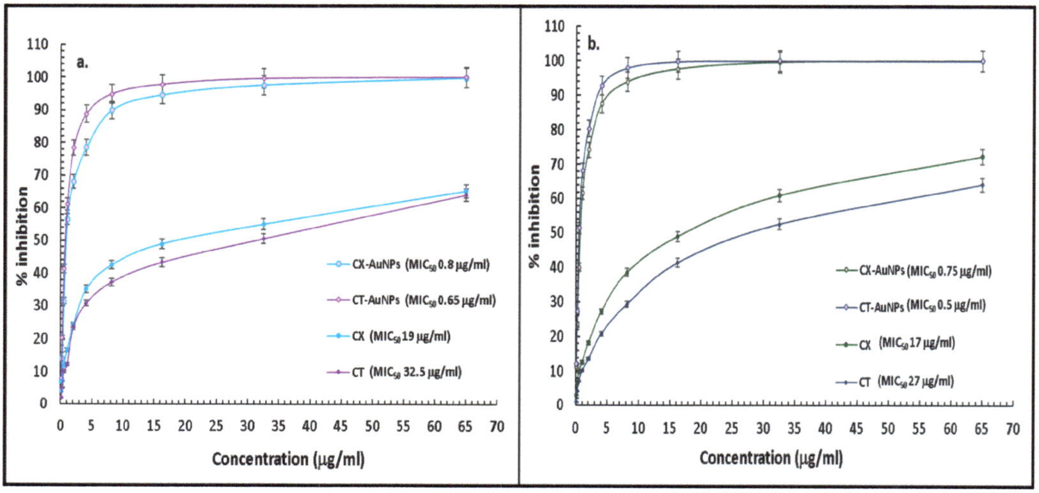

Figure 7. MIC graph of pure CT, pure CX, CT-AuNPs and CX-AuNPs against (**a**). *E. coli* and (**b**). *K. pneumoniae*.

Cephalosporin antibiotic(s) resistance in bacterial pathogens is generally associated with overexpression of extended-spectrum β-lactamase [6]. However, the close association of β-lactamase with the downregulation of porin channels and upregulation of efflux pump in bacterial pathogens significantly enhances their resistance towards cephalosporin [61–63]. In the present study, the bacterial strains tested were resistant to second and third-generation cephalosporins, and AuNPs convert them from ineffective antibiotics into effective antibiotic-nanoformulations. The enhanced antibacterial potential of cephalosporin-nanoformulations in the present investigation could be correlated with the ability of AuNPs to successfully deliver an ample amount of cephalosporins to resistant pathogens. It has to be noted that the active moiety i.e., β-lactam ring remains intact after conjugating/loading cephalosporins onto AuNPs, and due to the large surface-to-vol ratio significant amount of cephalosporin could be loaded on the AuNPs [20,24,25,47,64]. Thus, it could be proposed that β-lactamase might become saturated by the sufficient quantity of substrate (cephalosporin) received, meanwhile, the untouched cephalosporin could perform its usual action on the cell wall. In addition, AuNPs themselves have the ability to bind/inhibit the efflux pump, alter the permeability of the cell membrane and interact/disrupt the biomolecules of bacterial pathogens [17,24,65]. Therefore, it could be suggested that the antibacterial effect was due to the synergism of AuNPs and cephalosporins loaded onto them. Interestingly, AuNPs have no defined or single mechanism of action against bacterial pathogens, hence, developing resistance against them is quite a difficult task for the pathogens. The present investigation would like to emphasize one aspect: if an old generation cephalosporin could be resuscitated with AuNPs, "Why should the intellect, time and funds be spent on developing newer generations?". AuNPs appear to be a smarter alternative to overcoming resistance to bacterial pathogens. However, cost-effectiveness and safe applicability of AuNPs are still a question of debate.

It is noteworthy to mention that our team has recently worked on two third-generation cephalosporins gold nanoformulations (ceftriaxone and cefotaxime) [13,15] and one second-generation cephalosporin gold nanoformulations (cefoxitin) [11], and found that second-generation cephalosporin-AuNPs was equally effective antibacterial than third-generation AuNPs. However, the bacterial strains tested in each of these studies were different. These findings prompted our team to find a comparative analysis of the same resistant clinical strains to get a better insight into the antibacterial action of two different generations loaded onto AuNPs. The results suggested that an ineffective old-generation cephalosporin (CT) could be converted into an effective nanoformulation and show better antibacterial potential than the new-generation cephalosporin (CX) nanoformulations. However, it is too early to come to any conclusion as the human toxicity part and the fate of these AuNPs are still a point of debate, and our team has started working on toxicity aspects and getting deeper insights into the mechanism of their antibacterial action. Nevertheless, it is strongly believed that the scientific community could provide safer nano-antibiotic against resistant pathogens in the near future.

4. Conclusions

In the present study, the comparative analysis of second-generation cephalosporin (CT)-AuNPs and third-generation cephalosporin (CX)-AuNPs was performed against resistant clinical strains of *E. coli* and *K. pneumoniae*. A facile one-pot synthesis approach was used to successfully synthesize the CT-AuNPs and CX-AuNPs via using cephalosporins (CT and CX) as reducing and capping agents. The synthesized CT-AuNPs and CX-AuNPs were stable with ζ potential as -24.9 mV and -25.2 mV, and size as 45 ± 19 nm and 35 ± 17 nm, respectively. However, 79.84% of CT and 77.48% of CX were loaded onto the synthesized AuNPs. CT after loading to AuNPs becomes ~50 times more potent than the pure CT, while CX after loading to AuNPs becomes ~25 times more active than pure CX against the CT and CX-resistant tested strains. Hence, ineffective cephalosporins could be resuscitated into effective nano-antibiotic with the help of AuNPs. In addition, the idea of 'nano-conversion of old generation cephalosporin' appears to be more promising than spending efforts and intellects on developing a new generation of cephalosprorin. However, the toxicity and fate of AuNPs need to decipher in a planned manner. Nevertheless, the present investigation paved the way to design AuNPs-based nano-antibiotics to tackle resistance issues in bacterial pathogens.

Author Contributions: Conceptualization, S.M.D.R., A.M. and A.S.A.L.; methodology, T.H. and S.S.M.F.; software, A.S.A.L. and M.A.K.; validation, E.-S.K., F.A. and M.A.K.; formal analysis, H.S., E.-S.K. and F.A.; investigation, S.M.D.R. and T.H.; resources, S.S.M.F., N.A. and H.S.; writing—original draft preparation, S.M.D.R., N.A., A.M. and A.S.A.L.; writing—review and editing, T.H., N.A., E.-S.K. and F.A.; visualization, H.S., S.S.M.F. and M.A.K.; supervision, A.M. and A.S.A.L.; project administration, T.H.; funding acquisition, S.M.D.R. All authors have read and agreed to the published version of the manuscript.

Funding: This research has been funded by Scientific Research Deanship at the University of Ha'il-Saudi Arabia through project number MDR-22 008.

Institutional Review Board Statement: Not applicable.

Data Availability Statement: Not applicable.

Acknowledgments: This research has been funded by Scientific Research Deanship at the University of Ha'il-Saudi Arabia through project number MDR-22 008.

Conflicts of Interest: The authors declare no conflict of interest.

References

1. Murray, C.J.L.; Ikuta, K.S.; Sharara, F.; Swetschinski, L.R.; Robles Aguilar, G.; Gray, A.; Han, C.; Bisignano, C.; Rao, P.C.; Woo E.; et al. Global burden of bacterial antimicrobial resistance in 2019: A systematic analysis. *Lancet (Lond. Engl.)* **2022**, *399*, 629–655 [CrossRef]
2. Prestinaci, F.; Pezzotti, P.; Pantosti, A. Antimicrobial resistance: A global multifaceted phenomenon. *Pathog. Glob. Health* **2015** *109*, 309–318. [CrossRef] [PubMed]
3. Zhang, Z.; Zhang, Q.; Wang, T.; Xu, N.; Lu, T.; Hong, W.; Penuelas, J.; Gillings, M.; Wang, M.; Gao, W.; et al. Assessment of global health risk of antibiotic resistance genes. *Nat. Commun.* **2022**, *13*, 1553. [CrossRef] [PubMed]
4. Cameron, A.; Esiovwa, R.; Connolly, J.; Hursthouse, A.; Henriquez, F. Antimicrobial Resistance as a Global Health Threat: The Need to Learn Lessons from the COVID-19 Pandemic. *Glob. Policy* **2022**, *13*, 179–192. [CrossRef] [PubMed]
5. WHO. Available online: https://www.who.int/news-room/fact-sheets/detail/antimicrobial-resistance (accessed on 18 December 2022).
6. Shaikh, S.; Fatima, J.; Shakil, S.; Rizvi, S.M.; Kamal, M.A. Antibiotic resistance and extended spectrum beta-lactamases: Types epidemiology and treatment. *Saudi J. Biol. Sci.* **2015**, *22*, 90–101. [CrossRef]
7. Shaikh, S.; Fatima, J.; Shakil, S.; Danish Rizvi, S.M.; Kamal, M.A. Prevalence of multidrug resistant and extended spectrum beta-lactamase producing Pseudomonas aeruginosa in a tertiary care hospital. *Saudi J. Biol. Sci.* **2015**, *22*, 62–64. [CrossRef]
8. Shaikh, S.; Fatima, J.; Shakil, S.; Rizvi, S.M.; Kamal, M.A. Risk factors for acquisition of extended spectrum beta lactamase producing Escherichia coli and Klebsiella pneumoniae in North-Indian hospitals. *Saudi J. Biol. Sci.* **2015**, *22*, 37–41. [CrossRef] [PubMed]
9. Shaikh, S.; Rizvi, S.M.D.; Shakil, S.; Ahmad, A.; Pathak, N. Non-clonal Dissemination of Extended-Spectrum Beta-Lactamase Producing Pseudomonas aeruginosa Strains of Clinical Origin. *Iran. J. Sci. Technol. Trans. A Sci.* **2017**, *41*, 1011–1015. [CrossRef]
10. Reygaert, W.C. An overview of the antimicrobial resistance mechanisms of bacteria. *AIMS Microbiol.* **2018**, *4*, 482–501. [CrossRef]
11. Alafnan, A.; Rizvi, S.M.; Alshammari, A.S.; Faiyaz, S.S.; Lila, A.S.; Katamesh, A.A.; Khafagy, E.-S.; Alotaibi, H.F.; Ahmed A.B.F. Gold Nanoparticle-Based Resuscitation of Cefoxitin against Clinical Pathogens: A Nano-Antibiotic Strategy to Overcome Resistance. *Nanomaterials* **2022**, *12*, 3643. [CrossRef]
12. Abu Lila, A.S.; Huwaimel, B.; Alobaida, A.; Hussain, T.; Rafi, Z.; Mehmood, K.; Abdallah, M.H.; Hagbani, T.A.; Rizvi, S.M.D Moin, A.; et al. Delafloxacin-Capped Gold Nanoparticles (DFX-AuNPs): An Effective Antibacterial Nano-Formulation of Fluoroquinolone Antibiotic. *Materials* **2022**, *15*, 5709. [CrossRef]
13. Al Hagbani, T.; Rizvi, S.M.D.; Hussain, T.; Mehmood, K.; Rafi, Z.; Moin, A.; Abu Lila, A.S.; Alshammari, F.; Khafagy, E.S Rahamathulla, M.; et al. Cefotaxime Mediated Synthesis of Gold Nanoparticles: Characterization and Antibacterial Activity *Polymers* **2022**, *14*, 771. [CrossRef]
14. Hagbani, T.A.; Yadav, H.; Moin, A.; Lila, A.S.A.; Mehmood, K.; Alshammari, F.; Khan, S.; Khafagy, E.S.; Hussain, T.; Rizvi S.M.D.; et al. Enhancement of Vancomycin Potential against Pathogenic Bacterial Strains via Gold Nano-Formulations: A Nano-Antibiotic Approach. *Materials* **2022**, *15*, 1108. [CrossRef]
15. Alshammari, F.; Alshammari, B.; Moin, A.; Alamri, A.; Al Hagbani, T.; Alobaida, A.; Baker, A.; Khan, S.; Rizvi, S.M.D. Ceftriaxone Mediated Synthesized Gold Nanoparticles: A Nano-Therapeutic Tool to Target Bacterial Resistance. *Pharmaceutics* **2021**, *13*, 1896 [CrossRef]
16. Soliman, W.E.; Khan, S.; Rizvi, S.M.D.; Moin, A.; Elsewedy, H.S.; Abulila, A.S.; Shehata, T.M. Therapeutic Applications of Biostable Silver Nanoparticles Synthesized Using Peel Extract of Benincasa hispida: Antibacterial and Anticancer Activities *Nanomaterials* **2020**, *10*, 1954. [CrossRef] [PubMed]
17. Shaikh, S.; Nazam, N.; Rizvi, S.M.D.; Ahmad, K.; Baig, M.H.; Lee, E.J.; Choi, I. Mechanistic Insights into the Antimicrobial Actions of Metallic Nanoparticles and Their Implications for Multidrug Resistance. *Int. J. Mol. Sci.* **2019**, *20*, 2468. [CrossRef] [PubMed]
18. Rizvi, S.M.D.; Hussain, T.; Ahmed, A.B.F.; Alshammari, T.M.; Moin, A.; Ahmed, M.Q.; Barreto, G.E.; Kamal, M.A.; Ashraf, G.M Gold nanoparticles: A plausible tool to combat neurological bacterial infections in humans. *Biomed. Pharmacother.* **2018**, *107*, 7–18 [CrossRef]
19. Shaikh, S.; Rizvi, S.M.D.; Shakil, S.; Hussain, T.; Alshammari, T.M.; Ahmad, W.; Tabrez, S.; Al-Qahtani, M.H.; Abuzenadah A.M. Synthesis and Characterization of Cefotaxime Conjugated Gold Nanoparticles and Their Use to Target Drug-Resistant CTX-M-Producing Bacterial Pathogens. *J. Cell Biochem.* **2017**, *118*, 2802–2808. [CrossRef]
20. Khandelwal, P.; Singh, D.K.; Poddar, P. Advances in the Experimental and Theoretical Understandings of Antibiotic Conjugated Gold Nanoparticles for Antibacterial Applications. *ChemistrySelect* **2019**, *4*, 6719–6738. [CrossRef]
21. Allied Market Research. Available online: https://www.alliedmarketresearch.com/gold-nanoparticles-market-A08997 (accessed on 18 December 2022).
22. Globenewswire. Available online: https://www.globenewswire.com/news-release/2022/10/13/2534150/0/en/Global-Gold Nanoparticles-Market-to-Reach-7-6-Billion-by-2027.html (accessed on 18 December 2022).
23. IMARCGROUP. Available online: https://www.imarcgroup.com/gold-nanoparticles-market (accessed on 18 December 2022).
24. Rai, A.; Prabhune, A.; Perry, C.C. Antibiotic mediated synthesis of gold nanoparticles with potent antimicrobial activity and their application in antimicrobial coatings. *J. Mater. Chem.* **2010**, *20*, 6789–6798. [CrossRef]

25. Khandelwal, P.; Singh, D.K.; Sadhu, S.; Poddar, P. Study of the nucleation and growth of antibiotic labeled Au NPs and blue luminescent Au8 quantum clusters for Hg2+ ion sensing, cellular imaging and antibacterial applications. *Nanoscale* **2015**, *7*, 19985–20002. [CrossRef] [PubMed]
26. Fujimoto, M.; Maeda, T.; Okumura, K.; Uda, M.; Nakamura, M.; Kashiwagi, T.; Tsunoda, T. Process Development and Pilot-Scale Synthesis of Cefotetan. *Org. Process Res. Dev.* **2004**, *8*, 915–919. [CrossRef]
27. Ramadan, A.A.; Mandil, H.; Dahhan, M. UV-VIS spectrophotometric study for determination of cefixime in pure form and in pharmaceuticals through complex-ation with Cu(II) using acetate-NaOH buffer in water:methanol. *Int. J. Pharm. Phar. Sci.* **2013**, *51*, 428–433.
28. Gomes, M.J.; Martins, S.; Ferreira, D.; Segundo, M.A.; Reis, S. Lipid nanoparticles for topical and transdermal application for alopecia treatment: Development, physicochemical characterization, and in vitro release and penetration studies. *Int. J. Nanomed.* **2014**, *9*, 1231–1242. [CrossRef]
29. Balouiri, M.; Sadiki, M.; Ibnsouda, S.K. Methods for in vitro evaluating antimicrobial activity: A review. *J. Pharm. Anal.* **2016**, *6*, 71–79. [CrossRef]
30. Eloff, J.N. A sensitive and quick microplate method to determine the minimal inhibitory concentration of plant extracts for bacteria. *Planta Med.* **1998**, *64*, 711–713. [CrossRef]
31. Rusu, A.; Lungu, I.-A. The new fifth-generation cephalosporins–A balance between safety and efficacy. *Ro. J. Pharm. Pract.* **2020**, *13*, 121–126. [CrossRef]
32. Fu, Y.; Xu, X.; Zhang, L.; Xiong, Z.; Ma, Y.; Wei, Y.; Chen, Z.; Bai, J.; Liao, M.; Zhang, J. Fourth Generation Cephalosporin Resistance Among Salmonella enterica Serovar Enteritidis Isolates in Shanghai, China Conferred by bla (CTX-M-55) Harboring Plasmids. *Front. Microbiol.* **2020**, *11*, 910. [CrossRef] [PubMed]
33. Lester, R.; Musicha, P.; Kawaza, K.; Langton, J.; Mango, J.; Mangochi, H.; Bakali, W.; Pearse, O.; Mallewa, J.; Denis, B.; et al. Effect of resistance to third-generation cephalosporins on morbidity and mortality from bloodstream infections in Blantyre, Malawi: A prospective cohort study. *Lancet Microbe* **2022**, *3*, e922–e930. [CrossRef]
34. Chang, C.Y.; Huang, P.H.; Lu, P.L. The Resistance Mechanisms and Clinical Impact of Resistance to the Third Generation Cephalosporins in Species of Enterobacter cloacae Complex in Taiwan. *Antibiotics* **2022**, *11*, 1153. [CrossRef] [PubMed]
35. Watkins, R.R.; Papp-Wallace, K.M.; Drawz, S.M.; Bonomo, R.A. Novel β-lactamase inhibitors: A therapeutic hope against the scourge of multidrug resistance. *Front. Microbiol.* **2013**, *4*, 392. [CrossRef]
36. Ripoll, A.; Galán, J.C.; Rodríguez, C.; Tormo, N.; Gimeno, C.; Baquero, F.; Martínez-Martínez, L.; Cantón, R. Detection of resistance to beta-lactamase inhibitors in strains with CTX-M beta-lactamases: A multicenter external proficiency study using a well-defined collection of Escherichia coli strains. *J. Clin. Microbiol.* **2014**, *52*, 122–129. [CrossRef]
37. Helfand, M.S.; Bethel, C.R.; Hujer, A.M.; Hujer, K.M.; Anderson, V.E.; Bonomo, R.A. Understanding Resistance to β-Lactams and β-Lactamase Inhibitors in the SHV β-Lactamase: LESSONS FROM THE MUTAGENESIS OF SER-130. *J. Biol. Chem.* **2003**, *278*, 52724–52729. [CrossRef]
38. Suchomel, P.; Kvitek, L.; Prucek, R.; Panacek, A.; Halder, A.; Vajda, S.; Zboril, R. Simple size-controlled synthesis of Au nanoparticles and their size-dependent catalytic activity. *Sci. Rep.* **2018**, *8*, 4589. [CrossRef]
39. Piella, J.; Bastús, N.G.; Puntes, V. Size-Controlled Synthesis of Sub-10-nanometer Citrate-Stabilized Gold Nanoparticles and Related Optical Properties. *Chem. Mater.* **2016**, *28*, 1066–1075. [CrossRef]
40. Elia, P.; Zach, R.; Hazan, S.; Kolusheva, S.; Porat, Z.e.; Zeiri, Y. Green synthesis of gold nanoparticles using plant extracts as reducing agents. *Int. J. Nanomed.* **2014**, *9*, 4007–4021.
41. Muddapur, U.M.; Alshehri, S.; Ghoneim, M.M.; Mahnashi, M.H.; Alshahrani, M.A.; Khan, A.A.; Iqubal, S.M.S.; Bahafi, A.; More, S.S.; Shaikh, I.A.; et al. Plant-Based Synthesis of Gold Nanoparticles and Theranostic Applications: A Review. *Molecules* **2022**, *27*, 1391. [CrossRef]
42. Khan, S.; Danish Rizvi, S.M.; Avaish, M.; Arshad, M.; Bagga, P.; Khan, M.S. A novel process for size controlled biosynthesis of gold nanoparticles using bromelain. *Mater. Lett.* **2015**, *159*, 373–376. [CrossRef]
43. Amina, S.J.; Guo, B. A review on the synthesis and functionalization of gold nanoparticles as a drug delivery vehicle. *Int. J. Nanomed.* **2020**, *15*, 9823. [CrossRef] [PubMed]
44. Ghosh, N.S.; Pandey, E.; Kadian, J.P.; Chauhan, B. Green Synthesis of Gold Nanoparticles: A Novel, Environment-Friendly, Economic, Safe Approach. *Biomed. Pharmacol. J.* **2021**, *14*, 2041–2406. [CrossRef]
45. Sengani, M.; Grumezescu, A.M.; Rajeswari, V.D. Recent trends and methodologies in gold nanoparticle synthesis–A prospective review on drug delivery aspect. *OpenNano* **2017**, *2*, 37–46. [CrossRef]
46. Khan, S.; Haseeb, M.; Baig, M.H.; Bagga, P.S.; Siddiqui, H.H.; Kamal, M.A.; Khan, M.S. Improved efficiency and stability of secnidazole–An ideal delivery system. *Saudi J. Biol. Sci.* **2015**, *22*, 42–49. [CrossRef]
47. Khatoon, N.; Alam, H.; Khan, A.; Raza, K.; Sardar, M. Ampicillin Silver Nanoformulations against Multidrug resistant bacteria. *Sci. Rep.* **2019**, *9*, 6848. [CrossRef] [PubMed]
48. Maciejewska-Prończuk, J.; Morga, M.; Adamczyk, Z.; Oćwieja, M.; Zimowska, M. Homogeneous gold nanoparticle monolayers—QCM and electrokinetic characteristics. *Colloids Surf. A Physicochem. Eng. Asp.* **2017**, *514*, 226–235. [CrossRef]
49. Pourali, P.; Benada, O.; Pátek, M.; Neuhöferová, E.; Dzmitruk, V.; Benson, V. Response of Biological Gold Nanoparticles to Different pH Values: Is It Possible to Prepare Both Negatively and Positively Charged Nanoparticles? *Appl. Sci.* **2021**, *11*, 11559. [CrossRef]

50. Kato, H.; Nakamura, A.; Takahashi, K.; Kinugasa, S. Accurate Size and Size-Distribution Determination of Polystyrene Late Nanoparticles in Aqueous Medium Using Dynamic Light Scattering and Asymmetrical Flow Field Flow Fractionation wit Multi-Angle Light Scattering. *Nanomaterials* **2012**, *2*, 15–30. [CrossRef]
51. Rasmussen, M.K.; Pedersen, J.N.; Marie, R. Size and surface charge characterization of nanoparticles with a salt gradient. *Na Commun.* **2020**, *11*, 2337. [CrossRef]
52. Xu, R. Progress in nanoparticles characterization: Sizing and zeta potential measurement. *Particuology* **2008**, *6*, 112–115. [CrossRe
53. Zhou, W.; Apkarian, R.P.; Wang, Z.L.; Joy, D.C. Fundamentals of Scanning Electron Microscopy (SEM). In *Scanning Microscopy fc Nanotechnology*; Zhou, W., Wang, Z.L., Eds.; Springer: New York, NY, USA, 2006. [CrossRef]
54. Wilson, B.K.; Prud'homme, R.K. Nanoparticle size distribution quantification from transmission electron microscopy (TEM) c ruthenium tetroxide stained polymeric nanoparticles. *J. Colloid Interface Sci.* **2021**, *604*, 208–220. [CrossRef]
55. Souza, T.G.F.; Ciminelli, V.S.T.; Mohallem, N.D.S. A comparison of TEM and DLS methods to characterize size distribution c ceramic nanoparticles. *J. Phys. Conf. Ser.* **2016**, *733*, 012039. [CrossRef]
56. Goldstein, E. Rise in the prevalence of resistance to extended-spectrum cephalosporins in the USA, nursing homes and antibioti prescribing in outpatient and inpatient settings. *J. Antimicrob. Chemother.* **2021**, *76*, 2745–2747. [CrossRef]
57. Zamudio, R.; Boerlin, P.; Beyrouthy, R.; Madec, J.-Y.; Schwarz, S.; Mulvey, M.R.; Zhanel, G.G.; Cormier, A.; Chalmers, G.; Bonne R.; et al. Dynamics of extended-spectrum cephalosporin resistance genes in Escherichia coli from Europe and North Americ *Nat. Commun.* **2022**, *13*, 7490. [CrossRef]
58. Meyer, E.; Schwab, F.; Schroeren-Boersch, B.; Gastmeier, P. Dramatic increase of third-generation cephalosporin-resistant E. coli i German intensive care units: Secular trends in antibiotic drug use and bacterial resistance, 2001 to 2008. *Crit Care* **2010**, *14*, R11 [CrossRef] [PubMed]
59. van der Steen, M.; Leenstra, T.; Kluytmans, J.A.; van der Bij, A.K. Trends in Expanded-Spectrum Cephalosporin-Resistar Escherichia coli and Klebsiella pneumoniae among Dutch Clinical Isolates, from 2008 to 2012. *PLoS ONE* **2015**, *10*, e013808 [CrossRef]
60. Aracil-García, B.; Oteo-Iglesias, J.; Cuevas-Lobato, Ó.; Lara-Fuella, N.; Pérez-Grajera, I.; Fernández-Romero, S.; Pérez-Vázquez, M Campos, J. Rapid increase in resistance to third generation cephalosporins, imipenem and co-resistance in Klebsiella pneumonia from isolated from 7140 blood-cultures (2010–2014) using EARS-Net data in Spain. *Enferm. Infecc. Microbiol. Clin.* **2017**, *3* 480–486. [CrossRef]
61. Maurya, N.; Jangra, M.; Tambat, R.; Nandanwar, H. Alliance of Efflux Pumps with β-Lactamases in Multidrug-Resistant Klebsiell pneumoniae Isolates. *Microb. Drug Resist.* **2019**, *25*, 1155–1163. [CrossRef] [PubMed]
62. Xavier, D.E.; Picão, R.C.; Girardello, R.; Fehlberg, L.C.; Gales, A.C. Efflux pumps expression and its association with pori down-regulation and beta-lactamase production among Pseudomonas aeruginosa causing bloodstream infections in Brazil. *BM Microbiol.* **2010**, *10*, 217. [CrossRef]
63. Martínez-Martínez, L. Extended-spectrum beta-lactamases and the permeability barrier. *Clin. Microbiol. Infect.* **2008**, *14* (Supp S1), 82–89. [CrossRef] [PubMed]
64. Ghosh, P.; Han, G.; De, M.; Kim, C.K.; Rotello, V.M. Gold nanoparticles in delivery applications. *Adv. Drug Deliv. Rev.* **2008**, *6* 1307–1315. [CrossRef]
65. Khare, T.; Mahalunkar, S.; Shriram, V.; Gosavi, S.; Kumar, V. Embelin-loaded chitosan gold nanoparticles interact synergisticall with ciprofloxacin by inhibiting efflux pumps in multidrug-resistant Pseudomonas aeruginosa and Escherichia coli. *Environ. Re* **2021**, *199*, 111321. [CrossRef]

nanomaterials

Article

Synthesis of Metal/SU-8 Nanocomposites through Photoreduction on SU-8 Substrates

Yan-Jun Huang [1], Wen-Huei Chang [2,*], Yi-Jui Chen [1] and Chun-Hung Lin [1,*]

1 Department of Photonics, National Cheng Kung University, Tainan 70101, Taiwan
2 Department of Applied Chemistry, National Pingtung University, Pingtung 90003, Taiwan
* Correspondence: whchang@mail.nptu.edu.tw (W.-H.C.); chlin@ncku.edu.tw (C.-H.L.)

Abstract: The paper presents a simple, fast, and cost-effective method for creating metal/SU-8 nanocomposites by applying a metal precursor drop onto the surface or nanostructure of SU-8 and exposing it to UV light. No pre-mixing of the metal precursor with the SU-8 polymer or pre-synthesis of metal nanoparticles is required. A TEM analysis was conducted to confirm the composition and depth distribution of the silver nanoparticles, which penetrate the SU-8 film and uniformly form the Ag/SU-8 nanocomposites. The antibacterial properties of the nanocomposites were evaluated. Moreover, a composite surface with a top layer of gold nanodisks and a bottom layer of Ag/SU-8 nanocomposites was produced using the same photoreduction process with gold and silver precursors, respectively. The reduction parameters can be manipulated to customize the color and spectrum of various composite surfaces.

Keywords: nanocomposite; metal nanoparticles; photoreduction; plasmonic structure; nanofabrication; SU-8; antibacterial; color filter

1. Introduction

Nanoparticles of precious metals, such as silver and gold, have gained significant interest in the research community due to their unique properties [1]. These properties enable a wide range of applications across various scientific fields. For instance, adjusting the distribution, size, and shape of nanoparticles can manipulate the optical phenomena of localized surface plasmon resonance (LSPR), which differs from the characteristics of bulk metals [2] and has led to the development of applications in optical devices [3] and biomedical sensing [4]. Among the precious metal nanoparticles, silver nanoparticles (AgNPs) have shown great promise in the fields of sensing and catalysis due to their compatibility [5] and antibacterial properties. These properties make AgNPs ideal for use in biomedical and food packaging applications [6,7].

Incorporating nanoparticles into polymer media can offer diverse applications and uses, while maintaining their physical properties through packaging. Various standard patterning methods, including optical lithography [8], electron-beam lithography [9], and nanoimprinting [10], can be used to control the distribution and size range of nanoparticles, leading to the development of new patterning and material applications [11]. However, achieving an even distribution of nanoparticles in the polymer medium through the common metal deposition methods used in semiconductor processes can be challenging [12]. The recent research has focused on two mainstream production methods for mixing the composites. The first method involves pre-mixing the metal precursor with the polymer to form a stable colloid solution, followed by the reduction using electron-beam, UV exposure, or chemical reduction [9,13,14]. The second method involves mixing pre-synthesized nanoparticles with the polymer or growing them layer by layer [15,16]. The resulting nano-polymer composites can be used in various applications in the fields of electronics, optoelectronics, and biomedicine, including photopatterned electrodes [17], plasmonic sensor [8,18], and Raman sensing substrates [15]. However, these methods often require the preparation of nano-precursor solutions or the pre-synthesis of nanoparticles. The process can be cumbersome and time-consuming and may not ensure the long-term stability of storage properties or allow for post-adjustment. Additionally, care must be taken to prevent the aggregation of nanoparticles during growth [19]. Moreover, the use of electron beams and lasers for growth requires significant time and cost, especially in mass production.

SU-8 photoresist is a widely used negative-tone chemically amplified photoresist due to its heat and acid/base resistance [20], as well as its excellent mechanical properties. It finds extensive applications in the fields of microelectromechanical systems (MEMS) and microfluidics [21]. Its high compatibility with biological systems [22] also makes it suitable for use in biosensors. Additionally, its high transparency in the visible light range renders it applicable in optics [23]. SU-8 is typically spin-coated on wafers using standard processes and then undergoes patterning. Its characteristic of generating a large number of free radicals under UV exposure and elevated temperature also makes it an ideal material for reducing metal precursors within its structure via optical lithography [24–26]. For instance, Tan et al. made Au/SU-8 nanocomposites through the photoreduction of a gold precursor and SU-8 mixture for electrodes and grating applications [24]. Fischer et al. fabricated Ag/SU-8 nanocomposites through the photoreduction of a silver precursor and SU-8 mixture and explored their localized surface plasmon resonance (LSPR) response [25]. However, when mixing with the nanoparticles in solution form, the loading effect in the solution influences the spin-coating quality of the film and nanoparticle distribution inside the film.

In our previous study, we demonstrated the direct synthesis of a monolayer of gold nanoparticles (AuNPs) on the surface of SU-8 under UV exposure [26]. In this study, we would like to investigate the SU-8 photoreduction behavior on the silver precursor. A drop of the silver precursor was simply applied on the SU-8 surface, followed by UV exposure, without the need for mixing the precursor and SU-8 in advance. Additional photoinitiators (PIs) were added to SU-8 to test their ability to accelerate the reduction efficiency. The combination of the photoreduction of the gold and silver precursors on the same SU-8 surface, respectively, was also studied. AgNPs and AuNPs have distinct LSPR ranges in the visible spectrum, providing significant potential for color tuning in color filter applications through tailoring of their LSPR. Furthermore, the antibacterial ability of the reduced AgNPs was tested.

2. Materials and Methods

2.1. Chemicals and Materials

Perfluoropolyether (PFPE)-urethane dimethacrylate (Fluorolink MD700) was obtained from Solvay Specialty Polymers (Bollate, Italy). Additionally, 2,2-dimethoxy-2-phenylacetophenone and triarylsulfonium hexafluoroantimonate salts (mixed, 50 wt.% in propylene carbonate) were obtained from Sigma-Aldrich (St. Louis, MO, USA). SU-8 3025 was purchased from Kayaku Advanced Materials (Westborough, MA, USA). Poly(methyl methacrylate (PMMA, molecular weight: 35 k), silver nitrate ($AgNO_3$), gold (III) chloride trihydrate ($HAuCl_4 \cdot 3H_2O$), cyclopentanone, and 1H,1H,2H,2H-perfluorodecyltrichlorosilane (F_{13}-TCS) were purchased from Alfa Aesar (Ward Hill, MA, USA). Additionally, 1,1,2-Trichloro-1,2,2-trifluoroethane was purchased from Grand Chemical Co. (Miaoli, Taiwan). The solvents and chemicals were used without further purification.

2.2. Synthesis of Ag/SU-8 Nanocomposites by Photoreduction

A glass substrate (1.25 cm $\times$ 1.25 cm) was cleaned in acetone and isopropanol with an ultrasonic bath for 30 min and dried with a nitrogen flow. The SU-8 solution (20 µL SU-8 3025/cyclopentanone at a weight ratio of 1:8) was then spin-coated on the glass substrate at 500 rpm for 5 s and 2500 rpm for 25 s. The SU-8 film was then subjected to a softbake process at 95 °C for 10 min, resulting in a film thickness of approximately 300 nm. The SU-8 film was initially cured with an i-line UV mercury lamp (λ = 365 nm, 150 mW/cm^2) for 1 min. A drop of $AgNO_3$ aqueous solution (80 µL) was then dripped onto the SU-8 surface using a quantitative pipette. The same mercury lamp was employed again to expose the photoresist for 18 min, leading to the photochemical formation of AgNPs in the resist film. The process can be repeated for additional cycles after washing the sample with DI water.

2.3. Nanoimprint of SU-8 Nanopillar Arrays [26]

Nanoimprint lithography was employed to pattern SU-8 nanopillar arrays. An SU-8 resist film was prepared as described in Section 2.2. A PFPE working mold was replicated from a silicon master mold with anti-sticking treatment through the vapor deposition of F_{13}-TCS [27]. PFPE possesses a lower surface energy, making it easier to demold from the imprinted SU-8 polymers. A mixture of Fluorolink MD700 and its photoinitiator (2,2-dimethoxy-2-phenylacetophenone, 1 wt%) was prepared and poured onto the master mold. The solvent used for PI was 1,1,2-Trichloro-1,2,2-trifluoroethane. The sample was then placed in a vacuum and cured under UV light. After the curing, PFPE was released from the silicon master mold. Subsequently, the PFPE mold was placed on top of the SU-8 film, and the nanoimprint process was carried out using our home-built nanoimprint platform. The imprinting pressure was set at 3 bar for 10 min while maintaining a constant temperature of 80 °C. After cooling, the nanoimprinted SU-8 nanostructures were obtained by removing the mold from the resist. The process of the photochemical formation of AuNPs on SU-8 nanopillar arrays was the same as that used for an SU-8 film.

2.4. Growth of Gold Nanodisks on SU-8 through Photoreduction

The growth of gold nanodisks was achieved through the selective growth of patterned monolayer AuNPs via photoreduction [28]. To confine the area for AuNP production, a PMMA nanohole array mask was applied onto the SU-8 surface using a nanotransfer printing process. After exposing a drop of $HAuCl_4$ over the PMMA-masked SU-8 surface to UV light, AuNPs formed on the uncovered SU-8 surface. A gold nanodisk array was obtained after three cycles of photoreduction.

2.5. Characterization

Scanning electron microscopy (SEM) images were obtained using a field emission scanning electron microscope (JEOL 6340F, Tokyo, Japan), while transmission electron microscopy (TEM) samples were prepared using a focused ion beam system (FEI Helios G3CX, Thermo Fisher Scientific, Waltham, MA, USA). TEM brightfield images and energy scattering spectroscopy (EDS) were acquired using JEOL JEM-2100F CS STEM (Tokyo, Japan). The transmission spectra were measured using a miniature UV-VIS spectrometer (Model: BLK-CSR-SR, StellarNet Inc., Tampa, FL, USA), and a tungsten halogen light source (SL1-FILTER, StellarNet Inc., Tampa, FL, USA, and SLS301, Thorlabs Inc., Newton, NJ, USA). X-ray photoelectron spectroscopy (XPS) was carried out using PHI 5000 VersaProbe (ULVAC-PHI, Kanagawa, Japan). The CIE 1931 color space chromaticity diagrams were created by converting and mapping the measured spectra.

2.6. Antibacterial Screening Test

The experimental procedure for preparing the filter paper with the Ag/SU-8 nanocomposites used for the antibacterial screening test was carried out as follows: 1. apply 20 μL of SU-8 onto the filter paper (the sample). 2. Position the sample on a hotplate and set the temperature to 95 °C. Heat the sample to remove surface solvents and allow it to remain at this temperature for 10 min. 3. Transfer the sample to a mercury lamp exposure box for light pre-curing. Expose the sample to light for 5 min, then carefully remove it from the exposure box. 4. Apply 90 μL of $AgNO_3$ solution onto the surface of the sample. Place the sample back into the mercury lamp exposure box to initiate the photoreduction process for AgNPs. Allow the sample to be exposed to light for 18 min, then remove it from the exposure box. 5. Clean the surface of the sample by rinsing it with DI water. Use a nitrogen gun to dry the sample thoroughly. 6. Optionally, repeat steps 4 and 5 to increase the number of reduction cycles as desired. 7. Once again, position the sample on the hotplate and set the temperature to 95 °C. Heat the sample to eliminate any remaining surface moisture and allow it to stay at this temperature for 10 min. The Gram-positive Staphylococcus aureus (BCRC 10451) and Gram-negative Escherichia coli (BCRC 11634 and NCTC 11954) strains were utilized in this study. BCRC 10451 and NCTC 11954 are antibiotic-resistant strains that can withstand penicillin and streptomycin treatments. The strains were cultivated in nutrient broth medium (g/L: Tryptone 10; yeast extract 5; NaCl 10; pH 7.5) at 37 °C with continuous shaking at 120 rpm. The bacterial cell suspensions were diluted with sterile water to achieve a final concentration of 10^8 CFU/mL. For seeding the agar plates, 100 µL of this bacterial suspension was used. Filter paper disks with a diameter of 8 mm, incorporating Ag/SU-8 nanocomposites, were placed onto the surface of the seeded medium. After incubating at 37 °C for 24 h, the zones of inhibition, which indicate areas where bacterial growth is inhibited, were measured. As a control group in this study, filter paper disks without the application of SU-8, but following the previously described experimental procedure, were utilized.

3. Results and Discussion

3.1. Synthesis and Characterization of Ag/SU-8 Nanocomposites

The process of producing Ag/SU-8 nanocomposites is illustrated in Figure 1. A drop of silver precursor ($AgNO_3$) was placed on the SU-8 surface, and then, UV exposure was applied to initiate the reduction of AgNPs. The SU-8 film turned yellow and became progressively darker with repeated photoreduction, as seen in Figure 2a–c. The deepening color of the surface from the first to the third photoreduction suggests an increase in the concentration of the produced AgNPs. A TEM observation was conducted to confirm the composition and depth distribution of the AgNPs. Figure 2d,e display the TEM bright field images of the Ag/SU-8 nanocomposites reduced with the silver precursor (0.5 and 50 mM, respectively). In contrast to the reduced AuNPs that form a monolayer on the SU-8 film [26,28], the reduced AgNPs penetrated the SU-8 film and formed the Ag/SU-8 nanocomposites. The molecular weights of silver nitrate and chloroauric acid are approximately 169.87 and 339.79 g/mol, respectively. The large size of chloroauric acid may prevent it from penetrating into SU-8, which could explain why AuNPs grow on the SU-8 surface while AgNPs form inside the SU-8. The AgNPs were extremely small, with a size of less than 10 nm, due to the limited space available for growth within the SU-8 polymer. The AgNPs with 0.5 mM $AgNO_3$ tended to distribute towards the bottom side of the SU-8, while the AgNPs with 50 mM $AgNO_3$ were evenly and densely distributed throughout the entire area of the SU-8 film. Figure 2f presents the EDS elemental analysis of the same area in Figure 2e. The sample was coated with a platinum layer, which was identified and labeled with green points. The substrate was silicon and marked with red points. The layer on the substrate was SU-8 and the silver element (green points) was uniformly distributed within the SU-8 film.

The extinction spectra of Ag/SU-8 nanocomposites are displayed in Figure 3. A prominent absorption peak is observed at a wavelength of 440 nm, corresponding to the LSPR peak of reduced AgNPs. The effect of photoreduction parameters on AgNPs is evident from the extinction spectra. Figure 3a illustrates the impact of varying additional PI concentrations on photoreduction. The PI used for SU-8 polymerization is triarylsulfonium hexafluoroantimonate salt, a photoacid generator. Diluted SU-8 (SU-8 3025/cyclopentanone at a weight ratio of 1:8) was mixed with varying additional PI concentrations ranging from 0 to 10 wt%. The SU-8 without added PI (0 wt%) is the diluted commercial SU-8 that already contains a certain amount of PI. As the amount of added PI increases, the photoacid generation rate during UV exposure also increases, leading to the production of more free radicals required for reduction [26]. The most notable difference is observed between SU-8 without added PI and SU-8 with 1 wt% added PI. All samples with added PI exhibit a 440 nm absorption peak corresponding to AgNP LSPR after the first reduction, while the original SU-8 sample shows no significant absorption peak. The original SU-8 requires multiple reductions before its reduction effect approaches that of the PI-added sample (see Figure S1), indicating that adding PI improves reduction efficiency. As the amount of PI increases, the intensity of the absorption peak also increases, reaching its maximum at 5 wt% PI. To maintain consistency in subsequent experiments and facilitate observation of differences in other parameter effects, all subsequent experiments used SU-8 with 5 wt% added PI. Figure 3b shows reduction results at various $AgNO_3$ concentrations. As PI concentration increases, the LSPR absorption also increases, consistent with the increase in AgNP density shown in the TEM images of Figure 2a,b. Figure 3c displays the spectra of samples subjected to multiple photoreduction cycles. As the number of photoreduction cycles increases, the corresponding absorption peak intensity strengthens and the bandwidth widens. After the fourth reduction, however, the extinction peak does not exhibit a significant increase. The photoreduction process can also be performed directly on SU-8 nanostructures, resulting in the formation of Ag/SU-8 nanocomposites with a nanostructured shape. First, a nanopillar array with a diameter of 300 nm, a period of 600 nm, and a height of 300 nm was patterned through nanoimprint lithography [26]. Then, a drop of $AgNO_3$ solution was applied to the surface of the SU-8 nanopillar array and exposed to UV light. Figure 3d shows a cross-sectional TEM image of 50 mM $AgNO_3$ photoreduced in a SU-8 nanopillar array with 2 reduction cycles. The AgNPs within the SU-8 nanopillars were identified through EDS elemental analysis, as shown in the bottom inset of Figure 3d. The AgNPs were uniformly distributed within the SU-8 nanopillars.

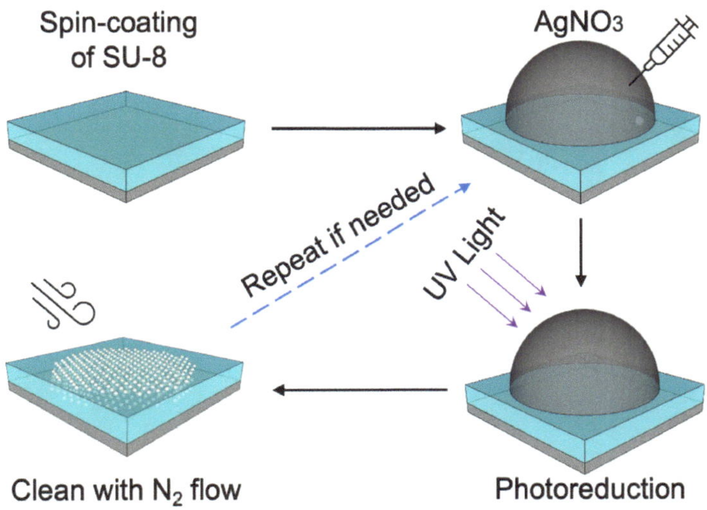

Figure 1. The process flow for synthesizing Ag/SU-8 nanocomposites.

3.2. Composite Surface Composed of Gold Nanodisks and Ag/SU-8 Nanocomposites

In our previous study, we demonstrated the selective growth of patterned monolayer AuNPs on SU-8 surfaces [28]. Here, we fabricated a composite surface composed of a top gold nanodisk array and bottom Ag/SU-8 nanocomposites, as shown in Figure 4a, using the same photoreduction process. To create the composite surface, we introduced a PMMA nanohole array mask on the SU-8 surface using a residual layer-free nanotransfer printing process. After three cycles of $HAuCl_4$ photoreduction, a gold nanodisk array was achieved. The same reduction process with 1 to 4 cycles was then applied on the same SU-8 surface with the silver precursor of $AgNO_3$. The SEM image of the composite surface is shown in Figure 4b. The top gold nanodisks had a diameter of approximately 150 nm, a period of 300 nm, and a height of approximately 100 nm. The AgNPs were inside the SU-8 film and were, therefore, invisible in this image. Figure 4c shows the measured extinction spectra of the composite surfaces with the reduction in AgNPs from 0 to 4 cycles. With the incorporation of AgNPs, the absorption at a wavelength of approximately 650 nm caused by the gold nanodisk array increased slightly, accompanied by slight blue shifts. The absorption at 435 nm wavelength caused by AgNPs increased with the reduction cycles. Figure 4d shows the CIE 1931 color space [29] calculated from the measured transmission spectra of the various Ag/SU-8 nanocomposites and composite surfaces. The triangle in the chromaticity diagram represents the area of the sRGB color space. For the composite surface (gold nanodisks + AgNPs), the color goes from light blue to light green (the red arrow trajectory with reduction cycles from one to four). For the AgNPs-only surface, the color goes from light purple to orange-red (the yellow arrow trajectory with 0.5 mM $AgNO_3$ from one to six reduction cycles, and then the green arrow trajectory with 50 mM $AgNO_3$ from one to six reduction cycles). By combining the reduction of AuNPs and AgNPs using the same photoreduction system, the composite surfaces can be fabricated in an inexpensive and convenient way. We can adjust the color and spectrum in three different dimensions: 1. adjusting AgNP photoreduction parameters; 2. adjusting geometric shapes of gold nanostructures; 3. the tuning between monolayer AuNPs and bulk gold nanostructures through photoreduction cycles. Compared to the fabrication of metasurface-based color filters [30–32], the current approach can achieve similar results with relatively few semiconductor equipment and process steps, avoiding a large amount of expensive and time-consuming e-beam writing time.

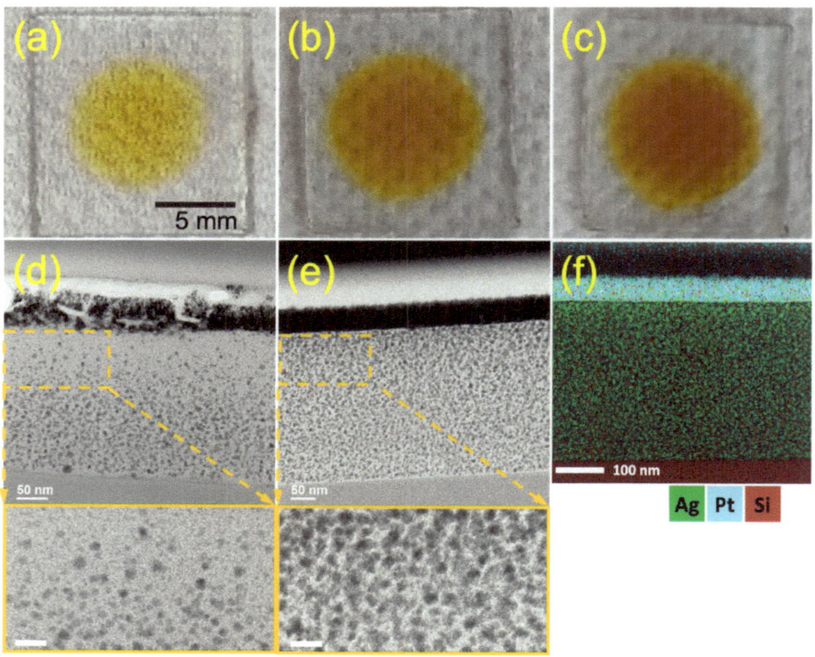

Figure 2. (**a**–**c**) Optical images of the Ag/SU-8 nanocomposites after one, two, and three photoreductions. The concentration of $AgNO_3$ was 0.5 mM. (**d**,**e**) Cross-sectional TEM images of the Ag/SU-8 nanocomposites photoreduced from 0.5 and 50 mM $AgNO_3$, respectively. (**f**) EDS elemental analysis of Ag/SU-8 nanocomposites photoreduced from 50 mM $AgNO_3$. The scale bars in the insets at the bottom of (**d**,**e**) measure 20 nm.

3.3. Antibacterial Ability of Ag/SU-8 Nanocomposites

AgNPs are shown to exert significant inhibitory activity against a broad spectrum of bacteria. The antibacterial activity of Ag/SU-8 nanocomposites was studied using the disk diffusion method. The control group consisted of standard antibiotics, such as Penicillin G and Streptomycin, as well as distilled water and SU-8. Our findings revealed that the Ag/SU-8 nanocomposites exhibited efficient antibacterial activity against both antibiotic-sensitive and antibiotic-resistant strains, as shown in Figure 5. The inhibition zones of Ag/SU-8 nanocomposite against antibiotic-sensitive *E. coli* were 9.3 ± 0.2, 10.3 ± 0.3, and 10.4 ± 0.1 mm at 0.5, 5, and 50 mM of $AgNO_3$ on SU-8 with one photoreduction cycle. Increasing the photoreduction cycles to two resulted in larger inhibition zones of 10.0 ± 0.3, 10.5 ± 0.4, and 11.6 ± 0.4 mm at 0.5, 5, and 50 mM of $AgNO_3$ on SU-8. Similar trends were observed in the other two antibiotic-resistant strains tested. The inhibition zones of Ag/SU-8 nanocomposite against antibiotic-resistant E. coli were 9.3 ± 0.2, 9.8 ± 0.2, and 11.0 ± 0.3 mm at 0.5, 5, and 50 mM of $AgNO_3$ on SU-8 with one photoreduction cycle. With two photoreduction cycles, the inhibition zones increased to 10.1 ± 0.2, 10.8 ± 0.3, and 11.1 ± 0.3 mm at 0.5, 5, and 50 mM of $AgNO_3$ on SU-8. For another antibiotic-resistant, Gram-positive Staphylococcus aureus strain, the inhibition zones of Ag/SU-8 nanocomposite were 10.5 ± 0.5, 11.3 ± 0.4, and 11.7 ± 0.4 mm at 0.5, 5, and 50 mM of $AgNO_3$ on SU-8 with one photoreduction cycle. Increasing the photoreduction cycles to two resulted in larger inhibition zones of 11.2 ± 0.4, 11.7 ± 0.3, and 12.4 ± 0.2 mm at 0.5, 5, and 50 mM of $AgNO_3$ on SU-8. The results indicated that the concentration of Ag/SU-8 nanocomposites and the number of photoreduction cycles increased the AgNP density and enlarged the inhibition zones. The control group, consisting of filter paper disks without SU-8 application but treated with $AgNO_3$ photoreduction, demonstrated an inhibition zone ranging from approximately 8.2 to 8.6 mm. This result signifies that SU-8 effectively enhanced the photoreduction efficiency of AgNPs and its antibacterial effect. The data representing this control group were presented as 0.5 mM paper on the rightmost side of Figure 5. There are several possible reasons why Ag/SU-8 nanocomposites may be more efficient against antibiotic-resistant strains compared to antibiotics: First, a different mechanism of action: antibiotics typically target specific cellular components or processes, such as cell wall synthesis or protein synthesis. Bacteria can develop resistance to antibiotics by evolving mechanisms to bypass or neutralize these targets. In contrast, AgNPs are believed to work through a combination of mechanisms, including disrupting the bacterial cell membrane, generating reactive oxygen species, and interfering with cellular signaling pathways. This makes it harder for bacteria to develop resistance to AgNPs, as they would need to evolve multiple mechanisms to counteract these different modes of action. Second, broad-spectrum activity: AgNPs were shown to have broad-spectrum antibacterial activity, meaning they can kill a wide range of bacterial species. This contrasts with antibiotics, which are often more specific to certain types of bacteria. Because AgNPs are effective against a broad range of bacteria, including antibiotic-resistant strains, they may be more useful in treating infections caused by multiple bacterial species or strains. Third, the lower likelihood of resistance: overuse and misuse of antibiotics has led to the emergence of antibiotic-resistant bacteria, which can render antibiotics ineffective. Because AgNPs have multiple modes of action, it may be harder for bacteria to develop resistance to them compared to antibiotics. Additionally, some studies suggest that prolonged exposure to AgNPs may select for resistant strains, but the likelihood of this happening may be lower than with antibiotics. Overall, AgNPs may be more efficient against antibiotic-resistant strains of bacteria because of their different mode of action, broad-spectrum activity, and lower likelihood of resistance. However, more research is needed to fully understand the mechanisms underlying the antibacterial activity of AgNPs and their potential for long-term use in treating bacterial infections.

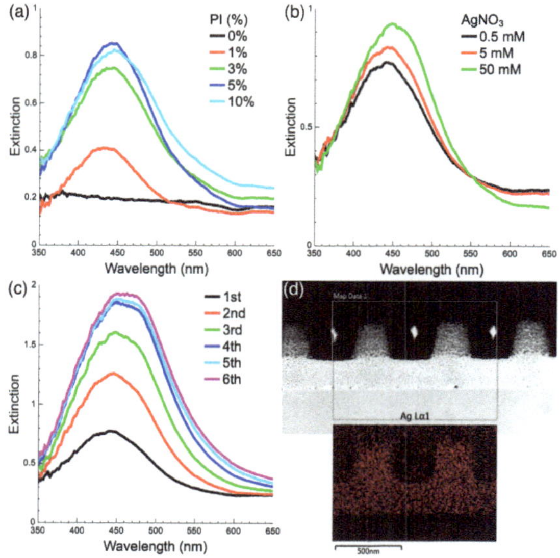

Figure 3. Extinction spectra of the Ag/SU-8 nanocomposites under various experimental conditions. (**a**) PI was added to SU-8 at concentrations ranging from 0 to 10 wt% with a fixed $AgNO_3$ concentration of 0.5 mM. (**b**) PI concentration was fixed at 5 wt% with $AgNO_3$ concentrations ranging from 0.5 to 50 mM. (**c**) Photoreduction was performed from one to six cycles with fixed concentrations of PI and $AgNO_3$ at 5 wt% and 0.5 mM, respectively. (**d**) Cross-sectional TEM image of the Ag/SU-8 nanocomposites photoreduced from 50 mM $AgNO_3$ on SU-8 nanopillars. The bottom inset shows EDS elemental analysis of the silver element.

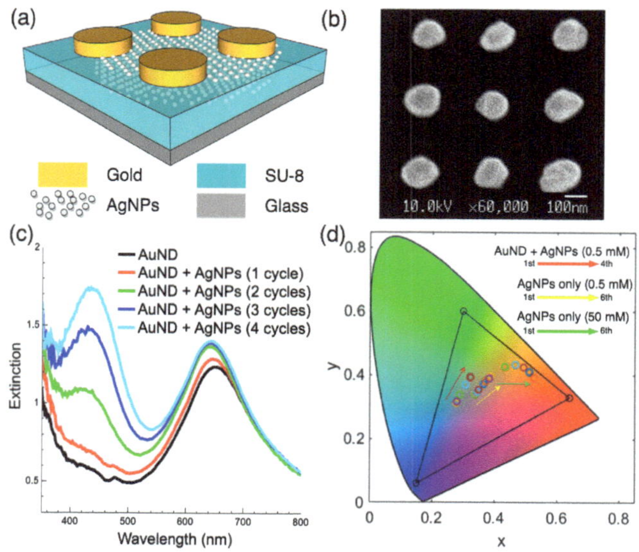

Figure 4. (**a**) The schematic of the composite surface made up of gold nanodisks and Ag/SU-8 nanocomposites. (**b**) The top-view SEM image of the composite surface. (**c**) The measured extinction spectra of the composite surface with the reduction in AgNPs from 0 to 4 cycles. (**d**) The CIE 1931 color space calculated from the measured spectra of the various Ag/SU-8 nanocomposites and composite surfaces.

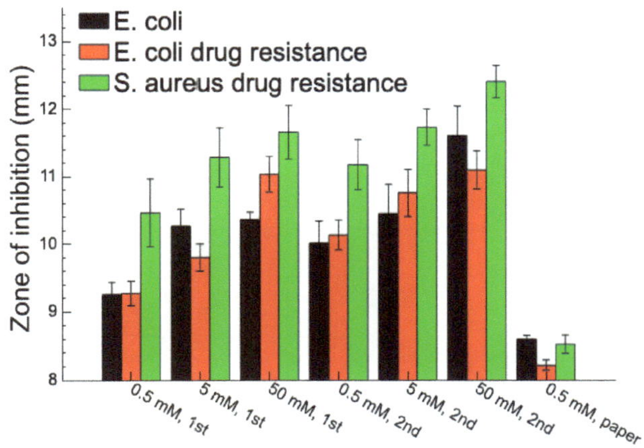

Figure 5. Antibacterial inhibition zones of various Ag/SU-8 nanocomposites against antibiotic-sensitive Escherichia coli, antibiotic-resistant Escherichia coli, and antibiotic-sensitive Staphylococcus aureus.

4. Conclusions

This study demonstrated the synthesis of AgNPs via the photoreduction of silver nitrate on SU-8. In contrast to the photoreduction of chloroauric acid, which produces monolayer AuNPs on the SU-8 surface, silver nitrate penetrates into SU-8 and forms Ag/SU-8 nanocomposites. Simply a drop of the silver precursor on the SU-8 surface is followed by UV exposure without the need of premixing the precursor and SU-8 or synthesizing AgNPs in advance. The impact of photoreduction parameters on AgNPs is evident from the measured extinction spectra, which vary with PI concentrations, $AgNO_3$ concentrations, and the number of photoreduction cycles. The photoreduction process can be directly applied to SU-8 nanostructures, resulting in the formation of Ag/SU-8 nanocomposites with a nanostructured shape, which is very appealing for nanodevice fabrication. Using the same photoreduction process, a composite surface consisting of a top layer of gold nanodisks and a bottom layer of Ag/SU-8 nanocomposites was fabricated. By adjusting the silver precursor concentration, photoreduction parameters, and geometry of the gold nanostructures, the transmission spectra of the composite surfaces can be customized, demonstrating their potential for color tuning in color filter applications. The Ag/SU-8 nanocomposites also exhibit antibacterial activity against two Gram-negative Escherichia coli and one Gram-positive Staphylococcus aureus. They were shown to be more effective against antibiotic-resistant strains. Increasing the AgNPs density in the nanocomposites can enhance their inhibitory activity against bacteria. The proposed method provides a simple, rapid, and cost-effective technique for producing AgNP nanocomposites.

Supplementary Materials: The following supporting information can be downloaded at: https://www.mdpi.com/article/10.3390/nano13111784/s1, Figure S1: Extinction spectra of Ag/SU-8 nanocomposites fabricated on the SU-8 film without the incorporation of additional PI, demonstrating the effects of 1 and 3 cycles of photoreduction.

Author Contributions: Conceptualization, W.-H.C. and C.-H.L.; methodology, Y.-J.C. and W.-H.C.; software, Y.-J.H. and C.-H.L.; validation, Y.-J.H. and W.-H.C.; formal analysis, Y.-J.H. and W.-H.C.; investigation, Y.-J.H. and Y.-J.C.; resources, C.-H.L.; data curation, Y.-J.H. and W.-H.C.; writing—original draft preparation, Y.-J.H. and C.-H.L.; writing—review and editing, W.-H.C. and C.-H.L.; visualization, W.-H.C. and C.-H.L.; supervision, W.-H.C. and C.-H.L.; project administration, C.-H.L.; funding acquisition, C.-H.L. All authors have read and agreed to the published version of the manuscript.

Funding: This research was funded by the Ministry of Science and Technology (MOST) of Taiwan, grant nos. MOST 108-2221-E-006-206-, MOST 109-2221-E-006-206-, MOST 109-2635-H-153-001-, and MOST 110-2221-E-006-159-.

Data Availability Statement: The data presented in this study is available upon request from the corresponding author.

Acknowledgments: The authors gratefully acknowledge the use of EM000800, EM025200 of MOST 110-2731-M-006-001 belonging to the Core Facility Center of National Cheng Kung University, and the shaped e-beam direct write system belonging to the Taiwan Semiconductor Research Institute.

Conflicts of Interest: The authors declare no conflict of interest.

References

1. Daniel, M.-C.; Astruc, D. Gold Nanoparticles: Assembly, Supramolecular Chemistry, Quantum-Size-Related Properties, an Applications toward Biology, Catalysis, and Nanotechnology. *Chem. Rev.* **2004**, *104*, 293–346. [CrossRef] [PubMed]
2. Liz-Marzán, L.M. Tailoring Surface Plasmons through the Morphology and Assembly of Metal Nanoparticles. *Langmuir* **2006**, *2* 32–41. [CrossRef] [PubMed]
3. Hwang, C.S.H.; Ahn, M.-S.; Lee, Y.; Chung, T.; Jeong, K.-H. Ag/Au Alloyed Nanoislands for Wafer-Level Plasmonic Color Filte Arrays. *Sci. Rep.* **2019**, *9*, 9082. [CrossRef]
4. Abu Hatab, N.A.; Oran, J.M.; Sepaniak, M.J. Surface-Enhanced Raman Spectroscopy Substrates Created Via Electron Bean Lithography and Nanotransfer Printing. *ACS Nano* **2008**, *2*, 377–385. [CrossRef]
5. Deshmukh, S.P.; Patil, S.M.; Mullani, S.B.; Delekar, S.D. Silver nanoparticles as an effective disinfectant: A review. *Mater. Sci. Eng C* **2019**, *97*, 954–965. [CrossRef]
6. Wu, Z.; Huang, X.; Li, Y.-C.; Xiao, H.; Wang, X. Novel chitosan films with laponite immobilized Ag nanoparticles for active foo packaging. *Carbohydr. Polym.* **2018**, *199*, 210–218. [CrossRef] [PubMed]
7. Sarwar, M.S.; Niazi, M.B.K.; Jahan, Z.; Ahmad, T.; Hussain, A. Preparation and characterization of PVA/nanocellulose/Ag nanocomposite films for antimicrobial food packaging. *Carbohydr. Polym.* **2018**, *184*, 453–464. [CrossRef]
8. Marqués-Hueso, J.; Abargues, R.; Valdés, J.L.; Martínez-Pastor, J.P. Ag and Au/DNQ-novolac nanocomposites patternable b ultraviolet lithography: A fast route to plasmonic sensor microfabrication. *J. Mater. Chem.* **2010**, *20*, 7436–7443. [CrossRef]
9. Abargues, R.; Marqués-Hueso, J.; Canet-Ferrer, J.; Pedrueza, E.; Valdés, J.L.; Jiménez, E.; Martínez-Pastor, J.P. High-resolutio electron-beam patternable nanocomposite containing metal nanoparticles for plasmonics. *Nanotechnology* **2008**, *19*, 35530 [CrossRef]
10. Kraus, T.; Malaquin, L.; Schmid, H.; Riess, W.; Spencer, N.D.; Wolf, H. Nanoparticle printing with single-particle resolution. *Na Nanotechnol.* **2007**, *2*, 570–576. [CrossRef]
11. Chen, L.; Dong, Y.; Tang, C.-Y.; Zhong, L.; Law, W.-C.; Tsui, G.C.P.; Yang, Y.; Xie, X. Development of Direct-Laser-Printabl Light-Powered Nanocomposites. *ACS Appl. Mater. Interfaces* **2019**, *11*, 19541–19553. [CrossRef]
12. Jiguet, S.; Bertsch, A.; Hofmann, H.; Renaud, P. SU8-Silver Photosensitive Nanocomposite. *Adv. Eng. Mater.* **2004**, *6*, 719–72 [CrossRef]
13. Yetisen, A.K.; Naydenova, I.; da Cruz Vasconcellos, F.; Blyth, J.; Lowe, C.R. Holographic Sensors: Three-Dimensional Analyte Sensitive Nanostructures and Their Applications. *Chem. Rev.* **2014**, *114*, 10654–10696. [CrossRef]
14. Ramesh, G.V.; Porel, S.; Radhakrishnan, T.P. Polymer thin films embedded with in situ grown metal nanoparticles. *Chem. Soc. Re* **2009**, *38*, 2646–2656. [CrossRef] [PubMed]
15. Martínez, E.D.; Urbano, R.R.; Rettori, C. Thermoplasmonic Maskless Lithography on Upconverting Nanocomposites Assisted b Gold Nanostars. *ACS Appl. Nano Mater.* **2019**, *2*, 6889–6897. [CrossRef]
16. Martínez, E.D.; Prado, A.; Gonzalez, M.; Anguiano, S.; Tosi, L.; Salazar Alarcón, L.; Pastoriza, H. Recent Advances on Nanocom posite Resists with Design Functionality for Lithographic Microfabrication. *Front. Mater.* **2021**, *8*, 629792. [CrossRef]
17. Jiguet, S.; Bertsch, A.; Hofmann, H.; Renaud, P. Conductive SU8 Photoresist for Microfabrication. *Adv. Funct. Mater.* **2005**, *15* 1511–1516. [CrossRef]
18. Pudlauskaitė, J.; Jankauskaitė, V.; Lazauskas, A.; Prosyčevas, I.; Narmontas, P. Ag/DNQ-novolac-based nanocomposite films fo controllable UV lithography morphological patterning. *Colloid Polym. Sci.* **2013**, *291*, 1787–1793. [CrossRef]
19. Gerardo, C.D.; Cretu, E.; Rohling, R. Fabrication of Circuits on Flexible Substrates Using Conductive SU-8 for Sensing Application *Sensors* **2017**, *17*, 1420. [CrossRef]
20. Lee, K.Y.; LaBianca, N.; Rishton, S.A.; Zolgharnain, S.; Gelorme, J.D.; Shaw, J.; Chang, T.P. Micromachining Applications of a High Resolution Ultrathick Photoresist. *J. Vac. Sci. Technol. B Microelectron. Nanometer Struct.* **1995**, *13*, 3012–3016. [CrossRef]
21. Tavakoli, H.; Zhou, W.; Ma, L.; Perez, S.; Ibarra, A.; Xu, F.; Zhan, S.; Li, X. Recent Advances in Microfluidic Platforms fo Single-Cell Analysis in Cancer Biology, Diagnosis and Therapy. *TrAC Trends Anal. Chem.* **2019**, *117*, 13–26. [CrossRef]
22. Márton, G.; Tóth, E.Z.; Wittner, L.; Fiáth, R.; Pinke, D.; Orbán, G.; Meszéna, D.; Pál, I.; Győri, E.L.; Bereczki, Z.; et al. The Neura Tissue around Su-8 Implants: A Quantitative in Vivo Biocompatibility Study. *Mater. Sci. Eng. C* **2020**, *112*, 110870. [CrossRef] [PubMed]

23. Serbin, J.; Ovsianikov, A.; Chichkov, B. Fabrication of woodpile structures by two-photon polymerization and investigation of their optical properties. *Opt. Express* **2004**, *12*, 5221–5228. [CrossRef] [PubMed]
24. Tan, E.K.; Shrestha, P.K.; Pansare, A.V.; Chakrabarti, S.; Li, S.; Chu, D.; Lowe, C.R.; Nagarkar, A.A. Density Modulation of Embedded Nanoparticles Via Spatial, Temporal, and Chemical Control Elements. *Adv. Mater.* **2019**, *31*, 1901802. [CrossRef] [PubMed]
25. Fischer, S.V.; Uthuppu, B.; Jakobsen, M.H. In Situ Su-8 Silver Nanocomposites. *Beilstein J. Nanotechnol.* **2015**, *6*, 1661–1665. [CrossRef]
26. Chen, Y.-J.; Chang, W.-H.; Li, C.-Y.; Chiu, Y.-C.; Huang, C.-C.; Lin, C.-H. Direct synthesis of monolayer gold nanoparticles on epoxy based photoresist by photoreduction and application to surface-enhanced Raman sensing. *Mater. Des.* **2021**, *197*, 109211. [CrossRef]
27. Liang, C.-C.; Lin, C.-H.; Cheng, T.-C.; Shieh, J.; Lin, H.-H. Nanoimprinting of Flexible Polycarbonate Sheets with a Flexible Polymer Mold and Application to Superhydrophobic Surfaces. *Adv. Mater. Interfaces* **2015**, *2*, 1500030. [CrossRef]
28. Chen, Y.-J.; Chang, W.-H.; Lin, C.-H. Selective Growth of Patterned Monolayer Gold Nanoparticles on SU-8 through Photoreduction for Plasmonic Applications. *ACS Appl. Nano Mater.* **2021**, *4*, 229–235. [CrossRef]
29. Smith, T.; Guild, J. The C.I.E. Colorimetric Standards and Their Use. *Trans. Opt. Soc.* **1931**, *33*, 73–134. [CrossRef]
30. Esposito, M.; Todisco, F.; Bakhti, S.; Passaseo, A.; Tarantini, I.; Cuscunà, M.; Destouches, N.; Tasco, V. Symmetry Breaking in Oligomer Surface Plasmon Lattice Resonances. *Nano Lett.* **2019**, *19*, 1922–1930. [CrossRef]
31. Song, S.; Ma, X.; Pu, M.; Li, X.; Liu, K.; Gao, P.; Zhao, Z.; Wang, Y.; Wang, C.; Luo, X. Actively Tunable Structural Color Rendering with Tensile Substrate. *Adv. Opt. Mater.* **2017**, *5*, 1600829. [CrossRef]
32. Xu, Z.; Li, N.; Dong, Y.; Fu, Y.H.; Hu, T.; Zhong, Q.; Zhou, Y.; Li, D.; Zhu, S.; Singh, N. Metasurface-based subtractive color filter fabricated on a 12-inch glass wafer using a CMOS platform. *Photon Res.* **2021**, *9*, 13–20. [CrossRef]

nanomaterials

Article

Reduction of Nitroaromatics by Gold Nanoparticles on Porous Silicon Fabricated Using Metal-Assisted Chemical Etching

Ling-Yi Liang [1], Yu-Han Kung [2], Vincent K. S. Hsiao [2,*] and Chih-Chien Chu [1,3,*]

1 Department of Medical Applied Chemistry, Chung Shan Medical University, Taichung 40201, Taiwan; s0964026@csmu.edu.tw
2 Department of Applied Materials and Optoelectronic Engineering, National Chi Nan University, Nantou 54561, Taiwan; s108328040@ncnu.edu.tw
3 Department of Medical Education, Chung Shan Medical University Hospital, Taichung 40201, Taiwan
* Correspondence: kshsiao@ncnu.edu.tw (V.K.S.H.); jrchu@csmu.edu.tw (C.-C.C.)

Abstract: In this study, we investigated the use of porous silicon (PSi) fabricated using metal-assisted chemical etching (MACE) as a substrate for the deposition of Au nanoparticles (NPs) for the reduction of nitroaromatic compounds. PSi provides a high surface area for the deposition of Au NPs, and MACE allows for the fabrication of a well-defined porous structure in a single step. We used the reduction of *p*-nitroaniline as a model reaction to evaluate the catalytic activity of Au NPs on PSi. The results indicate that the Au NPs on the PSi exhibited excellent catalytic activity, which was affected by the etching time. Overall, our results highlighted the potential of PSi fabricated using MACE as a substrate for the deposition of metal NPs for catalytic applications.

Keywords: porous silicon; metal-assisted chemical etching; gold nanoparticle; catalyst; nitroaromatic; *p*-nitroaniline

1. Introduction

Noble metal nanoparticles (NPs) have attracted considerable interest because of their unique chemical and physical properties, including their electronic [1], optical [2], magnetic [3], and catalytic properties [4]. These properties distinguish noble metal NPs from bulk metals and make them highly useful for a wide range of applications, such as catalysis, sensing, optics, and fuel cells. In various oxidation and reduction reactions, Au NPs have demonstrated great potential as highly efficient catalysts [5]. However, the tendency of NPs to aggregate in solutions because of their high surface energy may reduce their catalytic activity. To solve this problem, different approaches have been explored for stabilizing NPs, including capping them with organic molecules [6] or polymers [7] or dispersing them onto solid supports [8]. However, although organic molecules or polymers can serve as capping agents to prevent aggregation, they may also reduce the catalytic activity of NPs. By contrast, solid supports offer higher stability but often require time-consuming separation procedures to isolate the catalysts from the reaction system. Two-dimensional (2D) graphene, which has unique properties, such as a high specific surface area, excellent electrical conductivity, high charge carrier mobility, and high mechanical strength, has emerged as a promising support for various types of NPs [9]. Considerable interest has also been directed toward constructing size- and shape-controlled noble metal NPs supported on 2D carbon materials [10]. However, at the nanoscale, metal particles with highly active centers are not at thermodynamic equilibrium and are prone to aggregation with solid supports. Therefore, strategies for stabilizing NPs, such as capping them with multifarious stabilizers, designing core–shell structures, or anchoring them onto specific supports, must be explored.

Aromatic amines are crucial building blocks in organic synthesis and in the pharmaceutical industry [11]. Their synthesis can be achieved by reducing corresponding nitroaromatics [12]. However, these reduction reactions require catalysts for efficient conversion [13]. Although hydrogen gas and $NaBH_4$ are commonly used as reducing agents, noble metal NPs, such as Pt, Au, Ag, and Pd NPs, have been reported to be effective catalysts in the presence of $NaBH_4$ [14–18]. However, the aggregation of these NPs in the reaction system limits their catalytic efficiency, necessitating the development of novel techniques for ensuring well-dispersed stabilization of the NPs. To this end, various materials have been explored for stabilizing nanosized noble metal catalysts for catalyzing nitroaromatics. Kim et al. [19] synthesized a core–satellite structure using poly(N-isopropylacrylamide-acrylamide) and Au NPs for the photothermal-mediated catalytic reduction of 4-nitrophenol. Dong et al. [20] prepared Ag NPs dispersed in a nano-silica nanocatalyst, which exhibited excellent catalytic activity in the reduction of 4-nitrophenol and 2-nitroaniline using $NaBH_4$ in water at room temperature. Importantly, the nanocatalyst could be easily recovered and reused for at least ten cycles in both reduction reactions, demonstrating its good stability. Pandey et al. [21] synthesized highly stable dispersions of Pt NPs in guar gum, a natural, non-toxic, and eco-friendly biopolymer, serving as both the reducing and capping agent precursor in an aqueous medium. The catalytic activity of biopolymer-supported Pt NPs was demonstrated in the liquid-phase reduction of p-nitrophenol and p-aminophenol. The catalytic reduction of nitroaromatics achieved a remarkable efficiency of 97% within a total reaction time of 320 s at room temperature. Graphene oxide (GO) and reduced graphene oxide (r-GO) were also utilized to stabilize the noble NPs [22,23]. Without the need for additional reductants, surfactants, or protecting ligands, metallic noble metals were deposited on partially r-GO mats through a simple redox reaction between noble metal precursors and GO in an aqueous solution. These GO- or r-GO-supported noble NPs exhibited excellent catalytic activity for the selective reduction of nitroaromatic compounds. Cai et al. [24] developed a novel nanostructured catalyst comprising small and uniform Au NPs with a diameter of approximately 5 nm and ceria nanotubes (CeO_2 NTs). The catalytic performance of the Au NPs/CeO_2 NT catalyst in the reduction of 4-nitrophenol to 4-aminophenol was significantly higher compared to similar catalysts composed of chemically prepared AuNPs or commercially available CeO_2 powder as the support. The superior catalytic activity can be attributed to the unique surface properties of the synthesized Au NPs/CeO_2 NT catalyst, as well as the interaction between the barrier-free surface of Au NPs and surface defects (oxygen vacancies) of CeO_2 NTs, leading to the presence of oxidized Au species. Chen et al. [25] conducted a comprehensive investigation of the reduction of p-nitrophenol by $NaBH_4$ in the presence of raspberry-like composite sub-microspheres composed of poly(allylamine hydrochloride)-modified polymer poly(glycidyl methacrylate) with tunable Au NPs. They systematically examined the effects of polyelectrolyte concentration, the ratio of polymer spheres to Au NPs, and solution pH during composite synthesis on various reaction parameters such as the induction period, reaction time, average reaction rate, and average turnover frequency. They also proposed a mechanism to explain the observed enhancement in catalytic activity, which involves the active epoxy groups present on the polymer spheres and the strong adsorption of p-nitrophenolate anions onto the positively charged spheres.

Metal-assisted chemical etching (MACE) is a simple and versatile method for fabricating porous silicon (PSi) structures without requiring electrochemically etched electrodes [26]. The main principle of MACE is to deposit a noble metal on the surface of a Si substrate and then immerse the substrate in an etching solution containing fluoride and an oxidizing agent to induce an etching reaction [27]. MACE can be used to produce various PSi structures for which the pore size, porosity, and surface morphology can be controlled by adjusting the composition and concentration of the etching solution as well as the type, thickness, and distribution of the metal catalyst [28]. In conventional MACE, a mixture of HF and H_2O_2 is commonly used as an etching solution [29]. HF reacts with the oxygen atoms on the surface of the Si substrate to form fluorosilicic acid, which further dissolves the Si surface. Simultaneously, the noble metal catalyst on the surface of the Si substrate serves as an active site for catalyzing the etching reaction by promoting the generation of holes (positive charges) in the Si substrate through the reduction of H_2O_2. These holes are then injected into the interface between the Si substrate and the metal catalyst, resulting in the oxidation and dissolution of the Si substrate in the etching solution [30]. During the etching process, the metal catalyst serves as a cathodic reaction zone, whereas the Si substrate serves as an anodic reaction zone [31].

Because of their high surface area and tunable pore size, PSi or PSi substrates have been widely used for the deposition of metal NPs for catalyzing nitroaromatic compounds [32–34] in a different fabrication technique. To our knowledge, this is the first study to explore the use of PSi fabricated using MACE as a substrate for the deposition of Au NPs for the reduction of nitroaromatic compounds. As a substrate, PSi provides a high surface area for the deposition of Au NPs, and MACE allows for the fabrication of a well-defined porous structure with a tunable pore size. In this study, we used scanning electron microscopy (SEM) and energy-dispersive X-ray spectroscopy (EDS) to characterize Au NPs on PSi. We used *p*-nitroaniline (PNA) reduction as a model reaction to evaluate the catalytic activity of Au NPs. Our results indicate that the Au NPs on PSi exhibited excellent catalytic activity toward the reduction of PNA. They also indicated that the catalytic activity of the PSi substrate was affected by the etching time. That is, as the etching time increased, the surface area of the PSi increased, which increased the atomic weight percentage of Au NPs immobilized on the surface. Because the surface area available for the catalytic reaction increased, the catalytic activity also increased. However, at a certain point, further increasing the etching time resulted in a decrease in catalytic activity, which may have been attributable to the aggregation of Au NPs. Overall, these findings may have major implications for the development of efficient and cost-effective catalysts for various organic transformations.

2. Materials and Methods

Briefly, N-type Si wafers with a resistivity of 1–10 $\Omega\cdot$cm and a crystal orientation of (100) were cut using a glass cutter into 1.5×1.5 cm^2 square pieces, ultrasonically cleaned with methanol, acetone, and deionized (DI) water for 15 min in each solution, and dried with a nitrogen gas gun. The cleaned Si substrates were then placed vertically in an acid-resistant Teflon cell of 20 mL in size. Subsequently, a MACE mixture containing HF (48%), H_2O_2 (30%), DI water, and $HAuCl_4$ (3 mM) at a ratio of 1:5:2:4 (volume ratio) of 12 mL was added to the Teflon cell. Etching was then conducted at room temperature without stirring for different durations. After the etching process was completed, the PSi substrate was removed from the etching solution and rinsed with anhydrous alcohol and DI water to remove any residual HF solution. Finally, the PSi substrate was dried using a nitrogen gas gun. The fabrication process of electrochemically etched PSi is described in detail in previous studies [35]. In detail, a $\pm$20 V, 40 W source measure unit (PXI 4130) was used as power supply for offering a constant voltage mode at a current density of 30 mA/cm^2 for 30 min. An etching solution containing HF, ethanol, and DI water at a ratio of 1:2:1 (volume ratio) of 12 mL was added. A catalytic solution containing 10 mL of PNA at various concentrations and 33 mg of $NaBH_4$ was then premixed for 1 h using a magnetic stirrer. The prepared PSi and catalytic solution were then placed together in a glass vial for absorbance measurement at various time intervals. Finally, surface morphological analysis, elemental analysis, and EDS mapping were performed using a multifunction environmental field emission scanning electron microscope equipped with an energy-dispersive X-ray spectrometer (Hitachi SU-5000, Hitachi, Tokyo, Japan).

3. Results and Discussion

The conversion of a nitroaromatic molecule to an aniline molecule involves the hydrogenation–dehydration of nitroaniline to form a nitrogen–oxygen double bond, followed by hydrogenation to produce hydroxylamine, and finally, further hydrogenation–dehydration to yield the product, p-phenylenediamine, as shown in Figure 1a. The hydrogenation–dehydration of nitroaniline: Nitroaniline undergoes hydrogenation–

dehydration under appropriate conditions and in the presence of a suitable catalyst such as nickel or platinum. This reaction leads to the removal of the nitro group (NO_2) and the formation of a nitrogen–oxygen double bond. In the compound formed in the previous step, the nitrogen–oxygen double bond reacts with hydrogen gas, resulting in the formation of hydroxylamine (NH_2OH). This step involves a hydrogenation reaction where the nitrogen–oxygen double bond is reduced to an amino group. The final step involves the hydrogenation–dehydration of hydroxylamine. Under suitable conditions, hydroxylamine reacts with hydrogen gas once again, undergoing dehydration. This reaction removes the hydroxyl group (OH) and ultimately yields p-phenylenediamine.

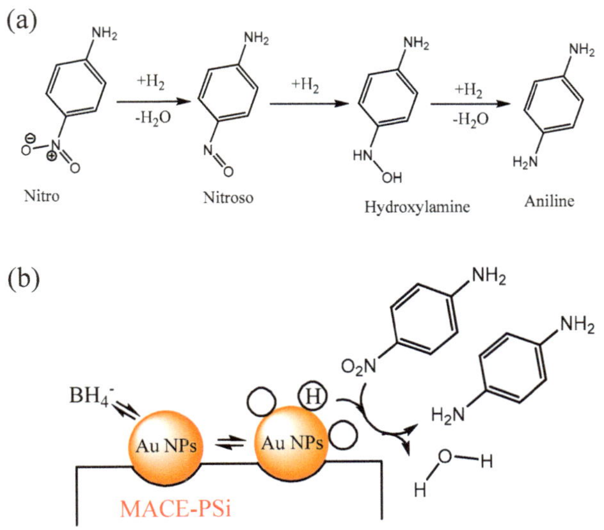

Figure 1. (**a**) Conversion of a nitroaromatic molecule to an aniline molecule. (**b**) Schematic of the catalytic reduction mechanism of PNA with Au NPs. Hydrogen and PNA adsorb on the surface of Au NPs, and the nitro group is reduced into a nitroso group. Further hydrogenation results in the formation of hydroxylamine, which undergoes dehydration to produce the final product, PPD.

Figure 1b depicts the reduction of PNA to *p*-phenylenediamine (PPD) in the presence of $NaBH_4$ through the catalytic reaction of Au NPs inside PSi fabricated using MACE. The mechanism of nitroaromatic reduction catalyzed by Au NPs involves four steps: adsorption, hydrogen atom generation, activation, and product formation. In the first step, a nitroaromatic molecule adsorbs, either chemically or physically, onto the Au surface, where it interacts with the surface electrons of Au NPs. In the second step, hydrogen gas molecules adsorb on the Au surface and dissociate into hydrogen atoms. This dissociation reaction is promoted by Au, which acts as a catalyst. In the third step, the nitroaromatic molecule is activated by the adsorbed hydrogen atoms, and the nitro group is reduced to an amino group, forming an intermediate aminoaromatic compound. In the fourth step, the intermediate aminoaromatic compound undergoes further reaction to form a corresponding reduced product. This product desorbs from the Au surface through different pathways.

In the presence of $NaBH_4$, the reaction between PNA and PPD was detected by monitoring the absorption spectra of the solution. In this reaction, the N–H bond of nitroaniline underwent a hydrogen transfer process with the hydrogen atoms of $NaBH_4$, resulting in the formation of PPD. Over time, the reduction of PNA lowered the degree of absorption, and a new absorption peak corresponding to PPD emerged. Consequently, the color of the solution changed from yellow, indicating the presence of PNA, to transparent, indicating the presence of PPD. This reaction was visible to the naked eye. As shown in Figure 2a, the beakers on the right and left contain identical mixtures of PNA and $NaBH_4$.

However, the beaker on the left contained MACE-PSi, whereas that on the right did not. After the mixture was allowed to stand for 1 h, the difference in color between the two solutions became evident to the naked eye. Specifically, the solution on the left, which contained PSi, changed from yellow to transparent, whereas the solution on the right, which lacked PSi, remained unchanged. During the catalytic reaction, absorption spectra were used to monitor the reduction of PNA to PPD (Figure 2b). In the ultraviolet–visible spectral region, the peak position of PNA was located between 300 and 400 nm, which differed from that of PPD. Therefore, whether a reaction occurred was determined by comparing the absorption spectra before and after the reaction. As the reaction proceeded, the characteristic peak of PNA at approximately 400 nm gradually decreased, whereas the peak of PPD at approximately 300 nm gradually increased. The disappearance of a peak at 400 nm and the appearance of a peak at 300 nm verified the successful reduction of PNA to PPD. As presented in Figure 2c, no changes in absorption were observed in a controlled experiment without the addition of PSi. The absorption spectra of the PNA/$NaBH_4$ solution remained unchanged, confirming the catalytic function of MACE-PSi. Therefore, electrochemically etched PSi was used as a catalytic substrate, and the same catalytic experiments were conducted. In addition, temporal changes in solution absorption were recorded. As indicated in Figure 2d, because PSi contains no metallic Au, the absorbance of the solution remained unchanged.

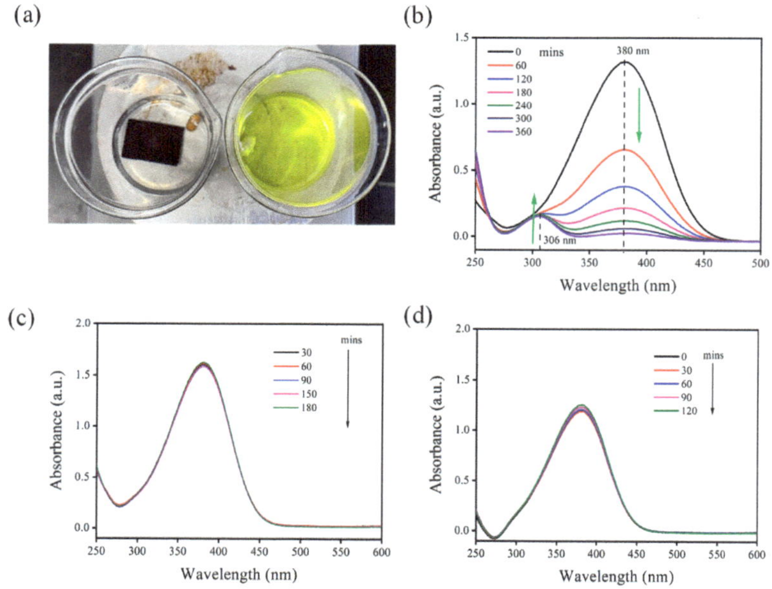

Figure 2. (**a**) PNA with (left) and without (right) PSi with the solution kept at room temperature for 60 min. Absorption spectra of PNA (**b**) with MACE-PSi, (**c**) without MACE-PSi, and (**d**) with electrochemically etched PSi.

Generally, the etching time of PSi is the most direct experimental parameter for modifying the surface morphology of PSi. Studies have indicated that the surface morphology of PSi can be used to determine changes in the deposition of Au thin films in surface-enhanced Raman scattering [35]. In the current study, PSi was fabricated over various etching times, and its catalytic effects on PNA were compared under the same catalytic conditions. The correlation between etching time and catalytic efficiency was also investigated by analyzing changes in the absorbance spectrum of PNA at a peak wavelength of 380 nm. Figure 3a presents the time-dependent absorbance peaks of PNA at 380 nm that were observed in the presence of PSi over various etching times. All PSi substrates prepared over various etching times and served as catalyst materials for PNA. However, the sample etched for 20 min exhibited the highest catalytic performance. After the initial absorbance peak intensity (A_0) was compared with the absorbance peak intensity after a certain period of time (A_t), a logarithmic calculation was performed to establish the correlation between the calculated values and elapsed time (Figure 3b). According to the results, the PSi sample etched for 20 min exhibited the highest catalytic performance for PNA. Therefore, the reduction rate and time constant (k) values were calculated using a pseudo first-order reaction [14], as shown below:

$$\ln(A_0/A_t) = kt \quad (1)$$

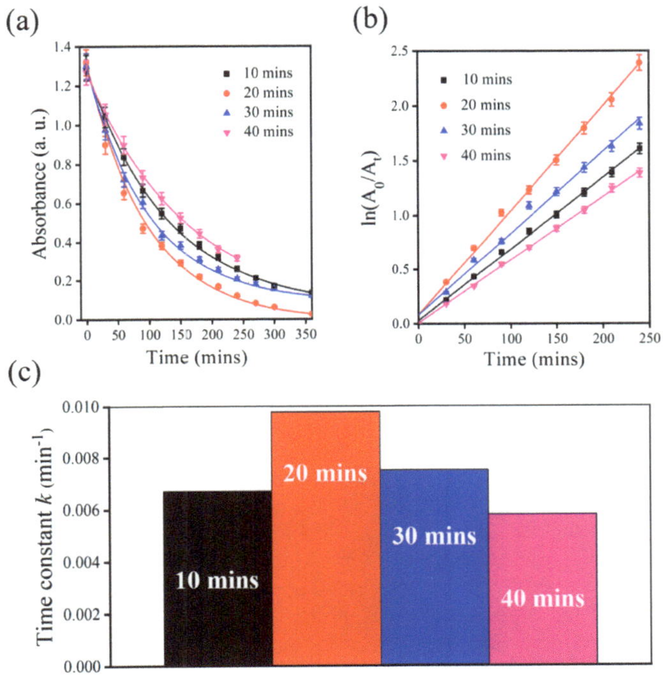

Figure 3. (**a**) Peak absorption (380 nm) changes in PNA with PSi over various etching times. (**b**) Calculated absorption values over time with MACE-PSi as a catalyst. (**c**) Calculated time constant values with PSi over various etching times.

The results are presented in Figure 3c. The PSi sample etched for 20 min exhibited the highest catalytic constant, followed by the sample etched for 30 min, whose catalytic constant was higher than that of the sample etched for 10 min but lower than that of the sample etched for 20 min. Of the samples, the PSi sample etched for 40 min exhibited the lowest catalytic performance for PNA.

Auric acid ($HAuCl_4$) is utilized as a metal catalyst in the MACE process to fabricate PSi, similar to the use of silver nitrate [28]. Auric acid, a metal salt containing gold (Au), acts as a catalyst in the HF etching solution during the MACE process. The gold ions (Au^{3+}) within auric acid react with silicon fluoride ($SiF_6{}^{2-}$) present in the etching solution. This reaction leads to the dissolution of Au species and the etching of the silicon material. By adjusting the concentration of auric acid, it is possible to control the formation of pores and the structural characteristics of porous silicon during the MACE process [29]. When auric acid is employed as the metal catalyst in MACE, the internal structure of PSi can contain gold nanoparticles [28]. This occurs due to the reduction of Au^{3+} from $HAuCl_4$ during the etching process, resulting in their deposition inside the pores of the silicon material. When the silicon material is immersed in the etching solution containing $HAuCl_4$, the Au^{3+} ions react with the silicon material present in the solution. Through this process, Au^{3+} ions are reduced to gold atoms, which then deposit inside the pores, forming gold nanoparticles within the PSi.

Figures 4–7 present both the surface morphological and elemental analysis results for the PSi prepared over various etching times and the EDS mapping results for the Au elements. Figure 4a is an SEM image of a PSi after 10 min of etching and reveals a porous structure. Figure 4b is an SEM image of the same sample but at a higher magnification and reveals a pore diameter of approximately 250 nm and fluff-like protrusions resembling Au structures at the locations at which the pores connect to one another. As indicated in Figure 4c, elemental analysis verified the presence of Au. Figure 4d is an EDS mapping image generated using Au elements, with green dots indicating the locations of Au. Both surface morphological analysis and elemental analysis confirmed the existence of a porous structure and Au in the PSi sample.

Using the same analytical techniques, we analyzed a PSi sample etched for 20 min (Figure 5). As shown in Figure 5a, etching for 20 min resulted in larger pores in the PSi, indicating a smaller diameter of Si. As indicated in Figure 5b, the pore size of the PSi was approximately 300 nm, which is larger than that of the PSi sample etched for 10 min. The figure also clearly reveals fluffy protrusions on the surface of the PSi, which are distributed more uniformly than those on the surface of the PSi sample etched for 10 min (Figure 4b). Figure 5c depicts the elemental analysis results for the PSi sample etched for 20 min, with the results indicating a higher Au content than that of the PSi sample etched for 10 min. The distribution of Au being uniform and the Au content being high may indirectly explain why the PSi sample etched for 20 min had a stronger PNA catalytic effect than that of the PSi sample etched for 10 min. Figure 5d depicts the EDS mapping results of the Au elements, with green dots indicating the locations of Au. Although the number of green dots in the sample etched for 20 min was not larger than that of the green dots in the sample etched for 10 min, the sample etched for 20 min was clearly brighter, confirming a difference in the amount of Au and the presence of fluffy protrusions on the surface of the PSi sample etched for 20 min.

Figure 6a depicts the surface morphology of a PSi sample etched for 30 min. A comparison with the samples etched for 10 and 20 min indicated a decrease in the pore size of the PSi sample etched for 30 min. As shown in Figure 6b, the pore diameter was approximately 150 nm, and the diameter of the PSi substantially increased. However, the hairy protrusions observed in the samples that were etched for 10 and 20 min were not observed on the surface of the PSi sample that was etched for 30 min. In addition, elemental analysis revealed a weight percentage of 7.5 wt% for Au in the PSi sample etched for 30 min, which was higher than that of the PSi sample etched for 10 min (6.7 wt%) but lower than that of the PSi sample etched for 20 min (8.2 wt%). These findings may explain why the PSi sample etched for 30 min was more effective than the PSi sample etched for 10 min at catalyzing PNA but not as effective as the PSi sample etched for 20 min. As shown in Figure 6d, Au element mapping revealed an uneven distribution of Au on the surface of the PSi, with some green dots exhibiting aggregation.

Figure 7 depicts the SEM and EDS results of a PSi sample etched for 40 min. As shown in Figure 7a, the surface morphology of the PSi changed, forming pore structures with an average size of approximately 1 μm and an irregular shape. Unlike the mesh pore structure of the PSi samples etched for 10 and 20 min, the diameter of the PSi was large (Figure 7b). However, similar to the PSi sample etched for 30 min, the surface of the PSi sample etched for 40 min did not exhibit fluffy protrusions. Elemental analysis indicated that the PSi sample etched for 40 min contained 7.9 wt% Au, which was higher than those of the PSi samples etched for 10 and 30 min. However, the catalytic effect of the PSi sample etched for 40 min did not exceed those of the PSi samples etched for 10 and 20 min. One possible reason for PNA being subject to a catalytic effect is Au content. In this case, the surface morphology of the Au may have also influenced the catalytic effect. Figure 7d depicts the EDS mapping results of Au for the PSi sample etched for 40 min and reveals a highly uneven distribution of Au and the presence of aggregation. These findings indirectly confirm our hypothesis that the structure and distribution of Au on porous surfaces influence the catalytic effect of PSi on PNA.

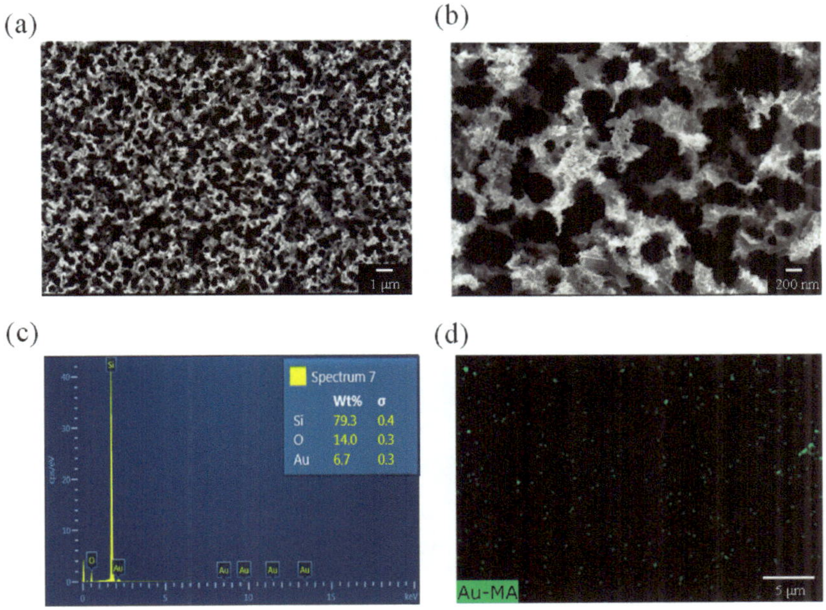

Figure 4. (**a**,**b**) SEM images, (**c**) EDS analysis, and (**d**) EDS mapping of a PSi sample etched for 10 min.

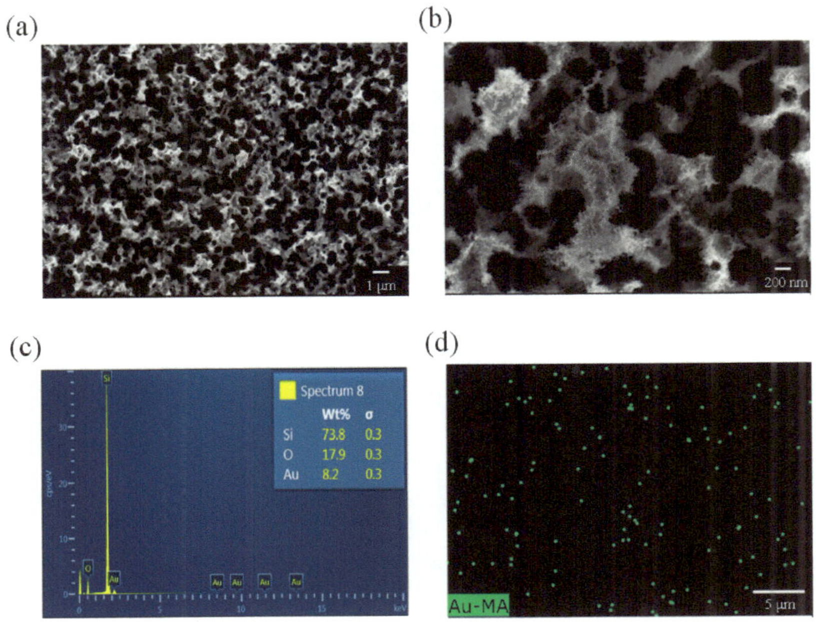

Figure 5. (**a**,**b**) SEM images, (**c**) EDS analysis, and (**d**) EDS mapping of a PSi sample etched for 20 min.

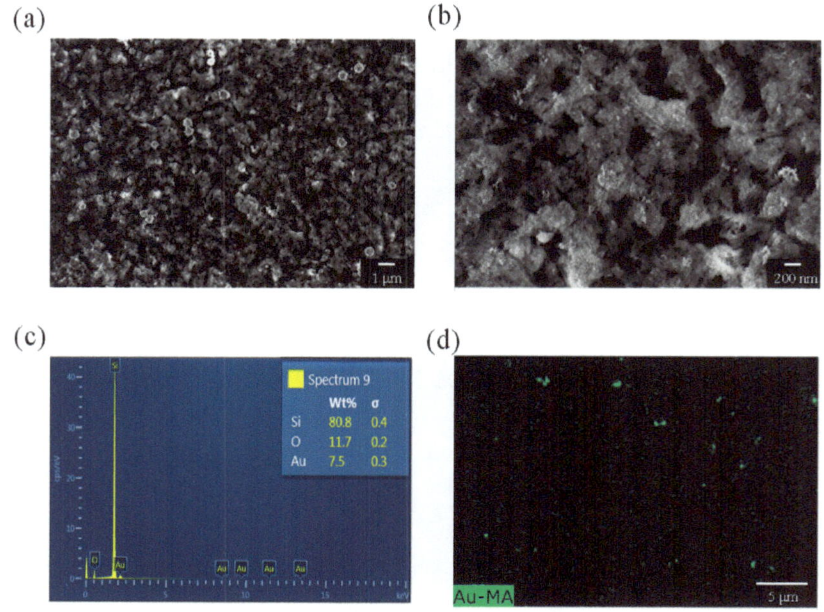

Figure 6. (**a**,**b**) SEM images, (**c**) EDS analysis, and (**d**) EDS mapping of a PSi sample etched for 30 min.

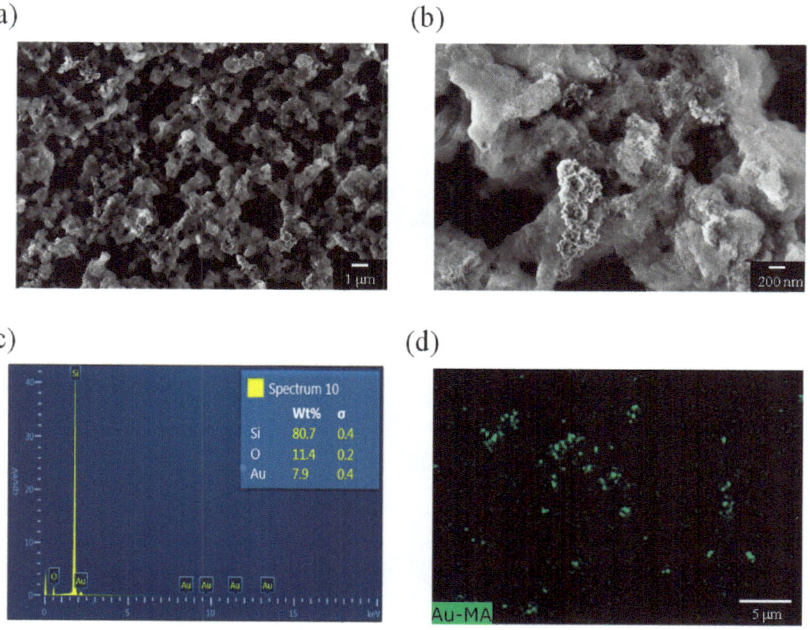

Figure 7. (**a**,**b**) SEM images, (**c**) EDS analysis, and (**d**) EDS mapping of a PSi sample etched for 40 min.

As a final step, PSi was etched for 20 min, and PNA was catalyzed at different concentrations. This testing method clarified the catalytic effect of PSi on different concentrations of PNA and thereby validated our experimental results. Because we used the same PSi for catalysis, our results confirmed that the produced sample played a role in repeated catalysis. The experimental results are presented in Figure 8. When the PNA concentration was low, the catalytic effect was relatively high. Therefore, we calculated the catalytic rate k by using the obtained PNA absorption spectrum change (Table 1). The results indicated that the k value was large when the concentration was low. However, when the PNA concentration exceeded 1.2 mM, the catalytic rate of PSi remained unchanged. Therefore, this concentration may be the catalysis and saturation concentration of PSi for PNA. However, by analyzing the catalytic usage times at the same concentration, we discovered that the catalytic efficiency decreased by approximately 35%. Hence, further studies are required to improve the catalytic efficiency and optimize the run time of the catalytic process by adjusting the concentration of auric acid.

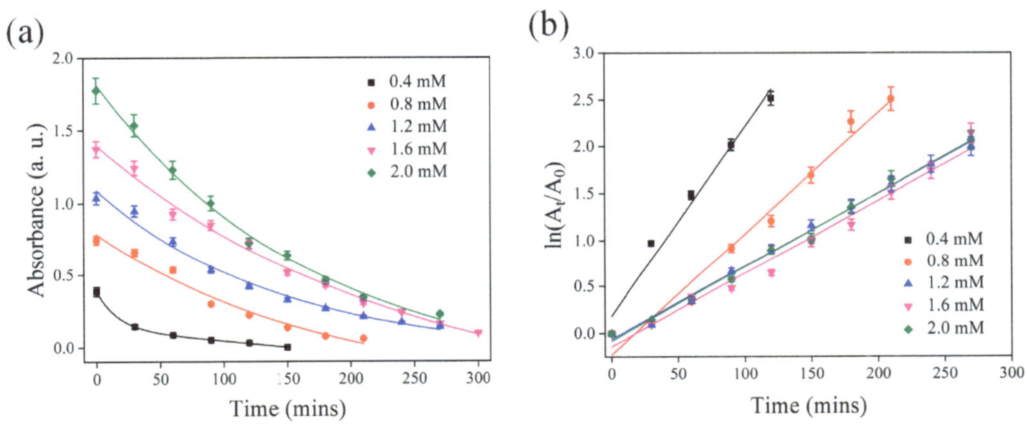

Figure 8. (**a**) Peak absorption (380 nm) change and (**b**) calculated absorption change with PSi etched for 20 min in the presence of different concentrations of PNA.

Table 1. Calculated time constants for PSi etched for 20 min in each cycle of the catalytic process.

Cycles	PNA Concentration (mM)	Time Constant (min^{-1})	Fitting Error (%)
1	1.2	0.01210	2
2	0.4	0.02024	3
3	0.8	0.01294	5
4	1.2	0.00780	4
5	1.6	0.00782	5
6	2.0	0.00791	6

4. Conclusions

To our knowledge, this is the first study to investigate the use of PSi fabricated by MACE as a carrier for Au NPs and to evaluate its catalytic performance in terms of nitroaromatic compound reduction. Overall, the PSi substrate provided a high surface area and tunable pore size, which facilitated the deposition and catalytic reaction of Au NPs. The Au NPs were then characterized using SEM and EDS, and their catalytic activity was evaluated with nitroaromatic reduction used as a model reaction. The results indicate that Au NPs on PSi exhibit excellent catalytic activity and that the catalytic activity of PSi substrates is affected by the etching time. In addition, a longer etching time results in a larger surface area of PSi and a higher atomic weight percentage of immobilized Au NPs, which leads to higher catalytic activity. However, an excessive etching time results in the aggregation of Au NPs and reduces catalytic activity. These findings have major implications for the development of efficient and cost-effective catalysts for different organic transformation reactions.

Author Contributions: Conceptualization, C.-C.C. and V.K.S.H.; methodology, L.-Y.L. and Y.-H.K.; formal analysis, L.-Y.L. and V.K.S.H.; writing—original draft preparation, L.-Y.L. and V.K.S.H.; writing—review and editing, C.-C.C. and V.K.S.H.; supervision, C.-C.C. and V.K.S.H. All authors have read and agreed to the published version of the manuscript.

Funding: This research was funded by National Science and Technology Council, grant number MOST 110-2221-E-260-008-MY3 and NSTC 111-2113-M-040-001.

Data Availability Statement: Not applicable.

Conflicts of Interest: The authors declare no conflict of interest.

References

1. Baig, N.; Kammakakam, I.; Falath, W. Nanomaterials: A review of synthesis methods, properties, recent progress, and challenge *Mater. Adv.* **2021**, *2*, 1821–1871. [CrossRef]
2. Zhang, Y.-X.; Wang, Y.-H. Nonlinear optical properties of metal nanoparticles: A review. *RSC Adv.* **2017**, *7*, 45129–4514 [CrossRef]
3. Saravanan, A.; Kumar, P.S.; Karishma, S.; Vo, D.V.N.; Jeevanantham, S.; Yaashikaa, P.R.; George, C.S. A review on bio-synthesis c metal nanoparticles and its environmental applications. *Chemosphere* **2021**, *264*, 128580. [CrossRef] [PubMed]
4. Gao, C.; Lyu, F.; Yin, Y. Encapsulated Metal Nanoparticles for Catalysis. *Chem. Rev.* **2020**, *121*, 834–881. [CrossRef]
5. Sankar, M.; He, Q.; Engel, R.V.; Sainna, M.A.; Logsdail, A.J.; Roldan, A.; Willock, D.J.; Agarwal, N.; Kiely, C.J.; Hutchings, G. Role of the Support in Gold-Containing Nanoparticles as Heterogeneous Catalysts. *Chem. Rev.* **2020**, *120*, 3890–3938. [CrossRef
6. Restrepo, C.V.; Villa, C.C. Synthesis of silver nanoparticles, influence of capping agents, and dependence on size and shape: A review. *Environ. Nanotechnol. Monit. Manag.* **2021**, *15*, 100428. [CrossRef]
7. Pedroso-Santana, S.; Fleitas-Salazar, N. The Use of Capping Agents in the Stabilization and Functionalization of Metalli Nanoparticles for Biomedical Applications. *Parti. Parti. Sys. Charact.* **2023**, *40*, 2200146. [CrossRef]
8. Hu, H.; Lu, S.; Li, T.; Zhang, Y.; Guo, C.; Zhu, H.; Jin, Y.; Du, M.; Zhang, W. Controlled growth of ultrafine metal nanoparticle mediated by solid supports. *Nanoscale Adv.* **2021**, *3*, 1865–1886. [CrossRef]
9. Koyappayil, A.; Yagati, A.K.; Lee, M.-H. Recent Trends in Metal Nanoparticles Decorated 2D Materials for Electrochemica Biomarker Detection. *Biosensors* **2023**, *13*, 91. [CrossRef]
10. Das, M.R.; Hussain, N.; Duarah, R.; Sharma, N.; Sarmah, P.; Thakur, A.; Bhattacharjee, P.; Bora, U.; Boukherroub, R. Meta nanoparticles decorated two-dimensional nanosheets as heterogeneous catalysts for coupling reactions. *Catal. Rev.* **2022**, 1–7 [CrossRef]
11. Berthold, D.; Haydl, A.M.; Leung, J.C.; Scholz, U.; Xiao, Q.; Zhu, Z. Recent Advances in the Synthesis of Arylamines in the Ligh of Application in Pharmaceutical and Chemical Industry. *Method. Amine Syn. Challen. Applicat.* **2021**, 377–444. [CrossRef]
12. Ju, K.S.; Parales, R.E. Nitroaromatic compounds, from synthesis to biodegradation. *Microbio. Mol. Bio. Rev.* **2010**, *74*, 250–27 [CrossRef] [PubMed]
13. Song, J.; Huang, Z.-F.; Pan, L.; Li, K.; Zhang, X.; Wang, L.; Zou, J.-J. Review on selective hydrogenation of nitroarene by catalyti photocatalytic and electrocatalytic reactions. *Appl. Catal. B Environ.* **2018**, *227*, 386–408. [CrossRef]
14. Mejía, Y.R.; Bogireddy, N.K.R. Reduction of 4-nitrophenol using green-fabricated metal nanoparticles. *RSC Adv.* **2022**, *1* 18661–18675. [CrossRef] [PubMed]
15. Gangula, A.; Podila, R.; Karanam, L.; Janardhana, C.; Rao, A.M. Catalytic Reduction of 4-Nitrophenol using Biogenic Gold an Silver Nanoparticles Derived from *Breynia rhamnoides*. *Langmuir* **2011**, *27*, 15268–15274. [CrossRef]
16. Ghosh, S.K.; Mandal, M.; Kundu, S.; Nath, S.; Pal, T. Bimetallic Pt–Ni nanoparticles can catalyze reduction of aromatic nitr compounds by sodium borohydride in aqueous solution. *Appl. Cat. A Gen.* **2004**, *268*, 61–66. [CrossRef]
17. Gu, Y.; Jiao, Y.; Zhou, X.; Wu, A.; Buhe, B.; Fu, H. Strongly coupled Ag/TiO_2 heterojunctions for effective and stable photo-therma catalytic reduction of 4-nitrophenol. *Nano Res.* **2018**, *11*, 126–141. [CrossRef]
18. Guo, M.; He, J.; Li, Y.; Ma, S.; Sun, X. One-step synthesis of hollow porous gold nanoparticles with tunable particle size for th reduction of 4-nitrophenol. *J. Hazard. Mater.* **2016**, *310*, 89–97. [CrossRef]
19. Moon, H.; Kim, Y. Photothermal-Mediated Catalytic Reduction of 4-Nitrophenol Using Poly(*N*-isopropylacrylamide-acrylamide and Hollow Gold Nanoparticles. *ACS Appl. Polym. Mater.* **2021**, *3*, 2768–2775. [CrossRef]
20. Dong, Z.; Le, X.; Li, X.; Zhang, W.; Dong, C.; Ma, J. Silver nanoparticles immobilized on fibrous nano-silica as highly efficient an recyclable heterogeneous catalyst for reduction of 4-nitrophenol and 2-nitroaniline. *Appl. Catal. B Environ.* **2014**, *158*, 129–13 [CrossRef]
21. Pandey, S.; Mishra, S.B. Catalytic reduction of p-nitrophenol by using platinum nanoparticles stabilised by guar gum. *Carbohyd Polym.* **2014**, *113*, 525–531. [CrossRef] [PubMed]

22. Ye, W.; Yu, J.; Zhou, Y.; Gao, D.; Wang, D.; Wang, C.; Xue, D. Green synthesis of Pt–Au dendrimer-like nanoparticles supported on polydopamine-functionalized graphene and their high performance toward 4- nitrophenol reduction. *Appl. Catal. B Environ.* **2016**, *181*, 371–378. [CrossRef]
23. Yang, M.-Q.; Pan, X.; Zhang, N.; Xu, Y.-J. A facile one-step way to anchor noble metal (Au, Ag, Pd) nanoparticles on a reduced graphene oxide mat with catalytic activity for selective reduction of nitroaromatic compounds. *Crystengcomm* **2013**, *15*, 6819–6828. [CrossRef]
24. Zhang, J.; Chen, G.; Chaker, M.; Rosei, F.; Ma, D. Gold nanoparticle decorated ceria nanotubes with significantly high catalytic activity for the reduction of nitrophenol and mechanism study. *Appl. Catal. B Environ.* **2012**, *132*, 107–115. [CrossRef]
25. Li, M.; Chen, G. Revisiting catalytic model reaction p-nitrophenol/NaBH 4 using metallic nanoparticles coated on polymeric spheres. *Nanoscale* **2013**, *5*, 11919–11927. [CrossRef]
26. Huang, Z.; Geyer, N.; Werner, P.; De Boor, J.; Gösele, U. Metal-assisted chemical etching of silicon: A review: In memory of Prof. Ulrich Gösele. *Adv. Mater.* **2011**, *23*, 285–308. [CrossRef]
27. Yae, S.; Nasu, N.; Matsumoto, K.; Hagihara, T.; Fukumuro, N.; Matsuda, H. Nucleation behavior in electroless displacement deposition of metals on silicon from hydrofluoric acid solutions. *Electrochim. Acta* **2007**, *53*, 35–41. [CrossRef]
28. Fan, Z.; Cui, D.; Zhang, Z.; Zhao, Z.; Chen, H.; Fan, Y.; Li, P.; Zhang, Z.; Xue, C.; Yan, S. Recent Progress of Black Silicon: From Fabrications to Applications. *Nanomaterials* **2020**, *11*, 41. [CrossRef]
29. Li, X.; Bohn, P.W. Metal-assisted chemical etching in HF/H2O2 produces porous silicon. *Appl. Phys. Lett.* **2000**, *77*, 2572–2574. [CrossRef]
30. Smith, R.L.; Collins, S.D. Porous silicon formation mechanisms. *J. Appl. Phys.* **1992**, *71*, R1–R22. [CrossRef]
31. Lee, C.-L.; Tsujino, K.; Kanda, Y.; Ikeda, S.; Matsumura, M. Pore formation in silicon by wet etching using micrometre-sized metal particles as catalysts. *J. Mater. Chem.* **2008**, *18*, 1015–1020. [CrossRef]
32. Abu Bakar, N.; Ridzwan, A.; Tan, W.; Abu Bakar, M.; Sabri, N. Facile preparation of porous silicon from cost-wise silicon powder as effective catalyst for reduction of p-nitrophenol. *Mater. Chem. Phys.* **2019**, *232*, 387–392. [CrossRef]
33. Liu, X.; Cheng, H.; Cui, P. Catalysis by silver nanoparticles/porous silicon for the reduction of nitroaromatics in the presence of sodium borohydride. *Appl. Surf. Sci.* **2014**, *292*, 695–701. [CrossRef]
34. Halim, M.Y.A.; Tan, W.L.; Abu Bakar, N.H.H.; Abu Bakar, M. Surface Characteristics and Catalytic Activity of Copper Deposited Porous Silicon Powder. *Materials* **2014**, *7*, 7737–7751. [CrossRef]
35. Khalil, I.; Chou, C.-M.; Tsai, K.-L.; Hsu, S.; Yehye, W.A.; Hsiao, V.K.S. Gold Nanofilm-Coated Porous Silicon as Surface-Enhanced Raman Scattering Substrate. *Appl. Sci.* **2019**, *9*, 4806. [CrossRef]

nanomaterials

MDPI

Article

Gold Nanorods as Radiopharmaceutical Carriers: Preparation and Preliminary Radiobiological In Vitro Tests

Ludovica Binelli [1,2,†], Valentina Dini [3,4,*,†], Simone Amatori [1], Teresa Scotognella [5], Alessandro Giordano [5,6], Barbara De Berardis [3], Federica Bertelà [1], Chiara Battocchio [1], Giovanna Iucci [1], Ilaria Fratoddi [7], Antonella Cartoni [7] and Iole Venditti [1,*]

1 Sciences Department, Roma Tre University, 00146 Rome, Italy; ludovica.binelli@uniroma3.it (L.B.); simone.amatori@uniroma3.it (S.A.); federica.bertela@uniroma3.it (F.B.); chiara.battocchio@uniroma3.it (C.B.); giovanna.iucci@uniroma3.it (G.I.)
2 Istituto Nazionale di Fisica Nucleare (INFN), Sezione di Roma3, Department of Sciences, Roma Tre University, 00146 Rome, Italy
3 National Center for Innovative Technologies in Public Health, Istituto Superiore di Sanità, 00161 Rome, Italy; barbara.deberardis@iss.it
4 Istituto Nazionale di Fisica Nucleare (INFN), Sezione di Roma1, Department of Physics, University La Sapienza, 00185 Rome, Italy
5 Nuclear Medicine Unit, Fondazione Policlinico Universitario A. Gemelli IRCCS, 00168 Rome, Italy; teresa.scotognella@policlinicogemelli.it (T.S.); alessandro.giordano@unicatt.it (A.G.)
6 Department of Radiological and Hematological Sciences, Università Cattolica del Sacro Cuore, 00168 Rome, Italy
7 Chemistry Department, Sapienza University, 00185 Rome, Italy; ilaria.fratoddi@uniroma1.it (I.F.); antonella.cartoni@uniroma1.it (A.C.)
* Correspondence: valentina.dini@iss.it (V.D.); iole.venditti@uniroma3.it (I.V.)
† These authors contributed equally to this work.

Abstract: Low-energy electrons (Auger electrons) can be produced via the interaction of photons with gold atoms in gold nanorods (AuNRs). These electrons are similar to those emitted during the decay of technetium-99m (^{99m}Tc), a radioactive nuclide widely used for diagnostics in nuclear medicine. Auger and internal conversion (IC) electron emitters appropriately targeted to the DNA of tumors cells may, therefore, represent a new radiotherapeutic approach. ^{99m}Tc radiopharmaceuticals, which are used for diagnosis, could indeed be used in theragnostic fields when loaded on AuNRs and delivered to a tumor site. This work aims to provide a proof of concept (i) to evaluate AuNRs as carriers of ^{99m}Tc-based radiopharmaceuticals, and (ii) to evaluate the efficacy of Auger electrons emitted by photon-irradiated AuNRs in inducing radio-induced damage in T98G cells, thus mimicking the effect of Auger electrons emitted during the decay of ^{99m}Tc used in clinical settings. Data are presented on AuNRs' chemical characterization (with an aspect ratio of 3.2 and Surface Plasmon Resonance bands at 520 and 680 nm) and the loading of pharmaceuticals (after ^{99m}Tc decay) on their surface. Spectroscopic characterizations, such as UV-Vis and synchrotron radiation-induced X-ray photoelectron (SR-XPS) spectroscopies, were performed to investigate the drug–AuNR interaction. Finally, preliminary radiobiological data on cell killing with AuNRs are presented.

Keywords: gold nanorods; technetium-99m; radiopharmaceuticals; theragnostic; nuclear medicine

1. Introduction

In the last decade, nanomaterials have found success in several fields, such as cosmetics, textiles, sensors, optoelectronics and medicine [1–4]. In the field of medicine, gold nanoparticles and gold nanorods (AuNRs) have found wide applications due to their peculiar chemical and physical features, such as the Localized Surface Plasmon Resonance (LSPR), which occurs when there is an interaction between a nanomaterial and electromagnetic radiation of the appropriate wavelength [5,6]. Another important outlook, which has favored the use of gold-based nanoparticles in biomedical applications, is their high biocompatibility. In fact, it is well known in the literature that gold nanoparticles are stable and biocompatible, even if these general characteristics also depend a lot on the specific dimension, shape and surface functionalization, and are studied on a case-by-case basis [7–9]. In particular, cetyltrimethylammonium bromide (CTAB)-functionalized AuNRs show different degrees of toxicity based on the types of performed tests. For example, Cornovale et al. found different cell viability for PC-3 cells (human prostate cancer cells) in free serum and supplemented serum, obtaining a range of 120–80% in viability using particle concentrations of 0.1–0.5 μg/mL in the substrate, respectively [10]. Guo et al. studied CTAB-functionalized gold nanoparticles' cytotoxicity using the MTT test in a range of human and murine cells, and their study evidenced different results: in the PC-3 cell line, GR5, GR7, GR8, GR9, GR11 and G12 samples displayed IC50 values of 8.2, 7.8, 8.2, 7.4, 8.5 and 8.0 μg/mL, respectively. In contrast, Au NP-CTAB samples (20 and 60 nm) caused a higher level of cytotoxicity with IC50 values of 2 and 3.5 μg/mL, respectively [11]. The general low cytotoxicity of AuNRs has largely allowed their implementation in the diagnostic field, using near-infrared (NIR) imaging, due to their two plasmonic peaks that are associated with different electron oscillations on the transverse and longitudinal side of the rod in an energy range in which biological tissues are not active [12–14]. Many techniques exploit AuNRs in diagnosis, one of which is two-photon luminescence imaging (TPL), which is capable of going deep and has sub-micron resolution. To go even deeper into a tissue, photoacoustic tomography (PAT) can be used [15]. The underlying principle of PAT is the ability of AuNRs to absorb a pulsed laser and emit an acoustic shock wave due to transient superheating and thermoelastic expansion. Other techniques that have been developed using AuNRs are optical coherence tomography (OCT), where AuNRs are used to enhance contrast, and X-ray computed tomography (XCT), where they are used as contrast agents instead of iodine molecules [16]. It is also well known that AuNRs are used for drug delivery due to their high surface/volume ratio: the surface can be functionalized for optimizing the interaction with drugs in view of the specific targets and the therapy used [17]. One of the main objectives is the fight against tumors. In these cases, it is possible to exploit the Enhanced Permeability and Retention (EPR) effect, a passive targeting, or create active targeting systems using surface functionalizations of AuNRs with ligands that increase their specificity to target cells [18].

In this context, the field of nuclear medicine is studying with interest the use of drug delivery systems based on radioisotopes that emit short-range charged particles, instead of common drugs; these radioisotopes could be an added value to common used therapy because they are able to deliver a therapeutic dose of ionizing radiation to a tumor, causing cellular damage due to their interaction with biological macromolecules [19]. Increasing the dose delivered to a tumor mass while simultaneously decreasing the dose delivered to healthy tissues is still a major challenge in radiotherapy, although several strategies have been proposed. Furthermore, if the radioisotope used is also a gamma emitter, it can be used in medical diagnostic procedures (e.g., scintigraphy), and in this way, it is possible to have a theragnostic system that improves personalized therapy [20].

Several new radiopharmaceuticals have been prepared in recent years due to the discovery of kits called *"shake and bake"*, which are very easy to use and optimize to ensure that the desired complex has a high labeling yield and stability. Indeed, these new drugs are influenced by several factors: the amount of reducing agent and ligand, pH and temperature. In addition, since the beginning of the 21st century, there has been significant growing interest in the field of so-called nanomedicine with the use of nanomaterials labeled with radionuclides and used for both diagnostic and therapeutic purposes [21].

Among different radionuclides, technetium 99m (^{99m}Tc), a gamma emitter, is widely used for diagnostic purposes in nuclear medicine, and several complexes labeled with it are available in clinical practice [22]. ^{99m}Tc, during its nuclear decay, also emits Auger electrons (AEs), most of which has a low energy (<25 keV) and can traverse tissues for very short distances on the order of a few micrometers, resulting in a high Linear Energy Transfer

(LET) between 1 and 23 keV/µm, which is very effective for producing clustered damage in the DNA and/or sensitive targets (e.g., cell membrane) of cancer cells. These types of damage are difficult to repair and generally lead to cell death [23]. The high lethality of AEs emitted near the nucleus is evident when observing their Relative Biological Effectiveness (RBE, defined as the ratio of the effectiveness of the radiation under investigation compared to X-rays or gamma rays used as the reference radiation); for AE radionuclides emitted from ^{99m}Tc in rat thyroid PC Cl3 cells, and assuming a cellular or nuclear target for dose calculation, the RBE increased from 0.75 to 2.18 [24]. Therefore, AEs emitted from ^{99m}Tc nuclear decay at the cellular level lead to a dense deposition of ionizing energy that is associated with increased radiobiological efficiency. If they are appropriately targeted to the DNA of tumor cells, they may represent an interesting new radiotherapy system: ^{99m}Tc loaded AuNRs, delivered to the tumor site, colud indeed be used as a theragnostic radiopharmaceutical [25,26]. Similar to those emitted by the decay of ^{99m}Tc, low-energy electrons (i.e., Auger electrons) are also produced by the interaction of photons with gold in AuNRs at energies below 1 MeV [27], thus providing a possible synergistic therapeutic effect on tumors if the therapeutic system is used in combination with conventional radiotherapy.

In this work, AuNRs were chosen as the drug delivery system (DDS) by exploiting their ease of synthesis and the possibility of subsequent surface engineering, as well as the presence of plasmonic absorption. This work is intended to be a proof of concept (i) to evaluate these new synthesized AuNRs as carriers of radiopharmaceuticals based on ^{99m}Tc, and (ii) to evaluate the effectiveness of Auger electrons emitted by photon-irradiated AuNRs in inducing radio-induced damage at the cellular level, thus mimicking the effect of Auger electrons emitted during the decay of ^{99m}Tc used in clinical settings. Preliminary data are presented on the chemical characterization of AuNRs (with typical Surface Plasmon Resonance bands in the visible range and an aspect ratio (A.R.) = 3.2) and the loading of radiopharmaceuticals based on long-lived ^{99}Tc. Working with the decayed radiopharmaceutical has allowed us, on the one hand, to work with a compound already in use and with all the excipients present in the commercial compound, while, on the other hand, not having the safety limits imposed by a radioactive compound, thereby optimizing the chemical characterization and loading studies on AuNRs. To study the drug–AuNR interaction, spectroscopic characterizations, such as UV-Vis, Fourier-transform infrared (FTIR) and synchrotron radiation-induced X-ray photoelectron (SR-XPS) spectroscopies were performed. Finally, preliminary radiobiological data on cell killing with AuNRs are presented.

2. Materials and Methods

2.1. Materials for AuNR Synthesis and Conjugation

Cetyltrimethylammonium bromide (CTAB) ($C_{19}H_{42}BrN$, ≥97% Merck, Rahway, NJ, USA), tetrachoroauric (III) acid trihydrate ($HAuCl_4 \cdot 3H_2O$, ≥99.9% Sigma-Aldrich, St. Louis, MO, USA), sodium borohydrate ($NaBH_4$, 99.99% Aldrich, St. Louis, MO, USA), silver nitrate ($AgNO_3$ 99.9%, Aldrich), L-ascorbic acid ($C_6H_8O_6$, AA, 99% Sigma, St. Louis, MO, USA) and bidistilled H_2O were used as received.

^{99m}Tc-sestaMIBI (chemical structure reported in Scheme 1) was chosen as the radiopharmaceutical: it is a methoxyisobutylisonitrile (MIBI) with an isonitrile group which together form a complex with ^{99m}Tc. The labeling procedure was performed according to the manufacturer's instructions (STAMICIS®, Curium Pharma, London, UK). The labeling procedure required the reconstitution of the vial with 3 mL (11.1 GBq) of fresh ^{99m}TcO4, which was eluted from a ^{99}Mo/^{99m}Tc generator (Ultratechnekow FW, Curium Pharma). The vial was heated to 100 °C and incubated at room temperature for 15 min. The percentage of radiochemical purity (%RP) was assessed using an aluminum oxide strip (Agilent Technologies, Santa Clara, CA, USA) and ethanol as the eluent system, and it was analyzed via autoradiochromatography (Cyclone Plus®, Perkin Elmer, Waltham, MA, USA). The OptiQuant® image analysis software was used to evaluate the %RP. After the quality control, ^{99m}Tc-sestaMIBI was stored at 4 °C until completed decay. After the decay, long-lived ^{99}Tc-sestaMIBI was used for loading on AuNRs.

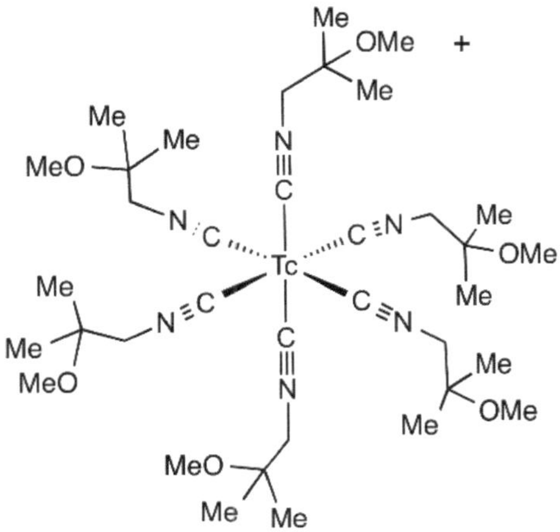

Scheme 1. Chemical structure of ^{99m}Tc-sestaMIBI.

2.2. Characterizations

The UV-Vis spectra were acquired in H_2O by using a quartz cell with a Shimadzu 2401 PC UV-Vis spectrophotometer in a wavelength range between 200 and 800 nm. The Energy-Dispersive X-ray Analysis (FESEM_EDX) images were acquired using a MIRA3 Tescan instrument (resolution 200 nm, SEM HV 30.0 kV). Nanoparticles dispersed in Milli-Q water and in the cell culture medium RPMI1640 at the final concentration of 0.1 mg/mL were characterized by using a Zetasizer Ultra instrument (Malvern Instrument, Malvern, UK), in order to determine the hydrodynamic diameter (Z-Average) and the polydispersity index (PDI). The equilibration step at 25 °C was set for 2 min. Three determinations were performed based on 1 mL of sample suspensions. The values of Z-Average and PDI were determined using the ZS Xplorer Software (Malvern Instruments, UK). AuNR sample stability and surface charge were assessed based on Zeta-potential measurements. The measurements were conducted in triplicate with 750 μL of suspensions using an automatic measurement protocol of Zetasizer Ultra. A Mini Spin Eppendorf centrifuge was used for the purification of the AuNR samples (13,000 rpm, 15 min, and two times with bidistilled water). The high-resolution X-ray Photoelectron Spectroscopy (SR-XPS) measurements were performed in situ using the SuperESCA beamline of the Elettra synchrotron radiation facility in Trieste, Italy. The experimental chamber is equipped with a 150 mm Phoibos hemispherical electron energy analyzer (SPECS GmbH), provided with a homemade delay line detector. The high-resolution core level spectra were measured in the normal emission configuration while keeping the sample at RT. C1s and N1s core levels were recorded at 550 eV of photon energy while Tc3d and Au4f were measured at 300 eV in order to maximize the intensity of signals. The overall resolution was always better than 100 meV. For each spectrum, the binding energy scale was calibrated using the aliphatic C1s component (285.00 eV) as an internal reference.

2.3. AuNR Synthesis

For AuNRs, a two steps synthesis was used, in analogy with the literature [18]. The first step was to prepare the seed solution according to the following protocol. In a reaction flask, 5 mL of CTAB at 0.2 M and 5 mL of $HAuCl_4$ at 0.0005 M were added. The solution was stirred and degassed with Argon for 3 min, and then 600 μL of $NaBH_4$ at 0.01 M was added. At this point, a color change was observed (the solution assumed the typical brownish-yellow coloration), and the solution was left in agitation for 5 min. In the second step, the growth solution was added 10 mL of CTAB at 0.2 M, 10 mL of $HAuCl_4$ at 0.001 M, and 400 µL of $AgNO_3$ at 0.004 M. The solution was stirred and degassed using Argon for 3 min, and then 70 µL of AA at 0.078 M and 24 µL of the seed solution were added. The solution was left in agitation for 20 min. For purification, the suspension was centrifugated at 13,000 rpm for 15 min, for two times. The AuNRs that were used for the biological test were synthesized following the experimental conditions shown in Table 1.

Table 1. Experimental conditions for the synthesis of AuNRs.

CTAB mL (M)	$HAuCl_4$ mL (M)	$AgNO_3$ µL (M)	AA µL (M)	Seed Solution, µL
10 (0.2)	10 (0.001)	400 (0.004)	70 (0.078)	24

2.4. Preparation of Conjugate Nanorods

To evaluate the loading of radiopharmaceutical on AuNRs, two tests were carried out, which were always performed in triplicate, following the protocol reported here. A total of 4 mL of the long-lived ^{99}Tc-sestaMIBI solution (0.0165 mg/mL) was placed in three vials with 1 mL of the AuNR solution (1 mg/mL). The solution was left in agitation for 24 h and, at the end, was centrifugated at 13,000 rpm for 20 min. The conjugate was stored at −20 °C, while the supernatant was used for the loading evaluation. From now on, the decayed, non-radioactive radiopharmaceutical will be indicated with ^{99}Tc and its conjugate on gold rods with AuNRs-99Tc.

2.5. Cell Culture

Human glioblastoma multiform cells (T98G cells) were purchased from the European Collection of Authenticated Cell Cultures (ECACC, UK Health Security Agency). The cells were grown in monolayer at 37 °C in a humidified atmosphere of 95% air and 5% CO_2, in RPMI-1640 medium (Euroclone S.p.A., Pero, Italy) supplemented with 10% fetal bovine serum (GIBCO®, Life Technologies, Waltham, MA, USA), 1 mM of glutamine (Euroclone S.p.A., Italy), and 50 U/dm^3 of penicillin and streptomycin (Euroclone S.p.A., Italy) with a doubling time of 27 h.

The flasks containing asynchronous non-confluent cells were gently rinsed with 10 mL of calcium and magnesium-free D-PBS (GIBCO®, Life Technologies, Waltham, MA, USA), and then detached using 1 mL of 1:1 *v/v* solution of 0.25% trypsin and 1×10^{-3} M EDTA. Trypsin was neutralized using a few mL of the fresh culture medium, and the cell solution was counted (using a Coulter Counter Z2 serie, Beckmann, Kristiansand, Norway). The T98G cells employed for the experiments were seeded on T-25 flasks at a concentration of 8×10^3 cells/cm^2 at 48 h before the experiment. At 24 h before the experiment, the cell culture medium was removed and replaced with a fresh medium or a medium containing AuNRs, at two different concentrations (0.1 µg/mL and 0.5 µg/mL), and placed in the incubator. On the day of the experiment, the cell culture medium was removed, and every flask was replaced with the fresh medium 1 h before irradiation. Some of the flasks were irradiated with doses of 1 Gy and 4 Gy. Then, the cells were detached from all flasks, counted and, through a series of successive dilutions, seeded in appropriate numbers in 4 Petri dishes at a final volume of 5 mL, as shown in Table 2.

Table 2. Experimental conditions utilized for the colony-forming assay.

Sample	Dose (Gy)	AuNRs (μg/mL)	Cells Plated
CN	-	-	400
0.1 μg/mL	-	0.1	400
0.5 μg/mL	-	0.5	400
1 Gy	1	-	400
4 Gy	4	-	1000
1a	1	0.1	900
2a	1	0.5	2000
3a	4	0.1	3000
4a	4	0.5	4000

Legend: CN: control, i.e., cells not treated; 0.1 μg/mL and 0.5 μg/mL: the AuNR concentration tested; 1 Gy and 4 Gy: the radiation dose used; 1a–4a: the combination treatment.

2.6. Irradiation

Irradiation with gamma rays (E = 0.662 MeV) was performed at the Istituto Superiore di Sanita' (ISS, Rome, Italy) with doses of 1 Gy and 4 Gy at a dose rate of 0.6 Gy/min using a ^{137}Cs source (Gammacell 40, Nordion Inc., Ottawa, ON, Canada). The doses were selected based on previous data [24]. All irradiations were performed at room temperature.

2.7. Colony-Forming Assay

To study the cytotoxic effect of AuNRs on T98G cells (in terms of reproductive cell death), the colony-forming assay developed by Puck and Marcus was used [28]. The assay allows the assessment of classical clonogenic survival. Briefly, after the treatment (i.e., with pristine AuNRs, upon γ-ray irradiation only and in combination), the cells were trypsinized, counted and plated into four 6 cm Petri dishes at the appropriate concentration to score the number of colonies ranging from 300 to 600 for each dose, as shown in Table 2. After about 12 days of growth at 37 °C under 5% CO_2 and 95% humidity, the colonies were stained with crystal violet (Figure S1 in Supplementary Materials). Colonies exceeding 50 cells were scored manually and represented surviving cells. The average colony count for the four Petri dishes was used to calculate plating efficiency (PE), which was defined as the number of colonies counted/number of cells plated. For each experiment, cell-surviving fractions (SF) were calculated as the ratio between the PE measured in the investigated sample and the measured PE of the corresponding control. All significance was calculated using Student's-*t*-test.

3. Results

3.1. AuNR Synthesis and Loading Studies

The choice of AuNRs as a drug carrier is strategic because it allows us to combine the high stability and biocompatibility of gold with the anisotropic form, which allows a versatile engineering possibility [6]. Furthermore, AuNRs possess specific plasmonic properties, presenting two plasmonic peaks in the visible and/or near-infrared spectral range.

In this work, AuNR synthesis was performed in accordance with recent literature [29] and consisted of two steps. In the first step, Au^{3+} was reduced by $NaBH_4$ in the presence of CTAB, and a brownish-yellow solution, i.e., the seed solution, was obtained. In the second step, which concerned the growth of nanorods, the reduction of gold by ascorbic acid was initiated, and this allowed the conversion of gold from Au^{3+} to Au^{1+} and then to Au^0. The AuNR spectrum in water (Figure 1a) shows the two typical plasmon bands at 520–680 nm, confirming the nanodimension of the material. FESEM EDX studies were performed, and average sizes of 39 ± 5 nm and 11 ± 2 nm and the respective A.R. = 3.2 were observed (Figure 1b).

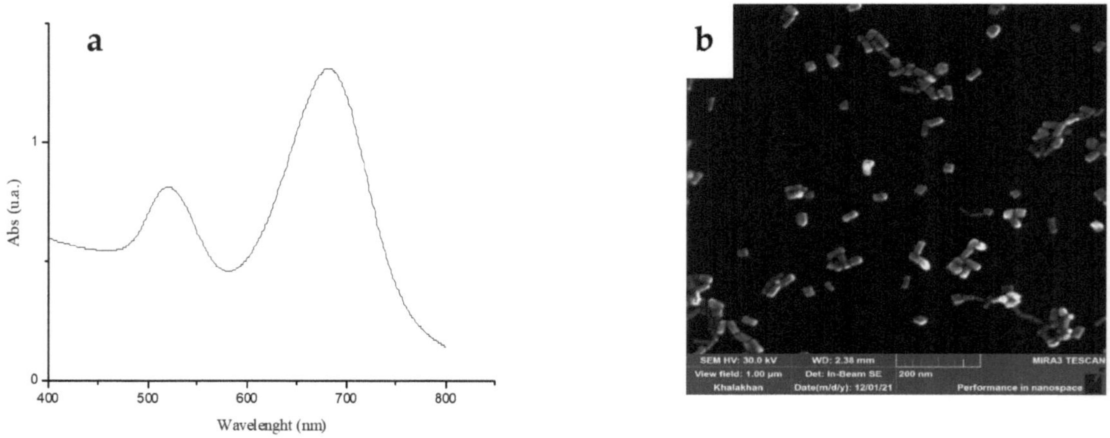

Figure 1. AuNRs: (**a**) visible spectrum in water suspension, with λ max = 520–680 nm, and (**b**) FESEM-EDX image (scale bar 200 nm), with rods having average dimensions of 39 ± 5 × 11 ± 2 nm (A.R. = 3.2).

Moreover, DLS studies were performed in water and in the cell culture medium RPMI1640. These data, presented in Table S1 in Supplementary Materials, show high values of PDI mainly due to the non-spherical shape of the particles, which leads to its incorrect determination and an overestimation of the hydrodynamic diameter. The differences observed between the AuNR characterization data obtained via the DLS and SEM analyses are related to the shape of nanoparticles and the different physical phenomena used to determine size distribution. SEM measures the true diameter, while DLS measures hydrodynamic diameter, leading to misleading results for non-spherical particles [30]. Both AuNRs and AuNRs-99Tc showed negative surface charge, and when AuNRs were suspended in the cell culture medium, the Zeta-potential values increased, suggesting adsorption of the protein corona. The AuNRs' stability was checked over time using UV-vis and DLS studies up to one month after their synthesis, confirming their dimension and polydispersity. For evaluating drug loading on the AuNR surface, a calibration curve was performed ($y = 55.409x$ and $R^2 = 0.999$, as reported in Figure S2 in Supplementary Materials) using different concentrations of the drug [31]. The protocol for loading ^{99}Tc-sestaMIBI onto the rods' surface was investigated via the simple contact between the radiopharmaceutical solution and the suspension of AuNRs for 24 h under stirring at room temperature. The obtained loading efficiency was η (%) = 5 ± 2%, which corresponds to 0.0033 mg of the drug for 1 mg of AuNRs.

3.2. Synchrotron Radiation-Induced X-ray Photoelectron Spectral (SR-XPS) Studies

SR-XPS measurements were performed on AuNRs-99Tc in order to ascertain the successful loading of the radiopharmaceutical onto AuNRs and to obtain a better insight into the stability of the ^{99}Tc molecular structure upon conjugation to the surface of AuNRs. For comparison, SR-XPS measurements were also carried out on ^{99}Tc. The SR-XPS spectra were acquired at the C1s, N1s and Tc3d core levels and, for the ^{99}Tc-sestaMIBI-AuNRs sample, at the Au4f core level as well. Complete SR-XPS data analysis results (binding energy (BE), full width-half maximum (FWHM), and atomic percentages and proposed assignments for all measured signal components) are summarized in Table S2 in Supplementary Materials. All signals appeared composite, and by applying a peak fitting procedure, several spectral components were individuated and assigned, based on a comparison with the literature [32], to the specific elements in the respective chemical groups. As shown in Figure 2, the measured spectra for the C1s and N1s core levels confirm the stability of the ^{99}Tc molecular structure, and for the AuNRs-99Tc system, the presence of CTAB, as expected by the nanorods' chemical composition.

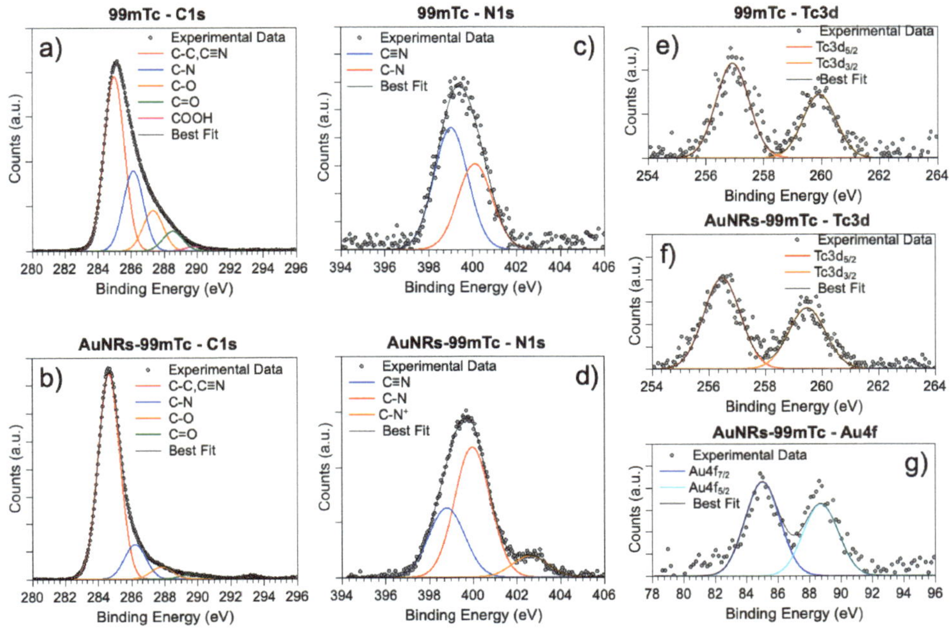

Figure 2. SR-XPS spectra: C1s (**a**,**b**), N1s (**c**,**d**) and Tc3d (**e**,**f**) of decayed and non-radioactive radiopharmaceutical ^{99}Tc and its conjugate on gold rods, AuNRs-99Tc. SR-XPS Au4f spectrum of AuNRs-99Tc (**g**).

In more detail, the intensity increment observed for the C1s component at 285.0 eV of BE (C-C) in AuNRs-99Tc, with respect to the analogous signal in pristine 99Tc, is associated with the aliphatic tail of CTAB (Figure 2a,b); an analogous effect is observed for the N1s spectral features associated with amine-like N (Figure 2c,d) with respect to the peak component assigned to the C≡N groups. On the other hand, the intensity of the third peak in the C1s spectra (287.3 eV for BE), arising due to the C-O groups of the -OMe moieties of ^{99}Tc, decreases in the AuNRs-99Tc sample, as expected. The C1s peaks at higher B.E. assigned to C=O (about 288 eV) and COOH (290 eV) reveal the presence of impurities, which are always observed in the samples prepared in air via deposition from aqueous solutions. To further prove the effectiveness of the ^{99m}Tc-sestaMIBI conjugation to the AuNR surface, Au4f and Tc3d signals were analyzed. For Au4f (Figure 2g), a single spin–orbit pair is observed, with the main Au4f$_{7/2}$ component being centered at 85.04 eV, as expected for gold atoms at the AuNRs surface interacting with ligands [33]. Finally, the reproducibility of Tc3d spectral position and shape [34,35] fully confirms the effectiveness of the loading process (Figure 2e,f).

3.3. Biological Studies

To evaluate the cytotoxic effect of AuNRs, alone or in combination with gamma ray (γ-ray) irradiation, the cellular reproductive death in T98G cells was studied using the colony-forming assay. Figure 3a shows that the SF of the samples treated only with AuNRs, in the concentration range of 0.1–1 µg/mL, decreases as the concentration increases, as expected based on a comparison with the literature [36–39]. For subsequent experiments with the combined treatment (i.e., AuNRs and γ-ray irradiation), we arbitrarily chose 0.1 µg/mL and 0.5 µg/mL because the former shows no toxicity whereas the latter reduces SF by approximately 50%. The results are shown in Figure 3b, along with the results of the samples treated with γ-rays only. In particular, the latter are consistent with the literature data [40–43].

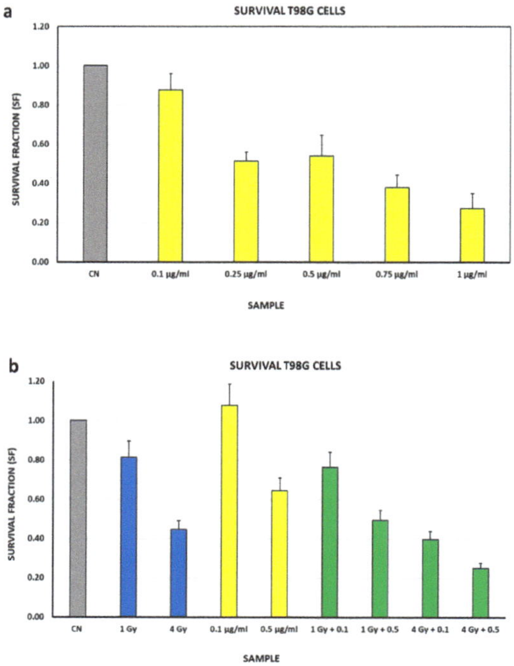

Figure 3. Bar plot for the results of the SF obtained from the colony-forming assay in glioblastoma T98G cells. Panel (**a**) shows the SF after treatment with different AuNR concentrations (yellow histograms). Panel (**b**) shows the SF after different treatments: blue histograms (only γ-rays), yellow histograms (only AuNRs), and green histograms (γ-rays + AuNrs). All results are significant compared to the control with a p value of <0.01, except for 0.1 µg/mL in panel (**b**). Significance was calculated using Student's t-test. The error bar represents the standard error of the mean (SEM) obtained from at least 3 independent experiments for each condition used.

In this analysis, all samples were normalized to the same control, (untreated cells—CN). Overall, it can be seen that the SF decreases as the dose and concentration of AuNRs increase (Figure 3b). In the samples with the combined treatment (i.e., AuNRs and γ-ray irradiation), the SF decreases at the same dose as the concentration of AuNRs increases, with an overall trend that appears to be linear. The decrease observed in the SF of the samples tested with the combined treatment can be due to three factors: the damage induced by γ-ray irradiation, the cytotoxicity of AuNRs, and the radiation damage due to the emission of Auger electrons from the irradiated gold.

To better understand which of the three factors most strongly affected cell survival in these samples, a more detailed analysis was conducted. The SF was calculated by considering the samples treated only with γ-rays as the control first (Figure 4a), and considering the samples treated only with AuNRs as the control afterward (Figure 4b).

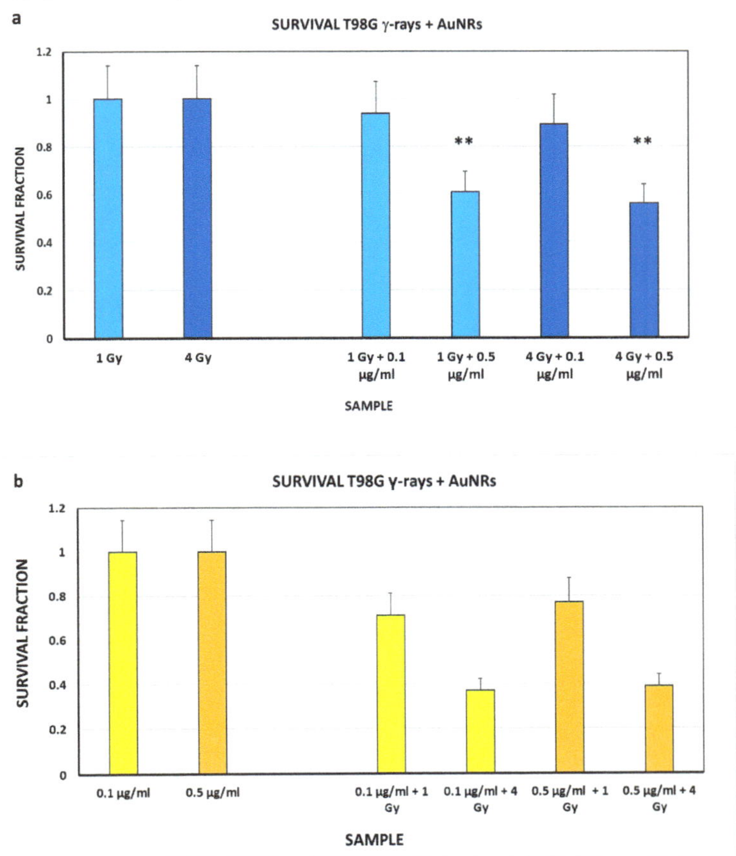

Figure 4. Bar plot of the SF obtained after normalizing to the dose used (**a**) or normalizing to the AuNR concentration (**b**). Significant difference with respect to the control was calculated using an ANOVA test (** $p < 0.01$). The error bar represents the propagation of the error.

In the former case (panel a), a similar decrease in the SF is observed for both doses as the AuNR concentration increases, but the decrease is significant with respect to the control only for the concentration of 0.5 µg/mL. The obtained data show that the observed effect is attributable to the presence of AuNRs and Auger electron emission from the irradiated gold.

In the latter case (panel b), the obtained data show that the SF decreases as the radiation dose increases, which is significant for all conditions with respect to the control. Furthermore, a significant decrease in SF is observed for the sample treated with 0.1 µg/mL or 0.5 µg/mL of AuNRs and irradiated at 4 Gy compared to the sample treated with 0.1 µg/mL or 0.5 µg/mL AuNRs and irradiated at 1 Gy. In this case, by normalizing the data to the AuNR concentrations used, the obtained results show that the decrease in SF is attributable to gamma rays and Auger electron emission from the irradiated gold.

The contribution of Auger electrons could be calculated as a first approximation by subtracting the SF values of the samples in panel b from the SF values of the samples treated with gamma rays only (see Figure 3b). The results are shown in Table 3.

Table 3. Contributions of the various components normalized to the radiation dose used.

Sample	SF (see Figure 3b)	Sample	SF (see Figure 4b)	Effect of Auger Electrons
CN	1 ± 0.14	0.1 μg/mL 0.5 μg/mL	1 ± 0.14 1 ± 0.14	- -
1 Gy	0.81 ± 0.14	0.1 μg/mL + 1 Gy 0.5 μg/mL + 1 Gy	0.71 ± 0.10 0.77 ± 0.11	0.11 ± 0.13 0.05 ± 0.14
4 Gy	0.45 ± 0.14	0.1 μg/mL + 4 Gy 0.5 μg/mL + 4 Gy	0.37 ± 0.05 0.39 ± 0.05	0.08 ± 0.07 0.06 ± 0.07

The obtained results seem to indicate that emission of Auger electrons from the irradiated gold of AuNRs occurs, although the energy of incident photons is only slightly higher than the suitable energy, and under our experimental conditions, it seems to be dose dependent rather than AuNR concentration dependent.

4. Conclusions

In this study, AuNRs were synthesized (A.R. = 3.2), fully characterized and conjugated with long-lived ^{99}Tc-sestaMIBI (STAMICIS®), a tumor-seeking radiopharmaceutical which is currently used in diagnostic imaging. ^{99}Tc was used in the non-radioactive form after complete decay. The obtained AuNRs were studied as a proof of concept from a radiobiological point of view to obtain an overall view of theragnostic action. In fact, the presence of the radiopharmaceutical used in diagnosis could induce the emission of Auger electrons from AuNRs, which has a therapeutic effect at the diagnostic site. This effect was explored as a proof of concept through the irradiation of AuNRs with an external γ source. The cell killing tests were performed on T98G cells. It emerged that in the case of AuNRs alone without radiation, the lowest concentration of 0.1 μg/mL was not toxic for the cells, while a concentration of 0.5 μg/mL showed some degree of toxicity. For the samples treated with both irradiation and AuNRs, the trend of SF appears to be linear: as the radiation dose and AuNR concentration increase, cell survival decreases. The effect of the induced AE emission on cell survival does not appear to be predominant. In fact, from our preliminary data, it appears to be the dose of gamma rays with which the cells are irradiated that contributes most significantly and in an independent manner from the concentration of AuNRs used. Further studies are needed to optimize the loading of this radiopharmaceutical onto AuNRs, and further biological tests, with and without irradiation, will be necessary to evaluate the mechanism and efficiency of action. However, these preliminary investigations are fundamental to verify and optimize the loading of a decayed radiopharmaceutical, thereby minimizing the dangers for operators and the costs, and, above all, they allow the optimization of the protocols under consideration, by mimicking the action of radionuclide on gold, in view of the loading of the radioactive drug, which would lead to the realization of the theragnostic system.

Supplementary Materials: The following supporting information can be downloaded at https://www.mdpi.com/article/10.3390/nano13131898/s1. Figure S1: Photo images of Petri dishes containing the colonies, which have been stained with crystal violet and are representative of the respective biological samples being analyzed; Figure S2: Calibration curve for ^{99}Tc-sestaMIBI in water (error bar showing the standard deviation is not appreciable); Table S1: Values of Z-Average, PDI and Zeta potential; Table S2: XPS data collected on the radiopharmaceutical before and after conjugation to AuNRs.

Author Contributions: Conceptualization, V.D. and I.V.; data curation and investigation, L.B., S.A., I.V. and V.D.; DLS data curation and formal analysis, B.D.B.; SR-XPS measurements and elaboration data, F.B., C.B. and G.I.; Supply of the nonradioactive radiopharmaceutical, T.S. and A.G.; Writing—original draft preparation, L.B., T.S., A.G., V.D. and I.V.; Writing—review and editing, I.F., A.C., I.V., V.D., L.B., C.B. and B.D.B. All authors have read and agreed to the published version of the manuscript.

Funding: This research received no external funding.

Data Availability Statement: Data supporting are in Supporting Materials.

Acknowledgments: The authors of department of Sciences of Rome Tre Unversity thank the Grant of Excellence Departments, MIUR (ARTICOLO 1, COMMI 314–337 LEGGE 232/2016) and the Rome Technopole Project CUP: F83B22000040006.

Conflicts of Interest: The authors declare no conflict of interest.

References

1. Shah, M.A.; Pirzada, B.M.; Price, G.; Shibiru, A.L.; Qurashi, A. Applications of nanotechnology in smart textile industry: A critical review. *J. Adv. Res.* **2022**, *38*, 55–75. [CrossRef]
2. Corsi, I.; Venditti, I.; Trotta, F.; Punta, C. Environmental safety of nanotechnologies: The eco-design of manufactured nanomaterials for environmental remediation. *Sci. Total Environ.* **2023**, *864*, 161181. [CrossRef]
3. Malik, S.; Muhammad, K.; Waheed, Y. Nanotechnology: A Revolution in Modern Industry. *Molecules* **2023**, *28*, 661. [CrossRef]
4. Jorfi, M.; Roberts, M.N.; Foster, E.J.; Weder, C. Physiologically Responsive, Mechanically Adaptive Bio-Nanocomposites for Biomedical Applications. *ACS Appl. Mater. Interfaces* **2013**, *5*, 1517–1526. [CrossRef] [PubMed]
5. Mabrouk, M.; Das, D.B.; Salem, Z.A.; Beherei, H.H. Nanomaterials for Biomedical Applications: Production, Characterisations, Recent Trends and Difficulties. *Molecules* **2021**, *26*, 1077. [CrossRef] [PubMed]
6. Venditti, I. Engineered gold-based nanomaterials: Morphologies and functionalities in biomedical applications. A mini review. *Bioengineering* **2019**, *6*, 53. [CrossRef] [PubMed]
7. Kus-Liśkiewicz, M.; Fickers, P.; Tahar, I.B. Biocompatibility and Cytotoxicity of Gold Nanoparticles: Recent Advances in Methodologies and Regulations. *Int. J. Mol. Sci.* **2021**, *22*, 10952. [CrossRef] [PubMed]
8. Fratoddi, I.; Venditti, I.; Battocchio, C.; Carlini, L.; Porchia, M.; Tisato, F.; Bondino, F.; Magnano, E.; Pellei, M.; Santini, C. Highly hydrophilic gold nanoparticles as carrier for anticancer copper(I) complexes: Loading and release studies for biomedical applications. *Nanomaterials* **2019**, *9*, 772. [CrossRef]
9. Ozcicek, I.; Aysit, N.; Cakici, C.; Aydeger, A. The effects of surface functionality and size of gold nanoparticles on neuronal toxicity, apoptosis, ROS production and cellular/suborgan biodistribution. *Mater. Sci. Eng. C* **2012**, *128*, 112308. [CrossRef]
10. Carnovale, C.; Bryant, G.; Shukla, R.; Bansal, V. Identifying Trends in Gold Nanoparticle Toxicity and Uptake: Size, Shape, Capping Ligand, and Biological Corona. *ACS Omega* **2019**, *4*, 242. [CrossRef]
11. Guo, J.; Armstrong, M.J.; O'Driscoll, C.M.; Holmes, J.D.; Rahme, K. Positively charged, surfactant-free gold nanoparticles for nucleic acid delivery. *RSC Adv.* **2015**, *5*, 17862. [CrossRef]
12. Sano, K.; Ishida, Y.; Tanaka, T.; Mizukami, T.; Nagayama, T.; Haratake, Y.; Munekane, M.; Yamasaki, T.; Mukai, T. Enhanced Delivery of Thermoresponsive Polymer-Based Medicine into Tumors by Using Heat Produced from Gold Nanorods Irradiated with Near-Infrared Light. *Cancers* **2021**, *13*, 5005. [CrossRef] [PubMed]
13. Maccora, D.; Dini, V.; Battocchio, C.; Fratoddi, I.; Cartoni, A.; Rotili, D.; Castagnola, M.; Faccini, R.; Bruno, I.; Scotognella, T.; et al. Gold nanoparticles and nanorods in nuclear medicine: A mini review. *Appl. Sci.* **2019**, *9*, 3232. [CrossRef]
14. Jelveh, S.; Chithrani, D.B. Gold Nanostructures as a Platform for Combinational Therapy in Future Cancer Therapeutics. *Cancers* **2011**, *3*, 1081–1110. [CrossRef] [PubMed]
15. Zhang, Z.; Wang, J.; Chen, C. Gold nanorods based platforms for light-mediated theranostics. *Theranostics* **2013**, *3*, 223–238. [CrossRef]
16. Zeng, H.; Du, X.; Singh, S.C.; Kulinich, S.A.; Yang, S.; He, J.; Cai, W. Nanomaterials via Laser Ablation/Irradiation in Liquid: A Review. *Adv. Funct. Mater.* **2012**, *22*, 1333–1353. [CrossRef]
17. Goddarda, Z.R.; Marín, M.J.; Russell, D.A.; Searcey, M. Active targeting of gold nanoparticles as cancer therapeutics. *Chem. Soc. Rev.* **2020**, *49*, 8774–8789. [CrossRef]
18. Nikoobakht, B.; El-Sayed, M.A. Preparation and Growth Mechanism of Gold Nanorods (NRs) Using Seed-Mediated Growth Method. *Chem. Mater.* **2003**, *15*, 1957–1962. [CrossRef]
19. Faure, G. *Principles of Isotope Geology*, 2nd ed.; John Wiley & Sons: Hoboken, NJ, USA, 1986; pp. 15–23.
20. González-Ruíz, A.; Ferro-Flores, G.; Jiménez-Mancilla, N.; Castellanos, A.E.; Ocampo-García, B.; Luna-Gutiérrez, M.; Santos-Cuevas, C.; Morales-Avila, E.; Isaac-Olivé, K. In vitro and in vivo synergistic effect of radiotherapy and plasmonic photothermal therapy on the viability of cancer cells using 177Lu–Au-NLS-RGD-Aptamer nanoparticles under laser irradiation. *J. Radioanal. Nucl. Chem.* **2018**, *318*, 1913–1921. [CrossRef]
21. Pijeira, M.S.O.; Viltres, H.; Kozempel, J.; Sakmár, M.; Vlk, M.; İlem-Özdemir, D.; Ekinci, M.; Srinivasan, S.; Rajabzadeh, A.R.; Ricci-Junior, E.; et al. Radiolabeled nanomaterials for biomedical applications: Radiopharmacy in the era of nanotechnology. *EJNMMI Radiopharm. Chem.* **2022**, *7*, 8. [CrossRef]
22. Saleh, T.B. *Basic Sciences of Nuclear Medicine*; Khalil, M., Ed.; Springer: Berlin/Heidelberg, Germany, 2010.
23. Burdak-Rothkamm, S.; Prise, K.M. New molecular targets in radiotherapy: DNA damage signalling and repair in targeted and non-targeted cells. *Eur. J. Pharmacol.* **2009**, *625*, 151–155. [CrossRef] [PubMed]

24. Freudenberg, R.; Runge, R.; Maucksch, U.; Berger, V.; Kotzerke, J. On the dose calculation at the cellular level and its implicatior for the RBE of 99mTc and 123I. *Med. Phys.* **2014**, *41*, 062503. [CrossRef]
25. Tavares, A.A.; Tavares, J.M. (99m)Tc Auger electrons for targeted tumourtherapy: A review. *Int. J. Radiat. Biol.* **2010**, *86*, 261–27 [CrossRef] [PubMed]
26. Ku, A.; Facca, V.J.; Cai, Z.; Reilly, R.M. Auger electrons for cancer therapy—A review. *EJNMMI Radiopharm. Chem.* **2019**, *4*, 2 [CrossRef] [PubMed]
27. Choppin, G.R.; Liljenzin, J.-O.; Rydberg, J. Absorption of Nuclear Radiation. *Radiochem. Nucl. Chem.* **2002**, *1*, 123–165.
28. Puck, T.T.; Marcus, P.I. Action of X-rays on mammalian cells. *J. Exp. Med.* **1956**, *103*, 653–666. [CrossRef]
29. Moreau, L.; Jones, M.; Roth, E.; Wu, J.; Kewalramani, S.; O'Brien, M.; Chen, B.-R.; Mirkin, C.A.; Bedzyk, M. The role of trace Ag i the synthesis of Au nanorods. *Nanoscale* **2019**, *11*, 11744–11754. [CrossRef]
30. Calzolai, L.; Gilliland, D.; Rossi, F. Measuring nanoparticles size distribution in food and consumer products: A review. *Foo Addit. Contam.* **2012**, *29*, 1183–1193. [CrossRef]
31. Venditti, I.; Iucci, G.; Fratoddi, I.; Cipolletti, M.; Montalesi, E.; Marino, M.; Secchi, V.; Battocchio, C. Directly Resveratro immobilization on hydrophilic charged gold nanoparticles: Structural investigations and cytotoxic studies. *Nanomaterials* **202** *10*, 1898. [CrossRef]
32. Venditti, I.; Cartoni, A.; Fontana, L.; Testa, G.; Scaramuzzo, F.A.; Faccini, R.; Terracciano, C.M.; Camillocci, E.S.; Morganti, S Giordano, A.; et al. Y3+ embedded in polymeric nanoparticles: Morphology, dimension and stability of composite colloida system. *Colloids Surf. A Physicochem. Eng. Asp.* **2017**, *532*, 125–131. [CrossRef]
33. *NIST X-ray Photoelectron Spectroscopy Database*; Version 4.1; National Institute of Standards and Technology: Gaithersburg, MI USA, 2012. Available online: http://srdata.nist.gov/xps/ (accessed on 1 April 2023).
34. Venditti, I.; Cartoni, A.; Cerra, S.; Fioravanti, R.; Salamone, T.A.; Sciubba, F.; Tabocchini, M.A.; Dini, V.; Battocchio, C.; Iucc G.; et al. Hydrophilic Gold Nanoparticles as anti-PD-L1 Antibody carriers: Synthesis and Interface Properties. *Part. Part. Sys Charact.* **2022**, *39*, 2100282. [CrossRef]
35. Stumpf, T.; Foerstendorf, H.; Bok, F.; Richter, A. *Annual Report 2018*; Institute of Resource Ecology: Dresden, Germany, 2019.
36. Vales, G.; Suhonen, S.; Siivola, K.M.; Savolainen, K.M.; Catalán, J.; Norppa, H. Genotoxicity and Cytotoxicity of Gold Nanoparticle In Vitro: Role of Surface Functionalization and Particle Size. *Nanomaterials* **2020**, *10*, 271. [CrossRef] [PubMed]
37. Abu-Dahab, R.; Mahmoud, N.N.; Abdallah, M.; Hamadneh, L.; Hikmat, S.; Zaza, R.; Abuarqoub, D.; Khalil, E.A. Cytotoxicity an Cellular Death Modality of Surface-Decorated Gold Nanorods against a Panel of Breast Cancer Cell Lines. *ACS Omega* **2021**, 15903. [CrossRef] [PubMed]
38. Xie, L.; Zhang, X.; Chu, C.; Dong, Y.; Zhang, T.; Li, X.; Liu, G.; Cai, W.; Han, S. Preparation, toxicity reduction and radiatio therapy application of gold nanorods. *J. Nanobiotechnol.* **2021**, *19*, 454. [CrossRef]
39. Xu, X.; Ding, Y.; Hadianamrei, R.; Lv, S.; You, R.; Pan, F.; Zhang, P.; Wang, N.; Zhao, X. Antimicrobial peptide functionalized gol nanorods combining near-infrared photothermal therapy for effective wound healing. *Colloids Surf. B Biointerfaces* **2022**, *22* 112887. [CrossRef] [PubMed]
40. Boglaienko, D.; Soltis, J.; Kukkadapu, R.; Du, Y.; Sweet, L.; Holfeltz, V.; Hall, G.; Buck, E.; Segre, C.; Emerson, H.; et al. Spontaneou redox continuum reveals sequestered technetium clusters and retarded mineral transformation of iron. *Commun. Chem.* **2020**, 87. [CrossRef]
41. Chew, M.T.; Bradley, D.A.; Suzuki, M.; Matsufuji, N.; Murakami, T.; Jones, B.; Nisbet, A. The radiobiological effects of He, C an Ne ions as a function of LET on various glioblastoma cell lines. *J. Radiat. Res.* **2019**, *60*, 178–188. [CrossRef]
42. Burdak-Rothkamm, S.; Smith, A.; Lobachevsky, P.; Martin, R.; Prise, K.M. Radioprotection of targeted and bystander cells b methylproamine. *Strahlenther. Onkol.* **2015**, *191*, 248–255. [CrossRef]
43. Short, S.; Mayes, C.; Woodcock, M.; Johns, H.; Joiner, M.C. Low dose hypersensitivity in the T98G human glioblastoma cell lin *Int. J. Radiat. Biol.* **1999**, *75*, 847–855. [CrossRef]

nanomaterials

MDPI

Article

Scanning Tunneling Microscopy Study of Lipoic Acid, Mannose, and cRGD@AuNPs Conjugates

Andrés Rodríguez-Galván [1,*], Mitzi Reyes [1], Marisol Ávila-Cruz [1], Margarita Rivera [2] and Vladimir A. Basiuk [3]

[1] Carrera de Biología, Unidad de Biomedicina, Facultad de Estudios Superiores Iztacala, Universidad Nacional Autónoma de México, Tlalnepantla 54090, Mexico; 316225425@iztacala.unam.mx (M.R.); biomar14sol@gmail.com (M.Á.-C.)

[2] Instituto de Física, Departamento de Materia Condensada, Universidad Nacional Autónoma de México, Coyoacán, Ciudad de México 04510, Mexico; mrivera@fisica.unam.mx

[3] Instituto de Ciencias Nucleares, Universidad Nacional Autónoma de México, Circuito Exterior C.U., Ciudad de México 04510, Mexico; basiuk@nucleares.unam.mx

* Correspondence: andres.rodriguez@comunidad.unam.mx

Abstract: The functionalization of AuNPs with different biological elements was achieved to investigate their possibility in biomedical applications such as drug delivery, vaccine development, sensing, and imaging. Biofunctionalized AuNPs are pursued for applications such as drug delivery, vaccine development, sensing, and imaging. In this study, AuNPs with diameters of 20 nm were functionalized with lipoic acid, mannose, or the cRGD peptide. By using UV-vis spectroscopy, Fourier transform infrared spectroscopy, dynamic light scattering, transmission electron microscopy, and scanning tunneling microscopy techniques, we showed that AuNPs can be functionalized by these biomolecules in a reliable way to obtain conjugates to explore potential biomedical applications. In particular, we demonstrate that the STM technique can be employed to analyze biofunctionalized AuNPs, and the obtained information can be valuable in the design of biomedical applications.

Keywords: biofunctionalization; gold nanoparticles; lipoic acid; mannose; RGD cyclic peptide; scanning tunneling microscopy

1. Introduction

Gold nanoparticles (AuNPs) are widely studied due to their interesting optical, thermal, and chemical properties that can be implemented for advanced biomedical applications [1]. In particular, they are promising platforms for molecular imaging, biomarker sensors, drug delivery, diagnostic, therapy (photon-induced), and theragnostics, among others. Several reports have shown the binding of peptides [2], proteins [3], nucleic acids [4], or carbohydrates [5] to AuNPs via ligand exchange, taking advantage of the strong thiol–gold interaction. The binding of biomolecules to AuNPs can improve their stability in physiological media and biodistribution and allows the design of conjugates with active targeting capability, which emerge due to the high specificity of biomolecules by specific biomarkers. These ligands help nanoparticles to target specific cells or tissues, improving the accumulation and biodistribution of loads (active drugs or contrast agents) compared with free loads [2,6].

Recently, the attachment of biomolecules on AuNP surfaces has been extensively studied, and several successful protocols have been reported [7]. In contrast, there is an open challenge over the orientation and assembly of the attached biomolecules. This is an important issue since the orientation, form (structure), and assembly of ligands on AuNPs can impact the biological activity and binding affinity of biomolecules; for example, the antigen-recognition Fv regions of immunoglobulin G should be correctly oriented on surfaces for effective antigen recognition [8]. Also, it is well known that the assembly of carbohydrates on cell surfaces can affect their interaction with proteins;

e.g., when carbohydrates are assembled into small clusters, their binding to proteins called lectins is more efficient than when they are isolated [9], because the number of simultaneous interactions formed between clusters of carbohydrates and recognition proteins are multiple and enhance the binding affinity compared with individual carbohydrates [9]. Liese and Netz [10] reported that the interaction of synthetic systems is most efficient when the size of the clusters of recognition of biomolecules is similar compared to the size of their receptor, and, also, that an excess of ligands does not improve the selectivity, i.e., there is not a linear correlation between the interaction and the number of ligands [11].

The direct visualization of the orientation and assembly of biological ligands on the surfaces of AuNPs can help to understand and tailor their biological behavior. Several techniques have been employed for the characterization of thiolated bounded ligands on AuNPs, such as inductively coupled plasma mass spectrometry (ICP–MS), that has been used for the quantification of ligands [12], and X-ray photoelectron spectroscopy (XPS), which has been used for density packing analysis [13], and the adsorption of thiol molecules has been extensively studied using UV–visible spectroscopy (UV–vis) and in silico studies [14]. In particular, to elucidate the assembly of thiolated molecules, electron spin resonance (EPR), small-angle neutron scattering (SANS), mass spectrometry (MS), and nuclear magnetic resonance (NMR) have been employed [15]. However, for the direct visualization of biomolecules on AuNPs, scanning tunneling microscopy (STM) represents a viable alternative; for example, it has been widely employed for the analysis of biomolecules on metallic surfaces and the surfaces of nanomaterials. In particular, STM has been employed in the analysis of thiolate molecules on AuNPs and Ag nanoclusters, showing that these molecules can form patterns on AuNPs [15,16]. Also, many biomolecules such as lipids, carbohydrates, proteins, and nucleic acids have been imaged near molecular resolution on flat surfaces of Cu, Au, and highly ordered pyrolytic graphite (HOPG) [17]. In previous works, we employed STM to obtain information of individual organic molecules and proteins deposited on HOPG and the surfaces of carbon nanomaterials [18–20]. Also, conventional techniques employed in the analysis of biomolecules, such as circular dichroism (CD) or X-ray crystallography, present several limitations for the rutinary analysis of biofunctionalized nanoparticles, for example, the necessity of high concentrations of conjugates or the compulsory requirement of diffracting crystals [17].

By considering the importance of the orientation and assembly of biomolecules on surfaces of AuNPs, in this work, we analyzed the surfaces of functionalized AuNPs with lipoic acid (ALA@AuNPs), mannose (MAN@AuNPs), and the Arg-Gly-Asp cyclic peptide (cRGD@AuNPs). These biomolecules were selected since they might exhibit potential biomedical applications: ALA is an antioxidant biomolecule that has been used as a linker molecule for the attachment of drugs and proteins to AuNPs [21,22]; MAN is a carbohydrate that has been employed in the design of contrast agents for molecular imaging in cancer [5]; and cRGD has been widely employed for the active targeting of cancer cells [23]. The resulting information from this study will be of benefit in the development of biomodified nanomaterials and their potential applications.

2. Materials and Methods

2.1. Materials

AuNPs of 20 nm diameter in citrate buffer (SC@AuNPs), lipoic acid (ALA), hydrochloric acid (HCl), sodium hydroxide (NaOH), and N-(3-dimethylaminopropyl)-N′-ethylcarbodiimide hydrochloride (EDC) were obtained from Sigma-Aldrich (Mexico City, Mexico). The cRGD peptide (Cyclo(Arg-Ala-Asp-D-Phe-Lys-cCys) was acquired from Peptides international (Louisville, KY, USA). 2-aminoethyl 2,3,4,6-tetra-O-acetyl-α-d-mannopyranoside hydrochloride (MAN) was obtained from Synthose Inc., (Concord, ON, Canada). Deionized water (18.2 MΩ cm) was used in all experiments.

2.2. ALA@AuNPs

Three milliliters (1 nM) of SC@AuNPs was washed three times at 11,500 rpm (30 min) and resuspended in deionized water (pH 11). Then, 2.5 mL (1 nM) of washed nanoparticles was incubated with 250 µL of ALA (10 mM) at constant stirring (3000 rpm, 48 h at room temperature). The excess of ALA was removed by centrifugation (11,500 rpm, 30 min) and then resuspended in deionized water at pH 7 and adjusted to 1 nM of AuNPs.

2.3. MAN@AuNPs

For the preparation of the MAN@AuNPs conjugates, 1 mL (1 nM) of ALA@AuNPs, 2 µL of MAN (100 mg/mL, in ethanol), and 10 µL of EDC (40 mM, pH 6.5) were mixed and incubated at constant stirring (3000 rpm, 48 h). Then, the conjugate was washed three times in deionized water at pH 11 and left in basic hydrolysis for 24 h.

2.4. cRGD@AuNPs

For the preparation of the cRGD@AuNPs conjugate, 1 mL of SC@AuNPs (1 nM, pH 7) was mixed with 2 µL of cRGD (100 mg/mL) and incubated at constant stirring (3000 rpm, 24 h). Then, cRGD@AuNPs were washed three times in deionized water and left in water at pH 7.

2.5. Characterization

UV-Vis spectra (300–700 nm) were recorded in a Beckman Coulter DU-530 (Life science) with quartz cuvettes (1 cm). The hydrodynamic diameter and Z–potential were recorded in a Zetasizer Nano ZSP analyzer (HeNe laser (633 nm, 5 mW), Malvern, Worcestershire, UK). Fourier transform infrared spectroscopy coupled to ATR analysis were recorded on a Perkin-Elmer Spectrum 100 spectrometer (Perkin Elmer, Akron, OH, USA). For this analysis, samples were deposited by drop casting on the ATR crystal. The microscopy analysis was performed in a JEM-2010F FASTEM JEOL coupled to a NORAN energy dispersive spectrophotometer (EDS) operating at 200 kV. For this analysis, the conjugates (SC-AuNPs, ALA@AuNPs, MAN@AuNPs, or cRGD@AuNPs) were prepared by depositing a drop of the samples onto carbon-coated copper grids (Ted Pella, Redding, CA, USA) and allowed to dry at room temperature overnight. The recorded images were analyzed using the software ImageJ version 1.54d (NIH, Wayne Rasband), and the statistical analysis was performed with the Origin Pro 2023b software (Northampton, MA, USA).

STM measurements were carried out in a NaioSTM system (Nanosurf AG, Liestal, Switzerland). The analysis was carried out at room temperature using mechanically cut Pt/Ir tips of 0.25 mm diameter (Nanosurf AG, Liestal, Switzerland). For the analyses, 50 µL of each sample was drop-casted onto a freshly cleaved HOPG substrate (5 × 5 × 2 mm, grade ZYB). The deposited samples were incubated for 30 min, washed with 2 mL of deionized water, and dried overnight in a vacuum. The STM images were analyzed by using the Nanotec Electronica WSxM 5.0 (Madrid, Spain) software [24].

3. Results

The UV–Vis spectroscopy results showed that SC@AuNPs and biofunctionalized samples ALA@AuNPs, MAN@AuNPs, and cRGD@AuNPs exhibited a strong absorbance band in the visible region at 523 nm without broadening, suggesting that the conjugates were stable after functionalization. The DLS measurement showed small differences; the diameters of SC@AuNPs and ALA@AuNPs were similar, 31.7 ± 0.11 nm and 31.9 ± 0.13 nm, respectively. After biofunctionalization with MAN, the size increased to 37.7 ± 0.81 nm, suggesting it was due to the covalent attachment of MAN. The size of the cRGD@AuNPs decreased to 29.5 ± 0.37 nm, a small change compared to the starting gold nanoparticles with sodium citrate. The size analysis showed that the conjugates and SC@AuNPs were moderately polydispersed (PDI 0.3–0.4); in particular, ALA@AuNPs and MAN@AuNPs showed a wider size distribution in comparison with SC@AuNPs and cRGD@AuNPs. Z potential analysis showed that the conjugates increased their negativity after biofunctionalization as follows: cRGD@AuNPs > MAN@AuNPs > ALA@AuNPs > SC@AuNPs (Figure 1, right column). These results suggest that the biomolecules contribute to the increased negativity. ALA can be attached to gold nanoparticles via the dithiol ring, with the carboxylic acid exposed to solution; the covalent binding of MAN to ALA increased the number of electronegative atoms on the AuNPs surface, and the binding of cRGD to AuNPs also increased the number of electronegative atoms on the surface.

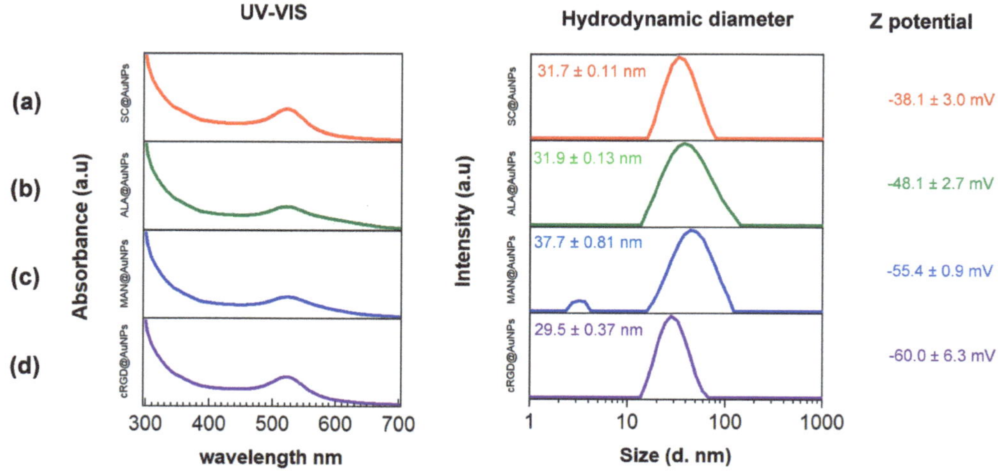

Figure 1. UV–Vis and DSL analysis of: (**a**) SC@AuNPs, (**b**) ALA@AuNPs, (**c**) MAN@AuNPs, and (**d**) cRGD@AuNPs. UV-VIS (left column), hydrodynamic diameter (middle column), and Z-potential (right column). All the measurements were conducted in deionized water at pH 11.0 by using NaOH (1 M). The color code of the surface coating is as follows: SC (red), ALA (green), MAN (blue), and cRGD (purple).

To confirm the presence of biomolecules on AuNPs, FT-IR experiments were performed. The SC@AuNPs (Figure 2a) showed a peak at 3300 cm^{-1} that was assigned to OH stretching, one peak at 2916 cm^{-1} that was assigned to CH_2, one peak at 1420 cm^{-1} assigned to carboxylic groups, and a peak at 1002 cm^{-1} assigned to CO; all these groups are present in sodium citrate [5]. The ALA@AuNPs (Figure 2b) showed one peak at 1730 cm^{-1} assigned to C=O double bond stretching, a peak at 2850 cm^{-1} assigned to CH_2 symmetric stretching, and a peak at 2916 cm^{-1} assigned to CH_2 asymmetric stretching; these peaks suggested the presence of ALA molecules [5]. After the covalent bond of MAN to ALA molecules attached to AuNPs, MAN@AuNPs (Figure 2c) showed a spectrum with three peaks at 1140, 1045, and 1082 cm^{-1} that suggested the presence of pyranose rings; meanwhile, the peak at 1738 cm^{-1} was assigned to the asymmetric vibration of C=O [25]. The peak at 1631 cm^{-1} was assigned to C=O and C-N stretching. Meanwhile, the peak at 1567 cm^{-1} suggested in-plane N-H bending and C-N stretching, and the C–O–C asymmetric frequency was placed between 1368 and 1045 cm^{-1}. Finally, the peak at 1221 cm^{-1} was assigned to C-N stretching, and the peak at 1045 cm^{-1} was related to C-O stretching. The cRGD@AuNPs showed a peak at 3295 that was assigned to the N-H stretching of Amide A, one peak at 1673 cm^{-1} was assigned to the stretching band of C=O of Amide I, and the peak at 1563 cm^{-1} was assigned to the C-N stretching and N-H bending of amide II; meanwhile, the peak at 777 cm^{-1} was assigned to the out-of-plane N-H bending. As a whole, the FTIR analysis suggested the presence of each one of the biomolecules employed for biofunctionalization [26].

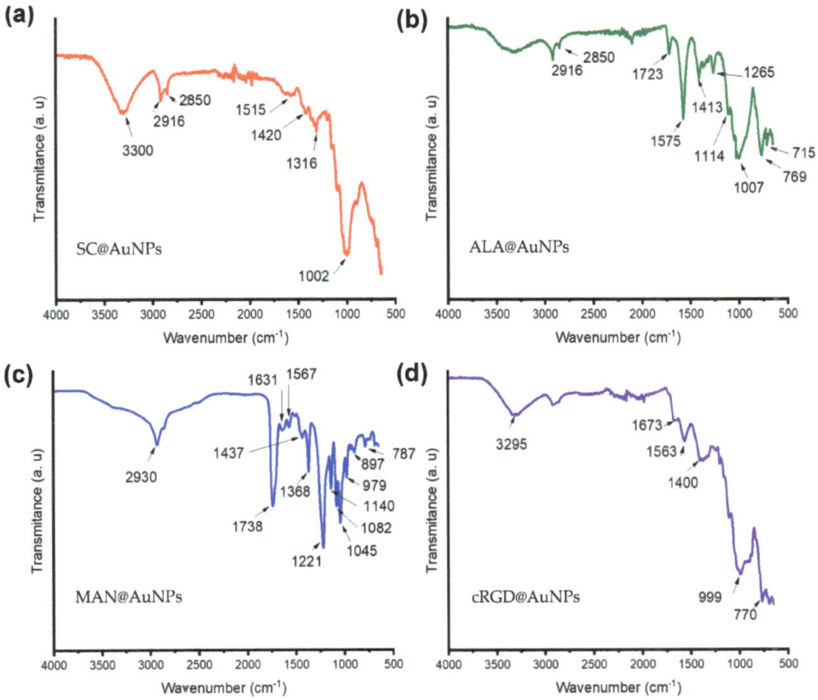

Figure 2. Attenuated total reflection Fourier transform infrared (ATR-FTIR) spectra of (**a**) SC@AuNPs, (**b**) ALA@AuNPs, (**c**) MAN@AuNPs, and (**d**) cRGD@AuNPs. The color code of the surface coating is as follows: SC (red), ALA (green), MAN (blue), and cRGD (purple).

Transmission electron microscopy images of the biofunctionalized nanoparticles are shown in Figure 3. All the conjugates were spherical, and no changes in size and morphology occurred after biofunctionalization. The histograms of size distribution are shown in Figure 3e. However, there was the presence of a low-electron-density halo around ALA@AuNPs, MAN@AuNPs, and RGDc@AuNPs (see Figure 3b–d). This material was associated with the presence of ALA, MAN, and cRGD on the surface of the AuNPs. The size of the coat for ALA was near 1.07 nm, for MAN near 3.72 nm, and for cRGD 4.8 nm. The elemental analysis of the conjugates confirmed the presence of gold as well as N and S, and the gold was related to AuNPs; meanwhile, the other elements were related to ALA, MAN, and cRGD.

Although TEM imaging suggested the adsorption of ALA, MAN, and cRGD molecules onto the surfaces of the AuNPs, it did not allow us to resolve finer structural details. For this purpose, the STM technique was explored to analyze the conjugates. We used HOPG as a substrate for the analysis since it may have advantages over metallic substrates. It was reported that the thiolated protecting coat of gold nanoparticles can redistribute on the Au(111) surface in minutes and form islands [27]. Also, HOPG consists of high-quality defect-free graphene layers that provide clean and atomically flat surfaces suitable for material depositions, it is easily cleaved [17], and the defects are easy to recognize and differentiate from gold nanoparticles.

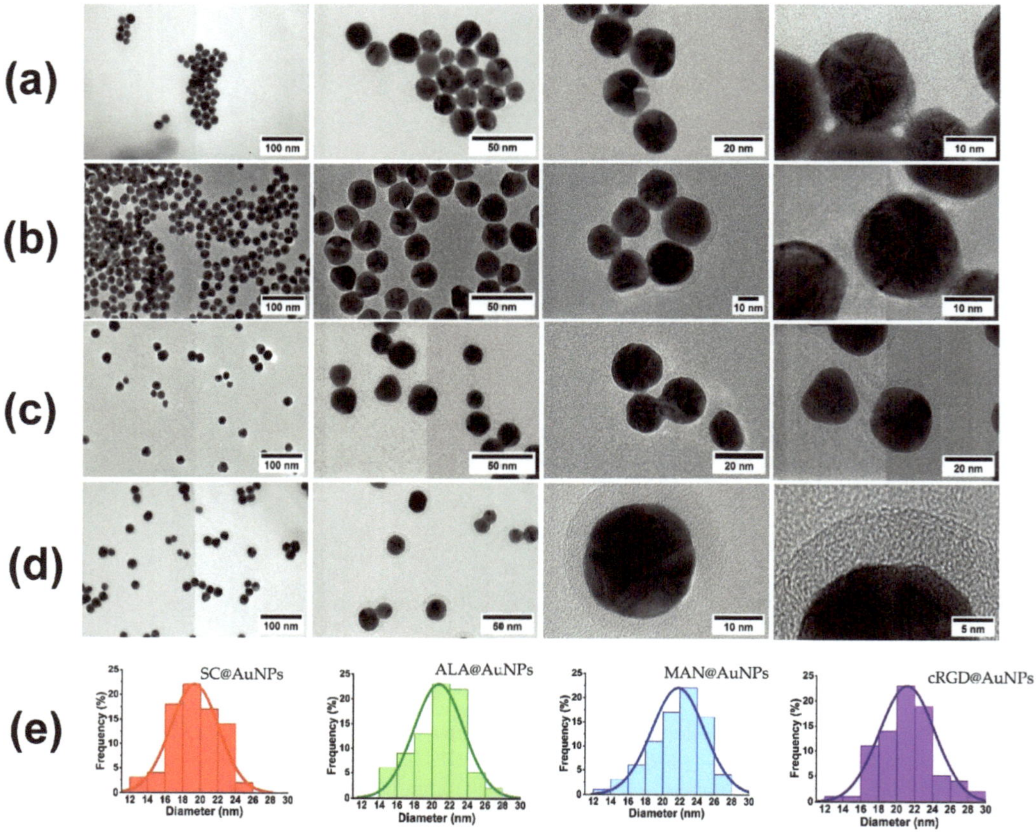

Figure 3. Sequences of transmission electron microscopy images and size histogram distribution for (**a**) SC@AuNPs, (**b**) ALA@AuNPs, (**c**) MAN@AuNPs, and (**d**) cRGD@AuNPs. (**e**) Histograms are indicated by the color code of the surface coating as follows: SC (red), ALA (green), MAN (blue), and cRGD (purple).

The SC@AuNPs showed large agglomerates of densely packed particles, in addition to sparse individual nanoparticles. Before STM analysis, these particles were washed several times, and throughout the process, the stabilizing citrate was washed off, decreasing the repulsive forces which maintain the stability of gold nanoparticles. Cross-sectional analysis of these samples showed diameters between 18 and 22 nm (Figure 4c,f,i) and 3 and 5 nm high. A smooth surface was observed after analyzing the conjugates at high resolution. The images of the ALA@AuNPs showed densely packed nanoparticles with evident borders between the nanoparticles as compared to SC@AuNPs. These particles are 25 to 28 nm long and 3 to 6 nm high (see Figure 5c,f). The analysis of individual nanoparticles also showed the presence of terraces. In addition, the cross-sectional analysis of particles under these conditions showed that the surface was not smooth (see Figure 5i). It is important to mention that ALA@AuNPs exhibited an ellipsoid shape that is characteristic of gold nanoparticles on flat surfaces, as seen with STM, which was not as evident for the SC@AuNPs. The images of MAN@AuNPs showed a pattern similar to ALA@AuNPs; however, it was more evident, even at high magnifications (Figure 6d). These particles showed a size between 20 and 25 nm long and 1 and 2 nm high. Finally, the cRGD@AuNPs showed the presence of a large number of individual nanoparticles with dimensions of 10 to 22 nm long and 3 to 8 nm high (Figure 7). The images showed the ellipsoid form despite the densely packed aggregation of the particles in contrast to the SC@AuNPs and MAN@AuNPs.

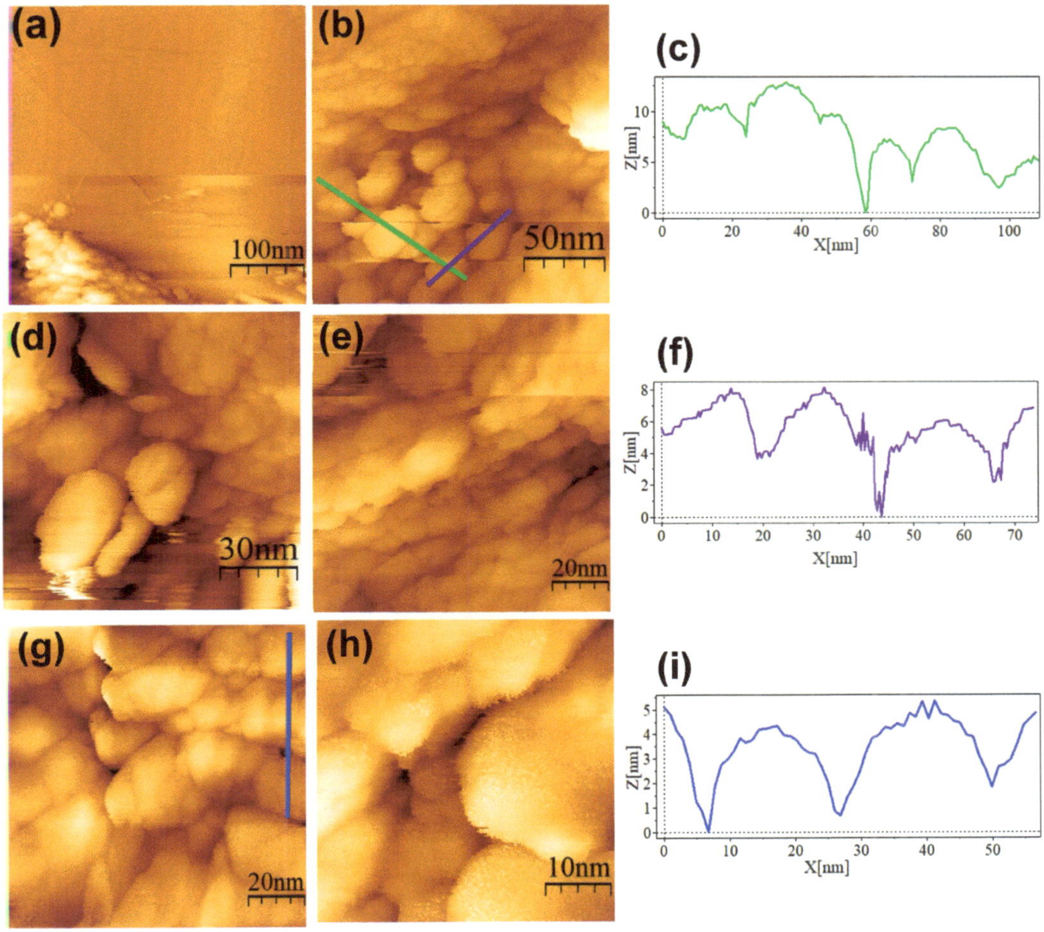

Figure 4. STM topographic images of SC@AuNPs deposited onto bare HOPG. STM imaging parameters: (**a**) (400×400 nm^2, V_s = 0.5 V, I = 1.0 nA, time/line: 0.382 s); (**b**) (181×181 nm^2, V_s = 0.5 V, I = 1.0 nA, time/line: 0.3 s); (**c**) profile corresponding to the blue line in (**b**); (**d**) (114×114 nm^2, V_s = 0.5 V, I = 1.0 nA, time/line: 0.3 s); (**e**) (103×103 nm^2, V_s = 0.5 V, I = 1.0 nA, time/line: 0.3 s); (**f**) profile corresponding to the green line in (**b**); (**g**) (103×103 nm^2, V_s = 0.5 V, I = 1.0 nA, time/line: 0.382 s); (**h**) (44.3×44.3 nm^2, V_s = 0.5 V, I = 1.0 nA, time/line: 0.3 s); (**i**) profile corresponding to the blue line in (**g**).

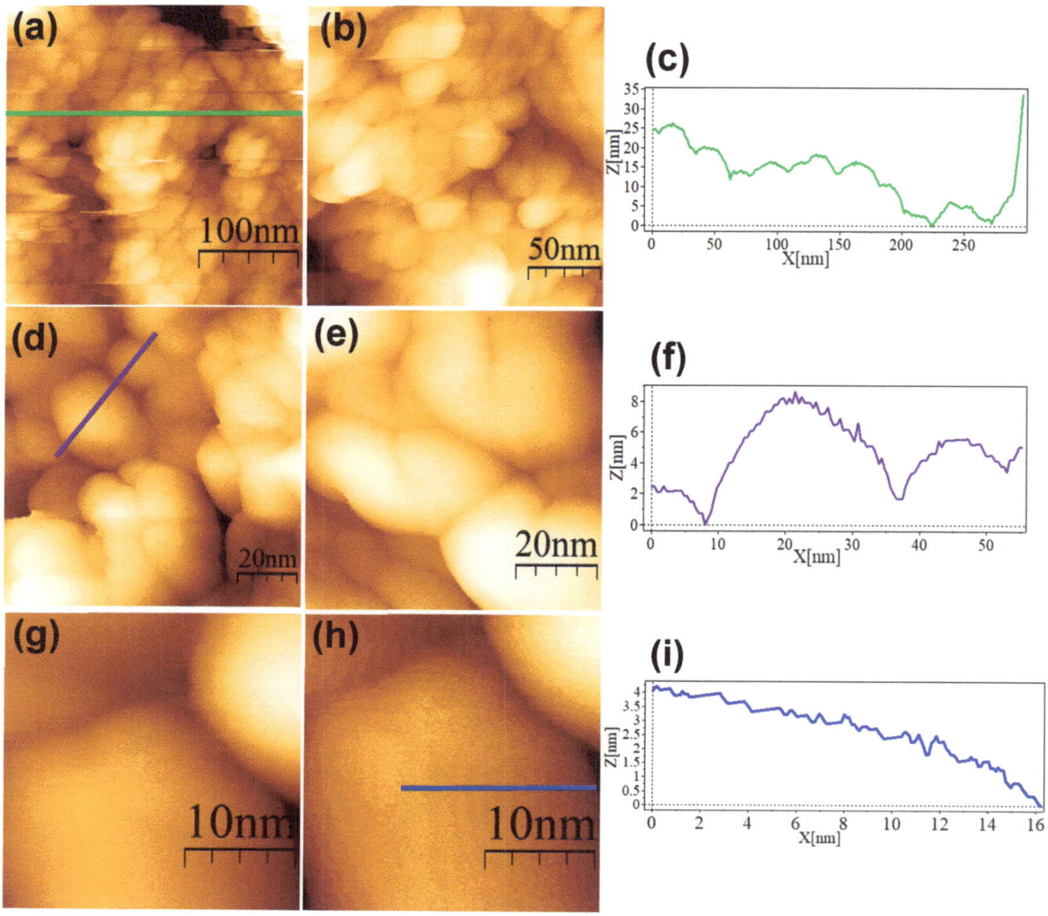

Figure 5. STM topographic images of ALA@AuNPs deposited onto bare HOPG. STM imaging parameters: (**a**) (300 × 300 nm^2, V_s = 0.5 V, I = 1.0 nA, time/line: 0.3 s); (**b**) (200 × 200 nm^2, V_s = 0.5 V, I = 1.0 nA, time/line: 0.3 s); (**c**) profile corresponding to the blue line in (**a**); (**d**) (97.3 × 97.3 nm^2, V_s = 0.5 V, I = 1.0 nA, time/line: 0.3 s); (**e**) (70.9 × 70.9 nm^2, V_s = 0.5 V, I = 1.0 nA, time/line: 0.3 s); (**f**) profile corresponding to the blue line in (**d**); (**g**) (26.6 × 26.6 nm^2, V_s = 0.5 V, I = 1.0 nA, time/line: 0.3 s); (**h**) (26.6 × 26.6 nm^2, V_s = 0.5 V, I = 1.0 nA, time/line: 0.3 s); (**i**) profile corresponding to the blue line in (**h**).

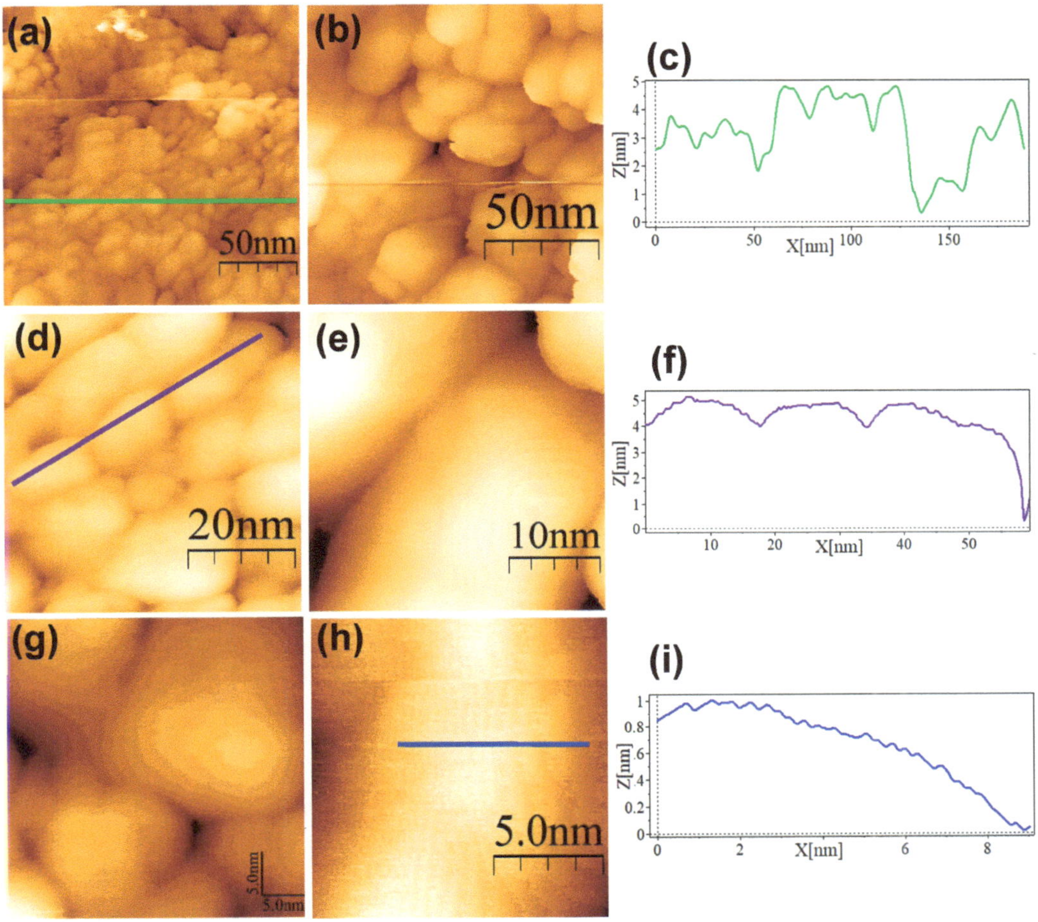

Figure 6. STM topographic images of MAN@AuNPs deposited onto bare HOPG. STM imaging parameters: (**a**) (189 × 189 nm^2, V_s = 0.5 V, I = 1.0 nA, time/line: 0.365 s); (**b**) (128 × 128 nm^2, V_s = 0.5 V, I = 1.0 nA, time/line: 0.365 s); (c) profile corresponding to the blue line in (**a**); (**d**) (54.7 × 54.7 nm^2, V_s = 0.5 V, I = 1.0 nA, time/line: 0.365 s); (**e**) (32 × 32 nm^2, V_s = 0.5 V, I = 1.0 nA, time/line: 0.365 s); (f) profile corresponding to the blue line in (**d**); (**g**) (35 × 35 nm^2, V_s = 0.5 V, I = 1.0 nA, time/line: 0.365 s); (**h**) (13.3 × 13.3 nm^2, V_s = 0.4 V, I = 1.0 nA, time/line: 0.365 s); (i) profile corresponding to the blue line in (**h**).

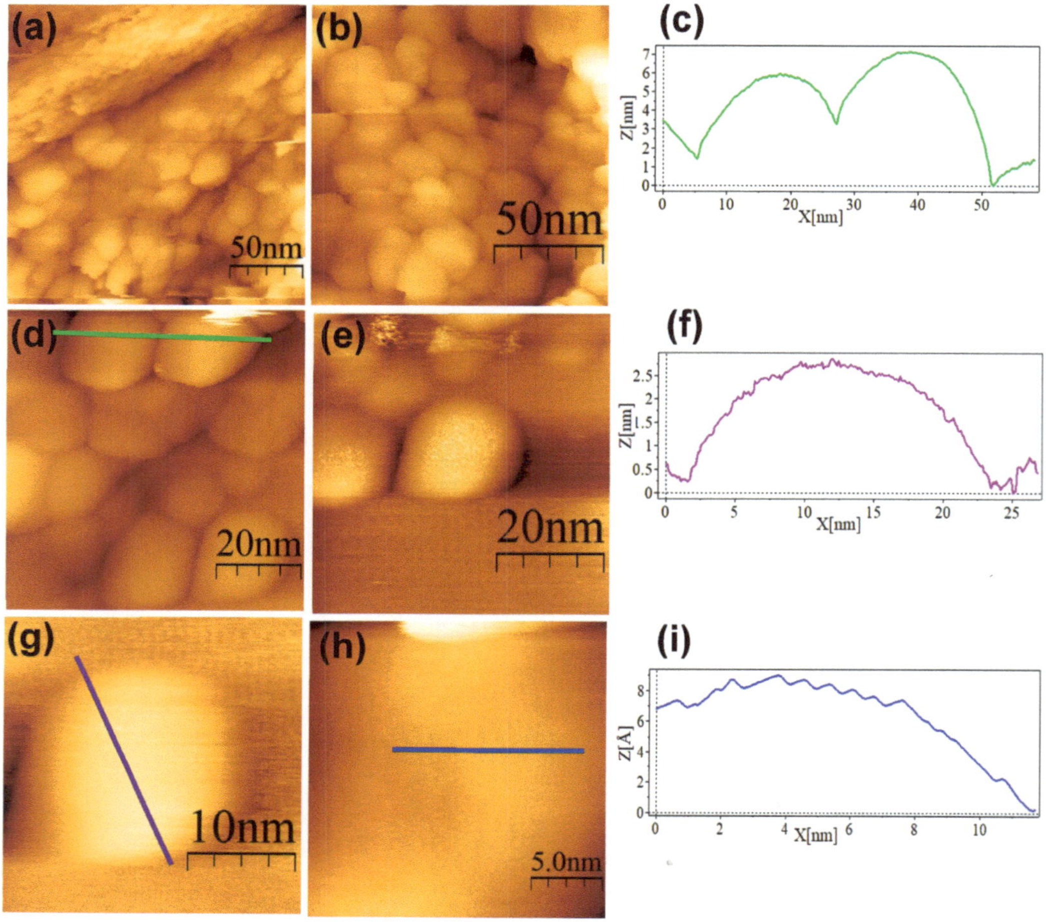

Figure 7. STM topographic images of cRGD@AuNPs deposited onto bare HOPG. STM imaging parameters: (**a**) (200 × 200 nm^2, V_s = 0.5 V, I = 1.0 nA, time/line: 0.4 s); (**b**) (134 × 134 nm^2, V_s = 0.5 V, I = 1.0 nA, time/line: 0.4 s); (**c**) profile corresponding to the blue line in (**d**); (**d**) (69.1 × 69.1 nm^2, V_s = 0.5 V, I = 1.0 nA, time/line: 0.399 s); (**e**) (54.6 × 54.6 nm^2, V_s = 0.5 V, I = 1.0 nA, time/line: 0.399 s); (**f**) profile corresponding to the blue line in (**g**); (**g**) (27.3 × 27.3 nm^2, V_s = 0.5 V, I = 1.0 nA, time/line: 0.399 s); (**h**) (20.4 × 20.4 nm^2, V_s = 0.5 V, I = 1.0 nA, time/line: 0.4 s); (**i**) profile corresponding to the blue line in (**h**).

4. Discussion

In this work, gold nanoparticles 20 nm in diameter were covered with thiolated biomolecules via S-Au bonds. This is a post synthesis methodology, well known as ligand exchange, where sodium citrate from the initial nanoparticles is exchanged by the thiolated molecules. It is important to mention that the exposed molecules at the interface determine the chemical and biological properties of the conjugates. Our results showed that the nanoparticles were stable after ALA, MAN, or cRGD modification, as confirmed by UV-Vis. The incorporation of functional groups provided by these molecules, as suggested by the FTIR analysis, can help to maintain the stability of these colloids in water, for example, allowing the formation of hydrogen bonds and electrostatic repulsion interactions. For biomedical applications, the stability of AuNPs is compulsory as it plays an important role in determining their toxicity [28]. Here, the stability was analyzed after synthesis and in water at pH 11. The size of the AuNPs after functionalization showed small changes due to the adsorption of the biomolecules compared to the initial AuNPs stabilized with sodium citrate (SC@AuNPs). In contrast, the Z potential showed that the conjugates increased their negativity after biofunctionalization: cRGD@AuNPs > MAN@AuNPs > ALA@AuNPs > SC@AuNPs. The change in the electronegativity of the conjugates can be attributed to the presence of aspartic acid amino acids (D) for cRGD, electronegative oxygens for MAN, and carboxylic acids for ALA. The size and charge of the nanoparticles are significant parameters as it was reported that these physicochemical parameters can influence the in vitro and in vivo behavior of functionalized gold nanoparticles, i.e., the in vitro uptake and the in vivo biodistribution [29].

Our TEM images showed the presence of a low-density coating around ALA@AuNPs, MAN@AuNPs, and cRGD@AuNPs, suggesting the presence of the biomolecules [30]. In addition, no aggregates were observed, which indicates that the nanoparticles' modification state diminished their aggregation behavior. On the other hand, our STM analysis showed that it is a viable microscopy technique for the characterization of biofunctionalized AuNPs. The STM images showed an aspherical morphology that was more evident for ALA@AuNPs and cRGD@AuNPs than for SC@AuNPs and MAN@AuNPs. The presence of free material could affect the morphology resolution in the MAN@AuNP case, the bonding reaction required mannose and EDC in excess, and even after several centrifugation washes, some residual material could be present, affecting the imaging of individual nanoparticles. Also, in the SC@AuNPs case, the presence of free molecules of sodium citrate could affect the resolution of individual molecules. The size of the nanoparticles measured by cross-sectional analyses showed particles with diameters longer than 20 nm, which can be attributed to the biomolecules attached to the AuNPs, as seen by TEM. A high-resolution analysis of the biomolecular coating of our as-synthetized conjugates was not achieved in our study. STM is a very sensitive technique, and several factors can limit its resolution power when analyzing biofunctionalized gold nanoparticles: the presence of free material can contaminate the tip, degrading its ability to produce images of high resolution; the mobility of the nanoparticles is another parameter that can affect an STM study. In our analysis, the mobility of individual nanoparticles on the HOPG substrate made it difficult to analyze their surfaces at a high resolution. Previous studies have reported the attachment of AuNPs on gold surfaces, which improved their analysis [15]. In addition, we employed the drop-cast technique for deposition, and we observed the formation of large agglomerates; in contrast, the deposition of monolayers can help to resolve individual AuNPs at a high resolution. In general, the previously mentioned parameters should be considered for the analysis of gold nanoparticles functionalized with biomolecules. Finally, the analysis of individual biomolecules as lipids, carbohydrates, proteins, or nucleic acids onto flat substrates is a complex task, and, moreover, the analysis of biomolecules on the surface of AuNPs represents an extreme challenge [17].

5. Conclusions

The biofunctionalization of AuNPs with lipoic acid, mannose, and the cRGD peptide via the formation of the Au-S bond via ligand exchange is a reliable method to obtain conjugates with potential biomedical applications. Combining several analytical techniques, the as-synthetized conjugates are described as stable without evidence of aggregation. Our preliminary studies using STM demonstrate that this technique can be employed to analyze biofunctionalized AuNPs; however, several parameters, such as the elimination of free ligands, the reduction in mobility, and the deposition of monolayers, should be optimized for the analysis of the surfaces of bioconjugates at a high resolution. The biofunctionalized nanoparticles with lipoic acid, mannose, and the cRGD peptide can also be conjugated with drugs and/or contrast agents. Also, it was shown that the STM can be used as a complementary technique for the characterization of biofunctionalized nanoparticles. Furthermore, the STM is a versatile technique that can be used for the electronic characterization of conjugates and the analyst of samples in physiological solutions.

Author Contributions: Conceptualization, A.R.-G.; methodology, M.R. (Mitzi Reyes), M.Á.-C. and A.R.-G.; validation, A.R.-G.; formal analysis, M.R. (Mitzi Reyes), M.Á.-C. and A.R.-G.; investigation, M.R. (Mitzi Reyes), M.Á.-C. and A.R.-G.; resources, A.R.-G.; data curation, A.R.-G.; writing—original draft preparation, A.R.-G.; writing—review and editing, M.R. (Margarita Rivera), A.R.-G. and V.A.B.; funding acquisition, A.R.-G. All authors have read and agreed to the published version of the manuscript.

Funding: Financial support from the National Autonomous University of Mexico (UNAM; Grant DGAPA-PAPIIT IA204521).

Data Availability Statement: Not applicable.

Acknowledgments: The authors thank Alejandro Heredia Barbero for his technical assistance with the FTIR analysis and Fís. Roberto Hernández for HRTEM imaging.

Conflicts of Interest: No conflicts of interest exist in the submission of this manuscript.

References

1. Anik, M.I.; Mahmud, N.; Al Masud, A.; Hasan, M. Gold Nanoparticles (GNPs) in Biomedical and Clinical Applications: A Revie *Nano Sel.* **2022**, *3*, 792–828. [CrossRef]
2. Amina, S.J.; Guo, B. A Review on the Synthesis and Functionalization of Gold Nanoparticles as a Drug Delivery Vehicle. *Int. Nanomed.* **2020**, *15*, 9823–9857. [CrossRef] [PubMed]
3. Fan, L.; Wang, W.; Wang, Z.; Zhao, M. Gold Nanoparticles Enhance Antibody Effect through Direct Cancer Cell Cytotoxicity b Differential Regulation of Phagocytosis. *Nat. Commun.* **2021**, *12*, 6371. [CrossRef] [PubMed]
4. Wang, C.; Zhang, H.; Zeng, D.; San, L.; Mi, X. DNA Nanotechnology Mediated Gold Nanoparticle Conjugates and The Applications in Biomedicine. *Chin. J. Chem.* **2016**, *34*, 299–307. [CrossRef]
5. Estudiante-Mariquez, O.J.; Rodríguez-Galván, A.; Ramírez-Hernández, D.; Contreras-Torres, F.F.; Medina, L.A. Technetiun Radiolabeled Mannose-Functionalized Gold Nanoparticles as Nanoprobes for Sentinel Lymph Node Detection. *Molecules* **202** *25*, 1982. [CrossRef] [PubMed]
6. Ghosh, P.; Han, G.; De, M.; Kim, C.K.; Rotello, V.M. Gold Nanoparticles in Delivery Applications. *Adv. Drug Deliv. Rev.* **200** *60*, 1307–1315. [CrossRef] [PubMed]
7. Lee, J.W.; Choi, S.-R.; Heo, J.H. Simultaneous Stabilization and Functionalization of Gold Nanoparticles via Biomolecul Conjugation: Progress and Perspectives. *ACS Appl. Mater. Interfaces* **2021**, *13*, 42311–42328. [CrossRef] [PubMed]
8. Iijima, M.; Somiya, M.; Yoshimoto, N.; Niimi, T.; Kuroda, S. Nano-Visualization of Oriented-Immobilized IgGs on Immunosenso by High-Speed Atomic Force Microscopy. *Sci. Rep.* **2012**, *2*, 790. [CrossRef]
9. Adak, A.K.; Lin, H.-J.; Lin, C.-C. Multivalent Glycosylated Nanoparticles for Studying Carbohydrate–Protein Interaction *Org. Biomol. Chem.* **2014**, *12*, 5563–5573. [CrossRef]
10. Liese, S.; Netz, R.R. Quantitative Prediction of Multivalent Ligand–Receptor Binding Affinities for Influenza, Cholera, an Anthrax Inhibition. *ACS Nano* **2018**, *12*, 4140–4147. [CrossRef]
11. Tjandra, K.C.; Thordarson, P. Multivalency in Drug Delivery—When Is It Too Much of a Good Thing? *Bioconjug. Chem.* **201** *30*, 503–514. [CrossRef] [PubMed]
12. Hinterwirth, H.; Kappel, S.; Waitz, T.; Prohaska, T.; Lindner, W.; Lämmerhofer, M. Quantifying Thiol Ligand Density of Sel Assembled Monolayers on Gold Nanoparticles by Inductively Coupled Plasma–Mass Spectrometry. *ACS Nano* **2013**, *7*, 1129–113 [CrossRef] [PubMed]
13. Volkert, A.A.; Subramaniam, V.; Ivanov, M.R.; Goodman, A.M.; Haes, A.J. Salt-Mediated Self-Assembly of Thioctic Acid on Gol Nanoparticles. *ACS Nano* **2011**, *5*, 4570–4580. [CrossRef] [PubMed]
14. Grönbeck, H.; Curioni, A.; Andreoni, W. Thiols and Disulfides on the Au(111) Surface: The Headgroup−Gold Interaction. *J. An Chem. Soc.* **2000**, *122*, 3839–3842. [CrossRef]
15. Ong, Q.; Luo, Z.; Stellacci, F. Characterization of Ligand Shell for Mixed-Ligand Coated Gold Nanoparticles. *Acc. Chem. Re* **2017**, *50*, 1911–1919. [CrossRef] [PubMed]
16. Zhou, Q.; Kaappa, S.; Malola, S.; Lu, H.; Guan, D.; Li, Y.; Wang, H.; Xie, Z.; Ma, Z.; Häkkinen, H.; et al. Real-Space Imagin with Pattern Recognition of a Ligand-Protected Ag 374 Nanocluster at Sub-Molecular Resolution. *Nat. Commun.* **2018**, *9*, 294 [CrossRef] [PubMed]
17. Rodríguez-Galván, A.; Contreras-Torres, F.F. Scanning Tunneling Microscopy of Biological Structures: An Elusive Goal for Man Years. *Nanomaterials* **2022**, *12*, 3013. [CrossRef] [PubMed]
18. Rodríguez-Galván, A.; Heredia, A.; Amelines-Sarria, O.; Rivera, M.; Medina, L.A.; Basiuk, V.A. Non-Covalent Attachment c Silver Nanoclusters onto Single-Walled Carbon Nanotubes with Human Serum Albumin as Linking Molecule. *Appl. Surf. Sc* **2015**, *331*, 271–277. [CrossRef]
19. Rodríguez-Galván, A.; Amelines-Sarria, O.; Rivera, M.; del Pilar Carreón-Castro, M.; Basiuk, V.A. Adsorption and Self-Assembl of Anticancer Antibiotic Doxorubicin on Single-Walled Carbon Nanotubes. *Nano* **2015**, *11*, 1650038. [CrossRef]

20. Rodríguez-Galván, A.; Heredia, A.; Plascencia-Villa, G.; Ramírez, O.T.; Palomares, L.A.; Basiuk, V.A. Scanning Tunneling Microscopy of Rotavirus VP6 Protein Self-Assembled into Nanotubes and Nanospheres. *J. Scanning Probe Microsc.* **2008**, *3*, 25–31. [CrossRef]
21. Dzwonek, M.; Załubiniak, D.; Piątek, P.; Cichowicz, G.; Męczynska-Wielgosz, S.; Stępkowski, T.; Kruszewski, M.; Więckowska, A.; Bilewicz, R. Towards Potent but Less Toxic Nanopharmaceuticals—Lipoic Acid Bioconjugates of Ultrasmall Gold Nanoparticles with an Anticancer Drug and Addressing Unit. *RSC Adv.* **2018**, *8*, 14947–14957. [CrossRef] [PubMed]
22. Abad, J.M.; Mertens, S.F.L.; Pita, M.; Fernández, V.M.; Schiffrin, D.J. Functionalization of Thioctic Acid-Capped Gold Nanoparticles for Specific Immobilization of Histidine-Tagged Proteins. *J. Am. Chem. Soc.* **2005**, *127*, 5689–5694. [CrossRef] [PubMed]
23. Gajbhiye, K.R.; Gajbhiye, V.; Siddiqui, I.A.; Gajbhiye, J.M. cRGD Functionalised Nanocarriers for Targeted Delivery of Bioactives. *J. Drug Target.* **2019**, *27*, 111–124. [CrossRef] [PubMed]
24. Horcas, I.; Fernández, R.; Gómez-Rodríguez, J.M.; Colchero, J.; Gómez-Herrero, J.; Baro, A.M. WSXM: A Software for Scanning Probe Microscopy and a Tool for Nanotechnology. *Rev. Sci. Instrum.* **2007**, *78*, 013705. [CrossRef] [PubMed]
25. Hong, T.; Yin, J.-Y.; Nie, S.-P.; Xie, M.-Y. Applications of Infrared Spectroscopy in Polysaccharide Structural Analysis: Progress, Challenge and Perspective. *Food Chem. X* **2021**, *12*, 100168. [CrossRef] [PubMed]
26. Krimm, S.; Bandekar, J. Vibrational Spectroscopy and Conformation of Peptides, Polypeptides, and Proteins. In *Advances in Protein Chemistry*; Anfinsen, C.B., Edsall, J.T., Richards, F.M., Eds.; Academic Press: Cambridge, MA, USA, 1986; Volume 38, pp. 181–364.
27. Pensa, E.; Albrecht, T. Controlling the Dynamic Instability of Capped Metal Nanoparticles on Metallic Surfaces. *J. Phys. Chem. Lett.* **2018**, *9*, 57–62. [CrossRef] [PubMed]
28. Wang, L.; Wu, W.-M.; Bolan, N.S.; Tsang, D.C.W.; Li, Y.; Qin, M.; Hou, D. Environmental Fate, Toxicity and Risk Management Strategies of Nanoplastics in the Environment: Current Status and Future Perspectives. *J. Hazard. Mater.* **2021**, *401*, 123415. [CrossRef]
29. Xu, M.; Soliman, M.G.; Sun, X.; Pelaz, B.; Feliu, N.; Parak, W.J.; Liu, S. How Entanglement of Different Physicochemical Properties Complicates the Prediction of in Vitro and in Vivo Interactions of Gold Nanoparticles. *ACS Nano* **2018**, *12*, 10104–10113. [CrossRef]
30. Egorova, E.A.; van Rijt, M.M.J.; Sommerdijk, N.; Gooris, G.S.; Bouwstra, J.A.; Boyle, A.L.; Kros, A. One Peptide for Them All: Gold Nanoparticles of Different Sizes Are Stabilized by a Common Peptide Amphiphile. *ACS Nano* **2020**, *14*, 5874–5886. [CrossRef]

nanomaterials

MDPI

Article

In Situ Incorporation of Atomically Precise Au Nanoclusters within Zeolites for Ambient Temperature CO Oxidation

Siriluck Tesana [1,2,3], John V. Kennedy [2,3], Alex C. K. Yip [2,4,*] and Vladimir B. Golovko [1,2,*]

1 School of Physical and Chemical Sciences, University of Canterbury, Christchurch 8041, New Zealand; s.tesana@gns.cri.nz
2 The MacDiarmid Institute for Advanced Materials and Nanotechnology, Wellington 6140, New Zealand; j.kennedy@gns.cri.nz
3 National Isotope Centre, GNS Science, Lower Hutt 5010, New Zealand
4 Department of Chemical and Process Engineering, University of Canterbury, Christchurch 8041, New Zealand
* Correspondence: alex.yip@canterbury.ac.nz (A.C.K.Y.); vladimir.golovko@canterbury.ac.nz (V.B.G.); Tel.: +64-3-3694086 (A.C.K.Y.); +64-3-3695942 (V.B.G.)

Abstract: Preserving ultrasmall sizes of metal particles is a key challenge in the study of heterogeneous metal-based catalysis. Confining the ultrasmall metal clusters in a well-defined crystalline porous zeolite has emerged as a promising approach to stabilize these metal species. Successful encapsulation can be achieved by the addition of ligated metal complexes to zeolite synthesis gel before hydrothermal synthesis. However, controlling the metal particle size during post-reduction treatment remains a major challenge in this approach. Herein, an in situ incorporation strategy of pre-made atomically precise gold clusters within Na-LTA zeolite was established for the first time. With the assistance of mercaptosilane ligands, the gold clusters were successfully incorporated within the Na-LTA without premature precipitation and metal aggregation during the synthesis. We have demonstrated that the confinement of gold clusters within the zeolite framework offers high stability against sintering, leading to superior CO oxidation catalytic performance (up to 12 h at 30 °C, with a space velocity of 3000 mL g^{-1} h^{-1}).

Keywords: gold nanoclusters; in situ encapsulation; LTA zeolite; CO oxidation

1. Introduction

The catalytic behavior of metal-based catalysts is strongly dependent on their particle size. This is especially true for metal nanoclusters (MNCs), whose unique properties are highly sensitive to the number of metal atoms in the core. The addition or subtraction of a single metal atom to the ultrasmall clusters can critically influence the catalytic activity [1–3]. Therefore, controlling and maintaining the sizes of MNCs is essential in the study of metal nanocluster-based catalysis. Capping ligands are generally used in the synthesis of metal colloids and MNCs to control the size and suppress the overgrowth of the metal particles. When such species are used to make catalysts, ligands can act as a barrier, prohibiting reactants from accessing active metal surfaces. Many strategies have been developed to preserve the ultrasmall sizes of metal particles, particularly during catalyst activation and catalytic reaction, where protecting ligands are removed [4]. Immobilization of the metal particles onto solid supports before removing the ligands is one common approach. However, depositing the metal particles onto the surface of the solid supports, in some cases, often cannot provide satisfactory stability to the metal particles [5–7]. Due to their high surface energy, naked ultrasmall metal particles tend to aggregate by Ostwald ripening or Smoluchowski ripening, hence, losing catalytically active surface [8]. Our team [9–15], as well as several other groups [8,16–18], have reported that controlling MNC aggregation can be both important and challenging.

In this respect, microporous aluminosilicate zeolites appear as a promising host for confining ultrasmall MNCs. A well-defined and rigid zeolite framework could offer better stability by accommodating MNCs inside their cavities, well-isolated from one another. In such cases, the ultrasmall MNCs would have a better chance to maintain their size and catalytically active surfaces. The great thermal stability of zeolites, their large specific surface area, and more importantly, the specific pore and cavity sizes could also facilitate unique reactant, product, and transition-state selectivity and suppress catalyst poisoning [19–23].

The challenging task is that the fabrication of such encapsulated MNCs within zeolite is not straightforward [24–26]. Common methods used for the preparation of zeolite-supported metal catalysts involve ion exchange and wet impregnation. In both cases, solvated metal complexes are introduced to the zeolite void space after the framework is formed. Therefore, the location and homogeneity of distribution of metal species and their sizes strongly depend on the diffusion of the metal precursor through the framework defined by the pore openings of a particular zeolite. It would be a major challenge to synthesize atomically precise metal particles using this approach. These post-synthetic protocols are generally restricted significantly for small-pore zeolites, where a majority of the metal species may not be able to access the micropores. Instead, they are likely to remain at the outer surface and sinter to form larger particles during the activation and reduction steps [27–31].

An alternative approach, an in situ encapsulation, where a ligated metal precursor is introduced to a zeolite synthesis gel before the framework formation may, therefore, be preferable [32–49]. Using this approach, zeolite building units can assemble around the metal species, occluding the metals within the resulting framework during crystallization. However, successful encapsulation can only be achieved if (1) the metal precursors possess adequate stability in a strongly alkaline media of zeolite synthesis, and (2) good interaction between the metal species and the zeolite building units is established. Therefore, choosing appropriate passivating ligands (e.g., amine-based [33–43] and mercaptosilane-based [44–49] ligands) that strongly coordinate with the metal precursors and promote the assembly of zeolite building units is a crucial step in this approach. Without the assistance of such ligands, the metal species tend to precipitate prematurely and severely agglomerate to form larger metal particles on the external surface of the zeolite or even bulk metal separated from zeolite hosts [33–36,44–46].

Bifunctional mercaptosilane-based ligands (such as 3-mercaptopropyl trimethoxysilane, MPTMS), among protecting ligands, have been reported to offer sufficient stability to simple metal precursors, such as H_2PtCl_6, $Pd(NH_3)_4Cl_2$, H_2IrCl_6, $(NH_4)_3RhCl_6$, $AgNO_3$, and $HAuCl_4$ [44,45]. Through a strong metal-S bond, the mercapto groups (-SH) of the ligands provide chemical protection to the metal precursors against reduction or hydrolysis, even in strong alkaline media. Meanwhile, the silane moieties of the ligand simultaneously induce condensation of the silicate oligomer around the metal precursors. The formation of covalent Si-O-Si or Si-O-Al bonds between the ligated metal precursors and the zeolite building units then encapsulates metal species. At the same time, the zeolite framework is being formed, allowing these simple metal species to be incorporated within the zeolite cavity without premature precipitation and/or severe aggregation.

This strategy, however, requires further reduction of these simple encapsulated ligated metal species using reductive thermal treatment to produce encapsulated metal nanoparticles (MNPs) or MNCs with clean surfaces. The conditions of such inevitable post-treatment have a prominent influence on the sizes and size distributions of the MNPs and MNCs [50]. The precise control of the metal particle size is challenging, especially for gold MNPs and MNCs, whose sizes are susceptible to the temperature and chemical atmosphere [51]. There exists a very narrow temperature range of reductive conditions that can achieve encapsulated, monodispersed Au MNCs with good control over the particle size [45].

In this study, we report the first successful attempt at in situ encapsulation of atomically precise gold nanoclusters (Au NCs) within the LTA zeolite void (Figure 1).

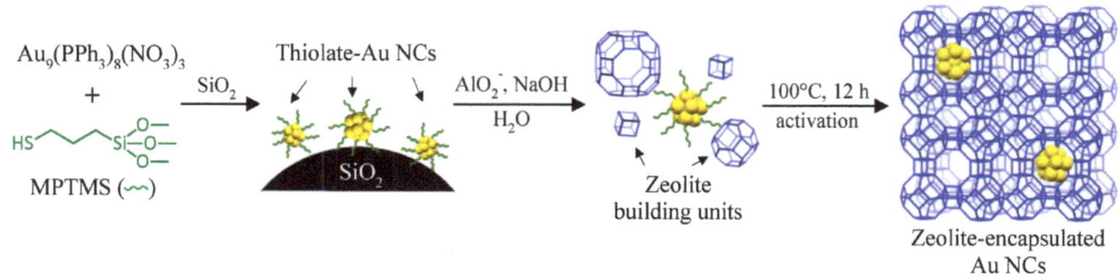

Figure 1. In situ encapsulation of atomically precise Au_9 within Na-LTA zeolite.

Instead of gold complexes, pre-made ligated Au_9 clusters, $Au_9(PPh_3)_8(NO_3)_3$, were synthesized and used as precursors to the metal active sites. It was important to carry out a ligand exchange of the phosphine ligands with bifunctional (3-mercaptopropyl) trimethoxysilane (MPTMS) ligands prior to the hydrothermal synthesis of zeolite to ensure the compatibility and stability of the metal precursor in the synthesis media. It was hypothesized that the ultra-small Au_9 clusters (~0.8 nm core size) with MPTMS ligands could match the LTA zeolite structure (~1.1 nm cavities with ~0.42 apertures). Indeed, we demonstrated that the zeolite-confined Au NCs maintained their ultra-small size and were resilient against sintering. Ambient temperature catalytic CO oxidation was used to investigate the catalytic reactivity and encapsulation efficiency of the LTA-encapsulated Au NCs, which were compared and contrasted with the post-impregnated Au NCs on the same zeolite.

2. Experimental Section

2.1. Materials Synthesis

2.1.1. Synthesis of $Au_9(PPh_3)_8(NO_3)_3$ Nanoclusters

Atomically precise $Au_9(PPh_3)_8(NO_3)_3$ nanoclusters, denoted as 'Au_9', were synthesized according to the protocol reported by Anderson et al. [14]. The ethanolic solution of $NaBH_4$ (0.02 M, 90 mL) was added to the ethanolic solution of $AuPPh_3NO_3$ (0.048 M, 160 mL). The mixture was stirred at 1000 rpm for 2 h in the absence of light to obtain a deep red solution. Subsequently, insoluble impurities were filtered off, and the solvent was evaporated under reduced pressure. The obtained solid was redissolved in 20 mL of CH_2Cl_2 before filtration. The dark green precipitate yielded after further solvent removal. The product was then washed with THF and hexanes before being dissolved in methanol and subjected to crystallization via vapor diffusion of diethyl ether as an anti-solvent at 4 °C in the dark for 5 d. The dark green crystals were washed with diethyl ether and dried in vacuo. The yield of the Au_9 was ca. 2.10 g, 60 ± 6 Au at%.

Details of Au_9 gold–phosphine cluster synthesis and characterization can be found in the Supporting Information, S3.1–S3.4.

2.1.2. Ligand Exchange of $Au_9(PPh_3)_8(NO_3)_3$ with (3-mercaptopropyl) Trimethoxysilane

The ligand exchange reaction of the Au_9 with (3-mercaptopropyl) trimethoxysilane (MPTMS) was performed following the procedure reported by Woehrle et al., with minor modifications [52]. Typically, 20 equivalents of MPTMS (262 μL, 1.34 mmol) were added to a dark brown methanolic solution of Au_9 (0.27 g, 0.067 mmol, 150 mL) while stirring. The mixture was stirred at 55 °C under an N_2 atmosphere for 18 h. A trace amount of insoluble solids was separated by centrifugation. The crude product from the ligand exchange reaction was named 'Au_9-MPTMS', and was characterized by NMR, MS, and UV-vis techniques (refer to the Supporting Information, S3.5) before using it in zeolite synthesis.

2.1.3. Pre-Mixing of SiO_2 with the Au_9-MPTMS

For the synthesis of Na-LTA encapsulated Au NCs, fumed SiO_2 (3.2 g, 0.053 mol) was suspended in a methanolic solution of the Au_9-MPTMS ligand exchange product under agitation. After 1 h, the solvent was removed in vacuo. The obtained fumed SiO_2 modified by Au_9-MPTMS (MPTMS:SiO_2 mol ratio of 1:40) was dried overnight under vacuum before being added to the aluminate solution for Na-LTA zeolite synthesis.

For the synthesis of Na-FAU encapsulated Au NCs, colloid SiO_2 LUDOX AM-30 (22.4 g, 30 wt% SiO_2, 0.1 mol) was added to a methanolic solution of the Au_9-MPTMS ligand exchange product under agitation. After the removal of methanol in vacuo, the colloidal SiO_2 modified by Au_9-MPTMS (MPTMS:SiO_2 mol ratio of 1:60) was added to the aluminate solution for Na-FAU zeolite synthesis.

2.1.4. Incorporation of Au_9-MPTMS within Na-LTA Zeolite

The encapsulation of Au NCs within LTA zeolite was performed using a strategy similar to that reported by Wu et al. and Otto et al. [34,45]. Instead of simple ligated metal complexes, the fumed SiO_2 modified by Au_9-MPTMS was added to the aluminate solution before hydrothermal crystallization. Typically, NaOH (4.8 g, 0.12 mol) and $NaAlO_2$ (6.0 g, 0.07 mol) were dissolved in 40 mL of Milli-Q water. While stirring (~750 rpm), the SiO_2 modified by Au_9-MPTMS and 22 mL of Milli-Q water were added to the mixture to form a homogeneous Au NCs-containing aluminosilicate gel with a molar ratio of 2.6 Na_2O: 1.0 Al_2O_3: 1.5 SiO_2: 93.6 H_2O: 0.02 Au. The gel was aged at 60 °C for 4 h before being transferred into 60 mL Teflon-lined stainless steel autoclaves (ca. 40 g of gel in each) and crystallized at 100 °C for 16 h. The product was collected by centrifugation at 12,000 rpm for 5 min and washed with Milli-Q water. After being suspended in methanol overnight, the as-made 'Au_9-MPTMS@Na-LTA' product was collected, dried overnight in vacuo, and further dried at 100 °C for 12 h in ambient air. The amount of Au clusters reported above (1.2 wt% Au loading, assuming 100% yield of Na-LTA) was adjusted to achieve 1.0 wt% Au loading in the final Au_9-MPTMS@Na-LTA product (refer to the Supporting Information, S3.6).

For comparison, in the absence of MPTMS, an 'Au_9-PPh_3@Na-LTA' sample was prepared by adding a methanolic solution of $Au_9(PPh_3)_8(NO_3)_3$ to the zeolite synthesis gel before the hydrothermal treatment. Through identical procedures to the Au_9-MPTMS@Na-LTA sample, the aged gel of Au_9-PPh_3@Na-LTA was subjected to hydrothermal treatment at 100 °C for 16 h. The obtained product was collected, washed, and suspended in methanol in the same fashion. The as-made Au_9-PPh_3@Na-LTA was dried overnight in vacuo and further treated at 100 °C for 12 h in ambient air.

The yield of the as-made Au_9-MPTMS@Na-LTA sample and Au_9-PPh_3@Na-LTA was ca. 9.42 g (96 ± 2% based on SiO_2). The addition of Au_9-MPTMS and $Au_9(PPh_3)_8(NO_3)_3$ did not significantly affect the yield of the final products.

2.1.5. Incorporation of Au_9-MPTMS within Na-FAU Zeolite

The FAU-encapsulated Au NCs were prepared following the methods reported by Chen et al. and Otto et al., with minor modifications [39,45]. In the typical synthesis, $NaAlO_2$ (1.82 g, 0.02 mol) and NaOH (6.24 g, 0.16 mol) were dissolved in 60 mL of Milli-Q water. The colloidal SiO_2 modified by Au_9-MPTMS was then added to the alumina solution while stirring (~750 rpm) to obtain the synthesis gel with the composition of 8.0 Na_2O: 1.0 Al_2O_3: 10.1 SiO_2: 376.5 H_2O: 0.07 Au. The mixture was stirred (~750 rpm) at room temperature for 24 h. The synthesis gel was transferred into 60 mL autoclaves (*ca.* 40 g of gel in each) and crystallized at 100 °C for 15 h. The 'Au_9-MPTMS@Na-FAU' product was collected, washed, and dried in the same manner as in the case of Au_9-MPTMS@Na-LTA.

The yield of the as-made Au_9-MPTMS@Na-FAU sample was ca. 4.30 g (65 ± 4%, based on Al_2O_3). Of note, the addition of Au_9-MPTMS did not significantly affect the yield of the FAU-based catalyst. Similar to Au_9-MPTMS@Na-LTA, the amount of Au NCs added was adjusted to achieve Au_9-MPTMS@Na-FAU with the targeted loading of 1.0 wt% Au (refer to the Supporting Information, S3.7).

2.2. Catalyst Activation

To remove the protecting ligands surrounding the Au NCs cores, the zeolite-encapsulated Au NCs were treated under ozone flow. The samples were exposed to O_3 (170 µg mL^{-1} of the target concentration) for 1 h at different temperatures under magnetic stirring (500 rpm). This was achieved using a Schlenk flask placed in an aluminum block on a hotplate stirrer. The ozone was produced by ozone generator OL100H1DS, Yanco Industries, Ltd., with an initial O_2 flow of 150 mL min^{-1}.

2.3. Ambient Temperature CO Oxidation for Testing of Activity and Encapsulation Efficiency

Catalytic CO oxidation was performed in a stainless steel continuous-flow fixed-bed reactor. The reactor was operated in differential mode between 30 and 200 °C at atmospheric pressure with a gas hourly space velocity (GHSV) ranging from 3000 to 30,000 mL g^{-1} h^{-1}. In a typical reaction, a gas mixture containing 1.0 vol% of CO, 10.5 vol% of O_2, balanced with Ar and N_2, was introduced into the reactor. Either 40 or 200 mg of the catalyst was loaded into the stainless steel reactor. The reactor was placed in a furnace controlled by a programmable temperature controller. The product gas was analyzed by an online GC (SRI Multiple Gas Analyzer, details in the Supporting Information, S4). All experiments were performed in duplicate.

3. Results and Discussion

3.1. Synthesis and Characterizations of LTA Zeolite-Encapsulated Au_9 Nanoclusters

Na-LTA zeolite was hydrothermally synthesized in the presence of the crude ligand exchange product, Au_9-MPTMS. Without further addition of organic structure-directing agents, the LTA zeolite framework could be formed under relatively mild conditions (100 °C, 16 h). Structural evidence of the LTA zeolite phase and its purity were verified by PXRD analysis. As shown in Figure 2, PXRD of Au_9-MPTMS@Na-LTA showed a characteristic peak pattern corresponding to the crystalline LTA zeolite, with approximately the same peak intensity as that of gold-free Na-LTA samples with and without the addition of MPTMS [53]. This result indicated that ligated Au_9 nanoclusters, Au_9-MPTMS, did not interfere with the hydrothermal crystallization process of the Na-LTA zeolite. The negligible difference in the PXRD peaks' intensity between the samples suggested that a similar zeolite crystallinity level was systematically achieved.

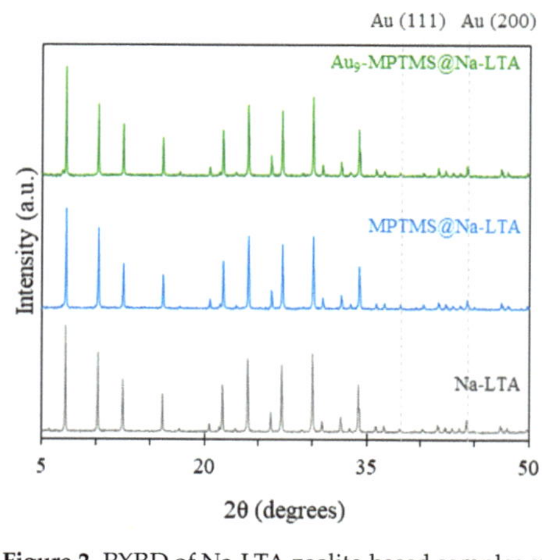

Figure 2. PXRD of Na-LTA zeolite-based samples synthesized in the presence of Au_9-MPTMS and MPTMS, compared to Na-LTA.

The total Au content in all samples was measured using MP-AES elemental analysis. By adding SiO_2-deposited Au_9-MPTMS to the aluminate solution before hydrothermal crystallization, ca. 93 wt% of the introduced Au NCs were incorporated within the Na-LTA zeolite. The Au_9-MPTMS@Na-LTA with 1.11 ± 0.02 wt% Au was achieved with excellent reproducibility, as confirmed by eight independent syntheses (Figure S3).

Without the MPTMS ligand exchange step, introducing $Au_9(PPh_3)_8(NO_3)_3$ to the zeolite synthesis gel led to a very low incorporation efficiency. Only ca. 10 wt% of the Au precursors were found in the final Au_9-PPh_3@Na-LTA sample (ca. 0.12 wt% Au). These results highlight a significant contribution of the MPTMS ligand in promoting Au NC incorporation into Na-LTA zeolite. As reported by Iglesia et al., the alkoxysilane moiety of MPTMS can be hydrolyzed in alkaline media to form covalent Si-O-Si or Si-O-Al bonds with the zeolite building units, which then enforce the encapsulation of the metal species during the growth of zeolite crystal [44–46].

It is worth mentioning that such contributions of MPTMS were observed only when the ligand exchange between the Au NCs and MPTMS was conducted beforehand. A direct co-addition of the Au_9 and MPTMS to the zeolite synthesis gel resulted in very low gold content (<0.2 wt%) in the final product, which is similar to the case of synthesis using only Au_9 (i.e., in the absence of MPTMS). This is in contrast to the case of zeolite containing Au NPs, as reported by Iglesia and co-workers, where Au NPs can be encapsulated within zeolite voids via the direct co-addition of $HAuCl_4$ and MPTMS to the similar zeolite synthesis gel, without a dedicated ligand exchange step beforehand [45]. This could be attributed to the properties of $HAuCl_4$, which are more soluble, stable, mobile, and active compared to phosphine-protected Au_9 NCs. The bonding between the Au^{3+} species of $HAuCl_4$ and the mercapto group of MPTMS can be quickly established after the direct addition. In contrast, the ligand exchange from phosphine to MPTMS in Au NCs can only occur under suitable conditions (methanol as a solvent, 55 °C temperatures, and prolonged periods of time—18 h). Thus, a dedicated ligand exchange step is necessary for the encapsulation of the Au NCs within the zeolite.

Determining the size of the incorporated Au species was one of the biggest challenges in this study, especially since they were just a few Au atoms in size and were incorporated within the zeolite framework [32]. The MNP sizes with larger diameters (e.g., >2 nm) can be determined using high-resolution TEM. For example, ~2 nm-encapsulated MNPs (Pt, Pd, Ru, and Rh) within GIS zeolite (8-MR) are marginally visible in HR-TEM images [33]. While the smaller MNPs (~1.0–1.5 nm) in ANA (8-MR) and SOD (6-MR) are not visible using conventional TEM but can be observed using the high-angle annular dark-field high-resolution scanning transmission electron microscopy (HAADF HR-STEM) [33]. With HAADF HR-STEM, the Pt NCs with a diameter of 0.2–0.7 nm or even individual metal atoms are distinguishable (as bright spots) within the MCM-22 zeolite crystallites [54]. Unfortunately, the size of Au NCs could not be directly verified by electron microscopy, as they fall below the detection limit of the conventional TEM used in this study. Apart from the ultrasmall size of Au species, the thickness of the zeolite crystallites (~200–700 nm, Figure S4a,b) was another factor that made size determination via electron microscopy impossible. However, the absence of large Au particles (>2 nm) in the as-made Au_9-MPTMS@Na-LTA sample containing 1.11 ± 0.02 wt% Au could be confirmed; Figure S4c,d.

In this work, the existence of Au NCs with a diameter smaller than 2 nm in the as-made Au_9-MPTMS@Na-LTA sample was indirectly evidenced by MP-AES analysis, along with PXRD and UV-vis DRS. The absence of bulk Au peaks at 2θ of 38.1 and 44.3 in the PXRD of Au_9-MPTMS@Na-LTA (Figure 2) suggested that the Au NCs retained their size and did not sinter to form larger Au crystallites (>10 nm) during the hydrothermal synthesis. Of note, the detection limits of PXRD are generally in the wide range of 1–10% by mass. However, these values vary significantly with the instrument and type of sample [55].

Apart from TEM and PXRD, UV-vis DRS was employed to monitor the formation of undesirably large Au NPs (>2 nm) due to cluster agglomeration. The size of the Au NPs could be roughly estimated from the position of the localized surface plasmonic resonance band (LSPR) in the UV-vis DRS. It is worth mentioning that such an estimation was not feasible for small Au NCs. Their optical properties are dominated by molecular-like single-electron transitions between quantized energy levels. As shown in Figure 3a, the Au_9-MPTMS@Na-LTA sample did not exhibit detectable LSPR-Au NPs absorption bands (LSPR, at 500–600 nm) [56,57]. or sharp ligand-to-metal charge transfer peaks (LMCT, at 350–500 nm) [58] in UV-vis DRS. These data suggest that Au NPs larger than 2 nm were not present in the sample. While the existing ligand-exchanged Au_9-MPTMS adducts did not exhibit LMCT peaks, the broad absorption feature at 350–500 nm matched well with the absorption feature of Au_9-MPTMS in methanol, as shown in Figure 3b. Based on the UV-vis DRS results, MPTMS-stabilized Au NCs were found to have sufficient stability against sintering at the pH and temperature conditions of the LTA hydrothermal synthesis, similar to the case of Au^{3+} complexes reported previously [45].

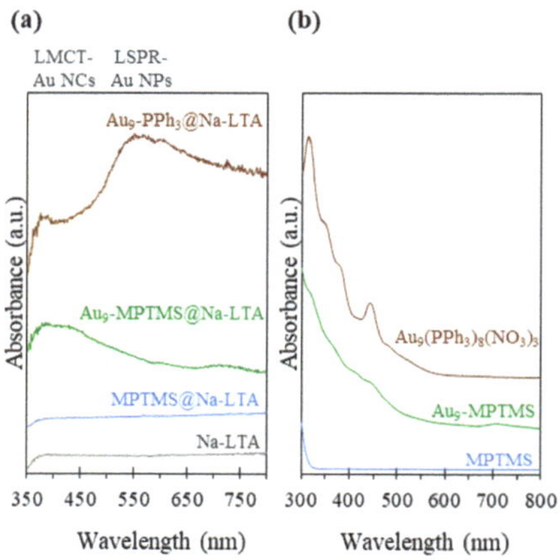

Figure 3. (**a**) UV-vis DRS of Na-LTA zeolite-based samples synthesized in the presence of Au_9-PPh_3, Au_9-MPTMS, and MPTMS, compared to Na-LTA. (**b**) UV-vis spectra of methanolic solutions of $Au_9(PPh_3)_8(NO_3)_3$, Au_9-MPTMS, and MPTMS.

Comparing UV-vis DRS of Au_9-MPTMS@Na-LTA with that of Au_9-PPh_3@Na-LTA, a prominent LSPR-Au NP band was found in the latter case, synthesized in the absence of MPTMS. These results emphasize the contribution of the mercapto group of the MPTMS ligand in offering high stability to the Au species through a strong Au-S bond [59], suppressing Au NCs from severe aggregation and premature precipitation under strong alkaline crystallization conditions of the zeolite. Of note, the results also highlight the high sensitivity of the UV-vis DRS method in detecting plasmonic particles, even with a significantly lower Au content (0.12 wt%) in the Au_9-PPh_3@Na-LTA sample.

Using HAADF HR-STEM studies via an aberration-corrected microscope, the position of embedded MNPs/MNCs in the internal space of zeolite crystallites has been directly confirmed [31,32,54]. In this work, information about the location of metal species was indirectly deduced from the catalytic behavior of zeolite-incorporated metal. The catalytic activity of zeolite-encapsulated metal species and their higher stability have been widely used as evidence of successful encapsulation [33–35,44–46]. The results using this approach are discussed in depth in the following section.

LTA zeolite-incorporated Au_9 nanoclusters were successfully synthesized via the in situ incorporation approach when the Au NCs were protected by MPTMS ligands. The product was obtained in significant yield with high zeolite crystallinity comparable to that of commercial zeolite. The sample showed great consistency in phase crystallinity and purity, the degree of incorporated Au, and absorption features across eight separately synthesized batches (Figure S3). More importantly, the sample exhibited high Au NC resilience against metal agglomeration and precipitation during the hydrothermal synthesis of zeolite.

3.2. Catalysts Activation

Since the pre-made Au clusters were employed for the synthesis of Au_9-MPTMS@Na-LTA catalysts, H_2 reduction treatment used to form Au NCs inside zeolite via the direct co-addition of $HAuCl_4$ and MPTMS to the similar zeolite synthesis gel was not required. However, post-synthesis treatment was necessary to remove ligands and open the metal active sites for the catalytic CO oxidation (Figure 1). Calcination (at 400 to 600 °C) has been reported to offer complete ligand removal for thiol-capped Au NCs and Au NPs [60–62]. In some cases, the full removal of protecting ligands did not result in greater catalytic performance but led to the inactivity of the catalyst [62]. Conversely, remarkable CO oxidation performance without the complete removal of ligands was reported in some catalytic systems (e.g., CeO_2-supported $Au_{25}(SR)_{18}$ [63] and $Au_{38}(SR)_{24}$ [64]). Therefore, the degree of ligand removal required to achieve high CO activity and good stability depends on the catalytic system.

The key challenge for the activation of Au_9-MPTMS@Na-LTA catalysts was to remove the MPTMS ligands of the confined Au NCs while suppressing the severe agglomeration of Au NCs. For noble metals, such as Pt, Pd, and Ir, the removal of MPTMS ligands could be achieved by H_2 treatment at 400 °C for 2 h [34]. Due to the weaker Au-S bonds, compared to the M-S bonds of the abovementioned noble metals (Au–S: 126 kJ mol^{-1}; Pt–S: 233 kJ mol^{-1}; Pd–S: 183 kJ mol^{-1}; Ir–S: 206 kJ mol^{-1}), it was expected that Au-bonded MPTMS can be removed under milder conditions [65,66]. Iglesia et al. reported the removal of MPTMS ligands from the Au^0 NP surface via treatment at 400 °C in air followed by treatment at 300 °C under H_2 [45]. However, under such conditions, we found that Au NCs sintered to form undesired larger Au NPs.

Thermogravimetric analysis (TGA) of pure $Au_9(PPh_3)_8(NO_3)_3$ clusters showed an onset temperature for MPTMS ligand removal of ~230 °C (under N_2 flow); Figure S5. However, there was no distinct difference in the TGA curves of Au_9-MPTMS@Na-LTA compared to those of pure Na-LTA or MPTMS@Na-LTA, according to the low sensitivity limit of TGA. Thus, the decomposition temperature of individual species in Au_9-MPTMS@Na-LTA cannot be identified. Moreover, the possibility of using temperature-programmed oxidation (TPO) to quantify the presence of ligands in the samples was investigated. Unfortunately, the ligand oxidation could not be quantified using the TPO experiment due to the CO_2 generated from oxidation interfering with the thermal conductivity detector (TCD) signals with O_2.

To establish the activation conditions of Au_9-MPTMS@Na-LTA, the sample was heated at temperatures varying from 200 to 400 °C for 1 h under static air (N.B.: b.p. of MPTMS is ~214 °C). The agglomeration of Au NCs after the calcination was monitored using UV-vis DRS. Activation conditions of Au_9-MPTMS@Na-LTA were established at an early stage of the study using the sample with 0.75 ± 0.01 wt% Au, which was prepared in the same fashion as discussed above. The optimal activation conditions were later applied to the most promising sample, Au_9-MPTMS@Na-LTA, with 1.11 ± 0.02 wt% Au in the following studies.

As shown in Figure 4a, the Au plasmonic band started to appear after calcination at 300 °C, indicating the onset of sintering of the Au species. Some ligand removal occurred during the heat treatment at 300 °C as well as undesirable Au agglomeration. There was no distinct difference between the UV-vis DRS of the 200 °C-treated sample and the untreated one, suggesting that severe Au sintering did not occur.

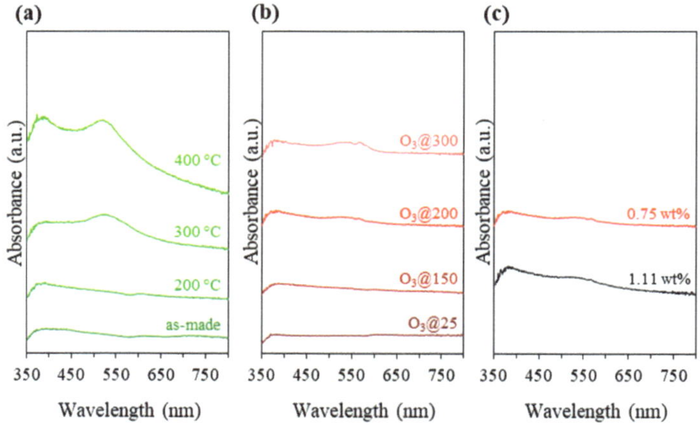

Figure 4. UV-vis DRS of Au_9-MPTMS@Na-LTA (0.75 ± 0.01 wt% Au) (**a**) before and after calcination under static air at different temperatures of 200, 300, and 400 °C and (**b**) after ozonolysis at different temperatures of 25, 150, 200, and 300 °C. (**c**) UV-vis DRS of Au_9-MPTMS@Na-LTA with 0.75 ± 0.01 and 1.11 ± 0.02 wt% Au after ozonolysis at 200 °C.

In the case of encapsulated noble MNP samples, such as Pt, Pd, Ru, Rh, and Ag, the degree of ligand removal could be inferred from the difference in metal surface areas derived from two approaches: (1) the mean diameter of metal particles obtained from electron micrographs and (2) H_2 or O_2 chemisorption by assuming spherical particles [33,34]. In the case of Au, the chemisorption of H_2 and O_2 is not feasible due to their high dissociation activation barriers [67]. Instead, CO chemisorption was used to obtain the surface area and derive the mean diameter of Au NPs within the zeolite framework [45]. However, the mean diameter of such Au NPs was not proportional to the total CO uptake by Au NPs encapsulated in zeolite. This is because the total CO uptake resulted from chemisorption on the Au species and the zeolite framework [68]. Therefore, FTIR spectra of the adsorbed CO must be recorded along with CO chemisorption to distinguish between CO adsorbed on Au NPs and zeolite, leading to a more complex monitoring approach.

In this work, the successful removal of the stabilizing ligands was inferred from the catalytic CO oxidation performance of the activated catalysts. The hypothesis was that if there was sufficient ligand removal, it would result in opened active sites, and the Au catalyst would show high CO oxidation activity at a relatively low reaction temperature. This hypothesis is based on earlier reports that ultra-small Au particles are active at much lower temperatures than larger Au species [69–75].

The CO oxidation was carried out in a fixed-bed flow reactor under atmospheric pressure at various temperatures, varying from 50 to 200 °C. As expected, without ligand removal, the Au_9-MPTMS@Na-LTA sample was inactive for CO oxidation in the temperature range studied. Surprisingly, after the conventional calcination under static air, all samples remained inactive. Insufficient ligand removal was possibly the main reason for the inactivity of the sample calcined at 200 °C, whereas the formation of larger agglomerated Au NPs was likely responsible for the inactivity of the samples treated at 300 and 400 °C. The inactivity in low-temperature CO oxidation due to the formation of large Au NPs generally agreed with the previous reports in the literature [69–72].

Since the removal of protecting ligands and the suppression of Au NC agglomeration could not be achieved simultaneously using conventional heat treatment, ozonolysis as a chemical treatment under milder conditions was explored. Ozone is the highly reactive oxidizing agent that allows oxidative removal of organic moieties under lower temperatures, reducing the chance of cluster agglomeration [13,76,77]. Adopting the ozonolysis approach, the Au_9-MPTMS@Na-LTA sample was treated under ozone flow at 25, 150, 200, and 300 °C for 1 h. A small change in the UV-vis DR spectra of all samples after ozone treatment is shown in Figure 4b. An O_3@300 sample featured the most pronounced plasmonic band, while the rest of the O_3-treated samples showed a negligible change in absorption intensity at the LSPR-Au NP absorption region of 500–600 nm.

The catalytic CO oxidation testing using O_3-treated samples was performed in the same manner as conventionally calcined samples. All O_3-treated samples showed their catalytic CO activity at a reaction temperature of 200 °C, except O_3@25, which was found inactive in the entire reaction temperature range (Figure S6). O_3@150 showed poor performance, especially at a low reaction temperature (≤100 °C). Even though UV-vis DRS and TEM of the post-reaction samples of O_3@25 and O_3@150 (Figure S7) showed a very low degree of Au agglomeration, these ozonolysis conditions did not allow for adequate ligand removal for low-temperature CO oxidation to take place. O_3@200 and O_3@300 showed a much better catalytic performance at low temperatures. O_3@200 gave more than 44% conversion and yield at 50 °C, and it was almost 50% higher than that of O_3@300 (Figure S6). The greater population of the larger LSPR-Au NPs in the O_3@300 sample (Figure 4b) was likely responsible for its lower activity.

According to the UV-vis DRS and TEM of the samples recovered after a catalytic test (Figure S7), the formation of larger LSPR-Au NPs during the catalytic test could be clearly confirmed in both O_3@200 and O_3@300 samples. As expected, the O_3@300 post-reaction sample showed a greater degree of cluster aggregation due to a higher MPTMS-removal efficiency during ozonolysis. The majority of sintered Au NPs in the post-reaction samples were in the range of 2–5 nm, while a minority >10 nm in diameter was found only in the O_3@300 post-reaction sample.

Ozonolysis combined with thermal treatment at 200 °C was shown to be the most promising activation approach for the Au_9-MPTMS@Na-LTA as it resulted in adequate ligand removal, as indicated by the best catalytic activity at ~50 °C. In addition, these results also indicated a relatively high ratio of surviving Au NCs to sintered Au NPs in the O_3@200 post-reaction sample. Ozonolysis at 200 °C was, therefore, applied to the most promising sample, Au_9-MPTMS@Na-LTA, with 1.11 ± 0.02 wt% Au. As shown in Figure 4c, there was no distinct difference in the UV-vis DRS of the sample compared to the one with 0.75 ± 0.01 wt% Au, suggesting that severe Au sintering did not occur in the sample with a higher Au content.

3.3. *Catalytic CO Oxidation*

The in situ incorporation approach aimed to encapsulate the Au NCs within the zeolite voids, enabling the greater stability of the clusters against sintering during catalytic reaction. Detailed catalytic CO oxidation studies discussed below confirm the efficiency of the Au NC encapsulation achieved using the in situ incorporation approach.

3.3.1. Effect of the Incorporation Approach—In Situ vs. Post-Incorporation

Since the catalytic performance of supported Au catalysts in low-temperature CO oxidation is known to strongly depend on the Au particle size [69–75] (as well as the support types [78,79]), sintering of the Au species was hypothesized to reduce the catalytic performance, as already discussed in Section 3.2 Catalyst Activation. With the higher stabilization given by the incorporation inside of the zeolite framework, the confined Au NCs were proposed to maintain their ultra-small size and be resilient against sintering and, hence, show a high catalytic CO oxidation activity. At the same time, it was hypothesized that if Au NCs were located at the external surface, their sintering would be more pronounced due to poorer stabilization. Thus, we compared the catalytic activity of the Au NCs confined within the zeolite framework, prepared using in situ incorporation (denoted as inst-Au_9-MPTMS@Na-LTA), and Au NCs immobilized on the external surface of the zeolite, prepared using post-impregnation (imp-Au_9-MPTMS@Na-LTA; synthesis details are described in Supporting Information, S3.10).

With the initial Au loading of 1.2 wt%, Au_9-MPTMS was introduced before and after the zeolite framework formation to give inst-Au_9-MPTMS@Na-LTA and imp-Au_9-MPTMS@Na-LTA, respectively. A much lower Au loading in the final product of imp-Au_9-MPTMS@Na-LTA (0.75 ± 0.03 wt%) compared to that of inst-Au_9-MPTMS@Na-LTA (1.11 ± 0.02 wt%) was confirmed by the MP-AES analysis. This result indicated that the in situ incorporation approach allowed Au NCs to incorporate with LTA zeolite better than in the case of post-impregnation. UV-vis DRS of as-made samples revealed much more apparent absorption features of Au_9-MPTMS at ~440 and ~720 nm in the case of imp-Au_9-MPTMS@Na-LTA in comparison to inst-Au_9-MPTMS@Na-LTA (Figure 5, as-made). This result correlated with the more intense red–brown color of the imp-Au_9-MPTMS@Na-LTA (Figure 6, as-made), implying that the majority of Au NCs were located on the external surface of Na-LTA. This is because the size of Au_9-MPTMS (>0.8 nm) was larger than the pore opening of Na-LTA (0.42 nm). The introduction of the clusters to Na-LTA after the framework formation has not allowed Au_9-MPTMS to access the zeolite pore.

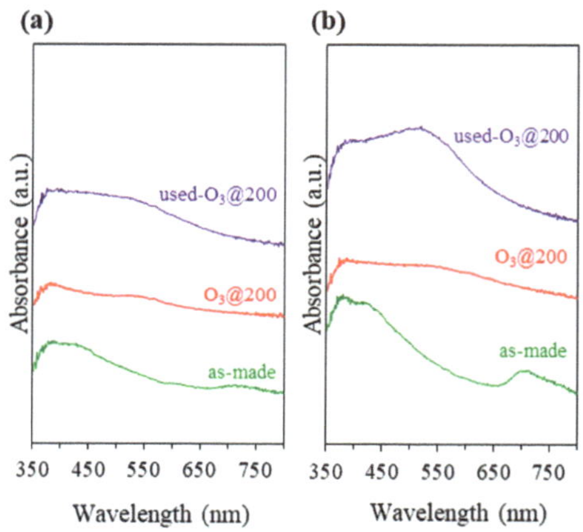

Figure 5. UV-vis DRS of as-made, O_3@200-treated, and post-reaction samples of (**a**) inst-Au_9-MPTMS@Na-LTA (1.11 ± 0.02 wt% Au) and (**b**) imp-Au_9-MPTMS@Na-LTA (0.75 ± 0.03 wt% Au). Reaction conditions: T_{max} of 100 °C, a total reaction time of 12 h.

Before the catalytic test, the samples were ozonolyzed at 200 °C for 1 h. The formation of Au NPs during the activation cannot be confirmed by UV-vis DRS (Figure 5, O_3@200). Nevertheless, TEM images (Figure 6, O_3@200) revealed numerous dark spots corresponding to Au NPs, especially in the case of imp-Au_9-MPTMS@Na-LTA. With the same total amount of gold per catalyst sample loaded into the reactor, the O_3@200-inst-Au_9-MPTMS@Na-LTA gave significantly higher CO conversion and CO_2 yield compared to the O_3@200-imp-Au_9-MPTMS@Na-LTA at all reaction temperatures (Figure 7). Impressively, using the O_3@200-inst-Au_9-MPTMS@Na-LTA, 100% conversion was achieved at 30 °C, the lowest temperature used in this study, and remained the same at 50 and 100 °C. Moreover, the catalyst exhibited 100% conversion even after 12 h of catalytic testing, indicating excellent stability of this catalyst (Figure S15). A higher degree of Au agglomeration during the catalytic test in the case of the O_3@200-imp-Au_9-MPTMS@Na-LTA was likely responsible for its lower catalytic activity compared to that of O_3@200-inst-Au_9-MPTMS@Na-LTA. This was confirmed by UV-vis DRS and TEM images of the post-reaction samples (Figures 5 and 6, used-O_3@200), where (1) a more pronounced LSPR band of sintered Au NPs and (2) a higher population of Au NPs visible in TEM, on the edge of zeolite or even dislodged Au NPs, were found in the case of the O_3@200-imp-Au_9-MPTMS@Na-LTA.

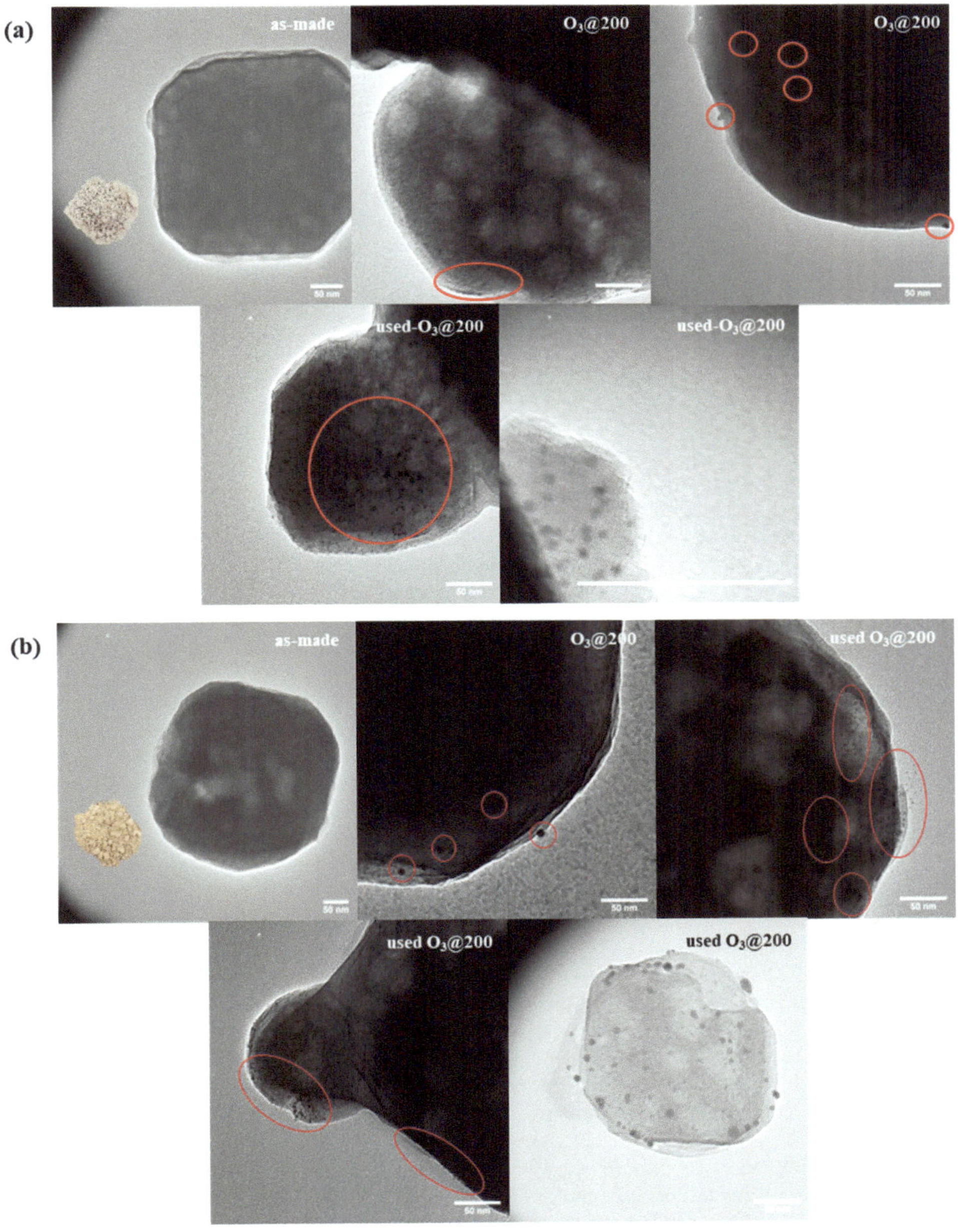

Figure 6. TEM images of as-made, O_3@200-treated, and post-reaction samples of (**a**) inst-Au_9-MPTMS@Na-LTA and (**b**) imp-Au_9-MPTMS@Na-LTA. Reaction conditions: T_{max} of 100 °C, a total reaction time of 12 h. Au NPs are highlighted in red circles. Scale bars are 50 nm. Photographs of the as made samples are shown as inserts in the as-made TEM images.

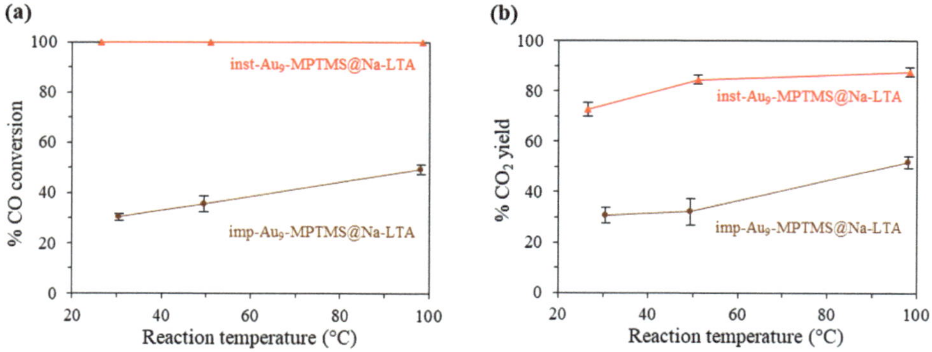

Figure 7. (**a**) % CO conversion and (**b**) % CO_2 yield in CO oxidation catalyzed by O_3@200-inst-Au_9-MPTMS@Na-LTA (1.11 ± 0.02 wt% Au) and O_3@200-imp-Au_9-MPTMS@Na-LTA (0.75 ± 0.03 wt% Au). Reaction conditions: GHSV of 3000- or 2030-mL g^{-1} h^{-1}, a total Au loading of 2.2 mg, 30–100 °C.

All these results suggested that Au NCs confined within the zeolite framework by the in situ incorporation have better stability during CO oxidation than surface-bound Au NCs prepared by post-impregnation. The excellent stability of the Au NCs led to superior catalytic performance, achieving 100% CO conversion and 73–88% CO_2 yield for up to 12 h time-on-stream. Hence, it could be concluded that the in situ incorporation approach offers Au NCs higher stability against agglomeration during the catalytic reaction in contrast to the post-impregnation approach.

3.3.2. Effect of Zeolite Framework—LTA vs. FAU

As shown in Figure 7, CO_2 yield was ~12–27% lower than CO conversion in the temperature range of 30–100 °C using the inst-Au_9-MPTMS@Na-LTA catalyst. A greater difference in CO conversion cf. CO_2 yield was found at lower temperatures (by 27% at 30 °C, and by 12% at 100 °C). The molecular adsorption of CO on zeolite counter ions and Au active sites in the zeolite-supported Au catalysts was previously reported as a reason for this carbon imbalance [45,68]. By conducting control experiments under the same conditions as the catalytic test but without O_2–passing 1.0% CO in Ar through the samples at 30–100 °C, the amount of CO adsorbed could be determined. Without gold (O_3@200-MPTMS@Na-LTA), a 4–7% loss of CO was found; in the presence of gold (O_3@200-Au_9-MPTMS@Na-LTA), a 10–14% loss was observed. Therefore, the carbon balance issue can be partially explained by the physisorption of CO within zeolite and the chemisorption of CO on Au [80]. Another possible explanation could be the adsorption of produced CO_2 on the zeolite surface; however, further experiments are required to confirm this hypothesis.

Apart from the molecular adsorption of CO and CO_2, the mass transfer rate of CO, O_2, and CO_2 within the LTA framework could also affect the catalytic reaction, especially when the conversion of CO to CO_2 takes place at the Au NC active sites encapsulated within the zeolite framework. Since access to the cages, cavities, or channels of a zeolite is controlled by the largest free path, CO, O_2, and CO_2 with kinetic diameters of ~0.38, 0.35, and 0.33 nm, respectively, are allowed to enter the alpha cavity of Na-LTA (maximum aperture 0.42 nm). By using zeolite with a larger cavity aperture, the mass transfer could be generally improved. Nevertheless, the zeolite cavity diameter itself is considered a critical factor for successful in situ encapsulation. Thus, for the study of the mass transfer rate effect, a Na-FAU zeolite was chosen as it possesses the same maximum cavity diameter (of ~1.1 nm) as Na-LTA but has larger apertures of 0.74 nm [53,81].

Au_9-MPTMS@Na-FAU was fabricated using the in situ incorporation approach, analogous to the Au_9-MPTMS@Na-LTA (see Supporting Information, S3.7). Both Au_9-MPTMS@Na-LTA and Au_9-MPTMS@Na-FAU were treated under O_3 at 200 °C before catalytic CO oxidation under the same conditions.

Synthesis optimization experiments showed that to achieve the target Au content of 1.0 wt%, twice the amount of Au was introduced to the Na-FAU zeolite synthesis gel (~2.4 wt% Au) compared to that of Na-LTA (~1.2 wt% Au). Interestingly, the Au_9-MPTMS@Na-FAU sample with 0.96 ± 0.05 wt% Au showed two prominent absorption bands at ~440 and ~740 nm (Figure 8b, as-made), which can be attributed to the Au_9-MPTMS, similar to the case of imp-Au_9-MPTMS@Na-LTA (Figure 5b, as-made). From the UV-vis DRS results, a higher density of Au NCs located at the outer surface of the FAU zeolite crystallite in comparison to that of LTA could be implied.

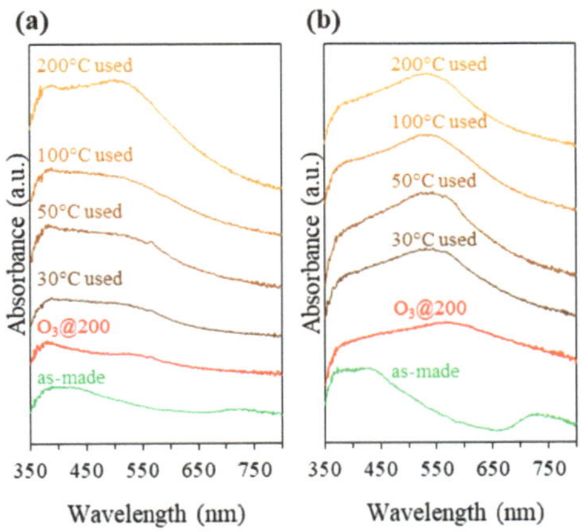

Figure 8. UV-vis DRS of as-made, O_3@200-treated, and post-reaction samples of (**a**) Au_9-MPTMS@Na-LTA (1.11 ± 0.02 wt% Au) and (**b**) Au_9-MPTMS@Na-FAU (0.96 ± 0.05 wt% Au). Reaction conditions: temperatures of 30, 50, 100, and 200 °C; reaction time of 2 h.

To compare the performance of the two catalysts, the gas hourly space velocity (GHSV) was adjusted to ~30,000 mL g^{-1} h^{-1} to ensure that the reaction was kinetically controlled. While the total Au loading for each catalytic test was maintained at 0.4 mg (~40 mg of the catalyst), each catalytic test was performed separately at 30, 50, 100, and 200 °C for 2 h. As shown in Figure 9, Au_9-MPTMS@Na-LTA showed a lower conversion and yield due to the lower total Au loading and shorter contact time (higher GHSV). The sample gave 31 ± 3% CO conversion and 33 ± 2% CO_2 yield at 30 °C. The conversion and yield increased with the increase in reaction temperature from 30 to 50 °C, where 44 ± 5% conversion and 43 ± 3% yield were observed. In comparison, a further increase in the reaction temperature from 50 to 100 and 200 °C did not significantly change CO oxidation activity. This consistency of the catalytic activity could be explained by the unchanged UV-vis DRS profiles of Au_9-MPTMS@Na-LTA before and after the catalytic test at 30, 50, and 100 °C, as shown in Figure 8a. This result highlights that LTA could offer great sintering resistance to the Au species during the catalytic reaction. TEM, however, revealed some degree of Au sintering in the post-reaction samples of Au_9-MPTMS@Na-LTA (used at 30, 50, and 100 °C). In comparison, the TEM of a post-reaction sample at 200 °C showed the highest numbers of sintered Au NPs among all used Au_9-MPTMS@Na-LTA catalysts (Figure S8).

Surprisingly, the Au_9-MPTMS@Na-FAU showed no catalytic activity in CO oxidation under the same conditions (Figure 9). The sample showed the highest CO loss (4–6%) at 30 °C, which was likely due to the adsorption rather than the conversion since no CO_2 product could be detected up to 200 °C. The severe sintering of Au NCs to form larger Au NPs was likely responsible for the inactivity of the activated Au_9-MPTMS@Na-FAU sample since the sample featured a broad plasmonic band even before the catalytic test (Figure 8b, O_3@200). Such an intense plasmonic band after the O_3 treatment was not found in any other samples reported in this work, not even in the imp-Au_9-MPTMS@Na-LTA sample, which was prepared by the post-impregnation method. Thus, the more accessible FAU zeolite framework must, indeed, have a strong influence on this pronounced sintering phenomenon.

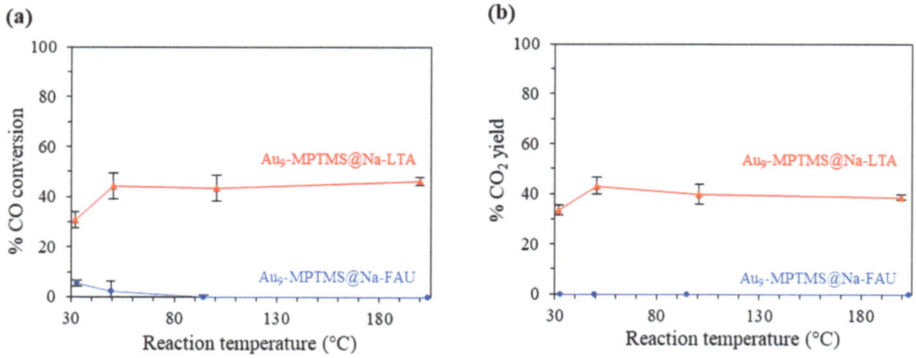

Figure 9. (**a**) % CO conversion and (**b**) % CO_2 yield in CO oxidation catalyzed by O_3@200-Au_9-MPTMS@Na-LTA (1.11 ± 0.02 wt% Au) (raw data are shown in Figure S16) and O_3@200-Au_9-MPTMS@Na-FAU (0.96 ± 0.05 wt% Au). Reaction conditions: GHSV of ~30,000 mL g^{-1} h^{-1}, catalyst loading of ~ 40 mg, total Au loading of 0.4 mg, 30–200 °C.

It is possible that the larger pore opening of FAU not only allows CO and O_2 to easily access the Au active sites but also allows Au NCs species to easily migrate out of the cavities and sinter with other Au particles, resulting in lower catalyst performance. The other hypothesis was that most Au species in the Au_9-MPTMS@Na-FAU sample were not encapsulated within the FAU framework but were instead located on the external surface zeolite crystallites (which would account for the more pronounced features in Figure 8b, as made). Therefore, Au NCs in such samples showed low stability, undergoing severe agglomeration during catalyst activation and the catalytic test, as evidenced by the rather pronounced plasmonic band. The intensity of plasmonic bands in the post-catalytic samples of Au_9-MPTMS@Na-FAU was even more significant than in the case of imp-Au_9-MPTMS@Na-LTA, implying that sintering is facilitated by the chemical nature of the FAU sample (e.g., Si/Al ratio).

3.3.3. Effect of Incorporated Au Species–Au NCs vs. Au NPs

Earlier, it was proposed that the formation of larger Au NPs due to the aggregation of Au NCs during the catalyst activation and/or the catalytic test was responsible for the lower catalytic activity or inactivity of the zeolite-supported Au catalysts. Under this hypothesis, the zeolite-supported ultra-small Au NCs were proposed to be the major active species responsible for CO oxidation at low temperatures, while it was hypothesized that the larger Au NPs were less active or inactive in CO oxidation under the same conditions. However, the size regime of the active Au species in this catalytic system cannot be verified due to the co-existence of Au NCs and Au NPs, as confirmed by the UV-vis DRS and TEM of all post-reaction samples. To directly compare the CO oxidation performance of the zeolite-supported Au catalyst with different sizes of incorporated Au species, AuNPs-MPTMS@Na-LTA was prepared following the procedure established by Iglesia et al. [45]. With a similar in situ incorporation strategy, the same Na-LTA zeolite host (Si/Al in the synthesis gel of ~0.8), and similar final Au content, the Au particle size was considered as the only different parameter influencing the catalytic performance of the two samples.

The AuNPs-MPTMS@Na-LTA was obtained with a 98 ± 2% yield based on SiO_2, comparable to the yield reported in the literature [45]. As confirmed by PXRD, AuNPs-MPTMS@Na-LTA showed a similar crystallinity to Au_9-MPTMS@Na-LTA (Figure S11). A slightly higher Au content was found in the former case (1.16 ± 0.02 wt% Au cf.

1.11 ± 0.02 wt% Au) due to a higher amount of introduced gold (1.3 wt% Au cf. 1.2 wt% Au). Noteworthy, the formation of Au NPs in the AuNPs-MPTMS@Na-LTA sample did not take place until post-synthesis thermal treatment under air and later under H_2, where MPTMS ligands were removed and, consequently, Au^{3+} was reduced to Au^0. Consistent with the previous report, H_2 reduction at 300 °C resulted in a sharp increase of the plasmonic band intensity with λ_{max} of 510 nm, indicating Au NPs (Figure 10b, H_2@300) [45]. Based on the position of the LSPR band maximum, the size of the Au NPs could be estimated to be between 2 and 2.5 nm [82], or close to 279 Au core atoms (2.25 nm) [83].

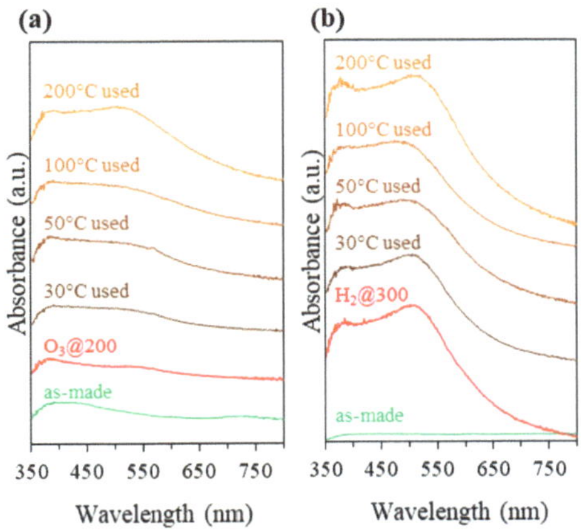

Figure 10. UV-vis DRS of as-made, activated, and post-reaction samples of (**a**) Au_9-MPTMS@Na-LTA (1.11 ± 0.02 wt% Au) and (**b**) AuNPs-MPTMS@Na-LTA (1.16 ± 0.02 wt% Au). Reaction conditions: temperature of 30, 50, 100, and 200 °C; reaction time of 2 h.

Comparing the UV-vis DRS of the two activated samples with a similar Au loading (Figure 10, O_3@200 cf. H_2@300), O_3@200-Au_9-MPTMS@Na-LTA exhibited a broad band across the spectrum with a barely detectable plasmonic band. This may suggest that most Au species were in the cluster form, whereas H_2@300-AuNPs-MPTMS@Na-LTA showed a prominent plasmonic peak, confirming the presence of larger Au NPs. However, further agglomeration of the Au species in both samples during the catalytic test was confirmed by a more intense and broader plasmonic band of the post-reaction samples.

Catalytic testing was performed similarly to the test discussed earlier. Each catalytic test was performed separately at 30, 50, 100, and 200 °C, for 2 h, with a GHSV of 30,000 mL g^{-1} h^{-1}. As shown in Figure 11, AuNPs-MPTMS@Na-LTA did not show any CO_2 production up to 200 °C. A 20–30% CO loss found across 30–200 °C could possibly be attributed to the molecular adsorption of CO, as discussed in the earlier section. These results, therefore, supported the hypothesis that Au NCs, not Au NPs, acted as the active species in the low-temperature CO oxidation using zeolite-supported Au catalysts.

A higher number of low-coordinated Au atoms in smaller supported Au particles was generally proposed as the origin of their excellent activity over larger Au particles due to their superior O_2 binding or dissociation sites [69–72,84–86]. Moreover, the adsorption energies of both CO and O atoms on Au were reported to depend strongly on the coordination number of the Au atom to which they bind. Specifically, the theoretical calculations showed that both O and CO adsorption energies are lowered by up to 1 eV when the coordination number of Au atom is reduced from 9 in the case of Au (1 1 1) to 4 in the case of Au_{10} clusters [70]. However, the fact that the LTA-supported Au NPs with an average diameter of ~1.3 nm (as reported by Iglesia et al. [45]) or 2.0–2.5 nm (as estimated from the LSPR band position) did not show any CO oxidation at 30–200 °C remains inexplicable [87]. The effect of support types on the size thresholds of the active Au species might be one reason as it was shown earlier that the specific size regime of the most active Au species in low-temperature CO oxidation varied, depending on the type of oxide support [63,78,79].

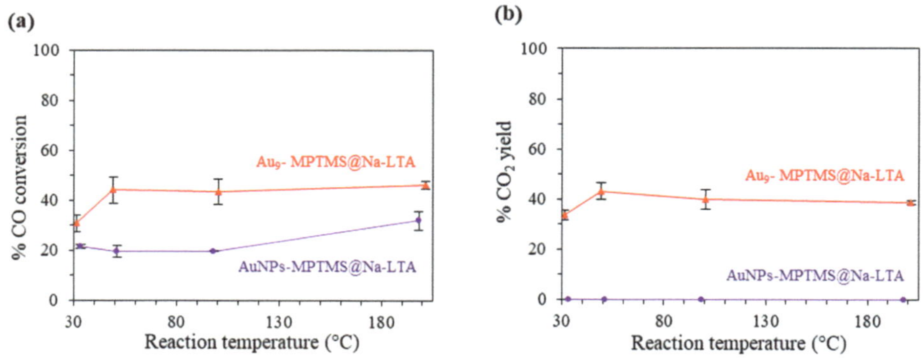

Figure 11. (**a**) % CO conversion and (**b**) % CO_2 yield in CO oxidation catalyzed by O_3@200-Au_9-MPTMS@Na-LTA (1.11 ± 0.02 wt% Au) and H_2@300-AuNPs-MPTMS@Na-LTA (1.16 ± 0.02 wt% Au). Reaction conditions: GHSV of ~30,000 mL g^{-1} h^{-1}, catalyst loading of ~40 mg, total Au loading of 0.4 mg, 30–200 °C.

4. Conclusions

In summary, we presented the fabrication of Au NC-based catalysts with clusters incorporated within the zeolite framework via the in situ incorporation of pre-made atomically precise Au_9 clusters during hydrothermal synthesis of Na-LTA zeolite. We demonstrated the importance of the ligand exchange (phosphine to mercaptosilane ligands) for the highly reproducible successful synthesis of zeolite-incorporated gold clusters, with high stability of the Au_9 clusters against sintering. The successful encapsulation of Au NCs was suggested by their superior catalytic CO oxidation performance. The catalyst illustrated good stability against sintering, maintaining a 100% CO conversion at 30 °C, up to 12 h, due to metal encapsulation. This result highlights the advantages of the restrictive framework of Na-LTA zeolite, which suppresses Au sintering, leading to the better performance of confined Au NCs in catalytic CO oxidation. The performance of this catalyst was superior to Au NCs immobilized on the external surface of LTA or encapsulated within the FAU framework with larger openings. Moreover, the unique catalytic activity of ultrasmall gold clusters compared to their larger gold nanoparticle counterparts was illustrated in this work. Ultra-small Au NCs have been shown to be major contributors to superior low-temperature catalytic CO oxidation performance. In contrast, even slightly larger Au NPs (2.0–2.5 nm) did not show any CO_2 yield across the temperatures studied.

Supplementary Materials: The following supporting information can be downloaded at https://www.mdpi.com/article/10.3390/nano13243120/s1. The supporting information is available free of charge. It includes detailed information on material synthesis and characterizations (NMR, PXRD, mass spectra, UV-vis DRS, SEM, TEM, MP-AES results), as well as the setup for the catalytic CO oxidation test, including a process flow diagram, GC settings, typical chromatograms, and calibration plots (PDFs). (Reference [88] is cited in the supplementary materials).

Author Contributions: Conceptualization, A.C.K.Y. and V.B.G.; Formal analysis, S.T.; Investigation, S.T.; Resources, J.V.K., A.C.K.Y. and V.B.G.; Data curation, A.C.K.Y. and V.B.G.; Writing–original draft, S.T.; Writing–review & editing, J.V.K., A.C.K.Y. and V.B.G.; Supervision, J.V.K., A.C.K.Y. and V.B.G. All authors have read and agreed to the published version of the manuscript.

Funding: This work was supported by the University of Canterbury (A.C.K.Y, V.G. S.T.). S.T. acknowledges the University of Canterbury for the UC Doctoral scholarship. All the co-authors thank the MacDiarmid Institute for Advanced Materials and Nanotechnology for generous financial support.

Data Availability Statement: The majority of data created during this study are available within this manuscript and its Supporting Information. All other data can be made available upon reasonable request to the corresponding authors.

Acknowledgments: We thank Shaun Mucalo from the School of Mechanical Engineering for SEM training and TEM imaging. We thank Marie Squire, Amanda Inglis, and Matthew Polson from the School of Physical and Chemical Sciences for their help with NMR, MS, and PXRD analysis.

Conflicts of Interest: The authors declare no competing financial interest.

References

1. Takano, S.; Tsukuda, T. Chapter 2—Controlled synthesis: Size control. In *Frontiers of Nanoscience*; Tsukuda, T., Häkkinen, H., Eds.; Elsevier: Amsterdam, The Netherlands, 2015; Volume 9, pp. 9–38.
2. Jin, R.; Zeng, C.; Zhou, M.; Chen, Y. Atomically precise colloidal metal nanoclusters and nanoparticles: Fundamentals and opportunities. *Chem. Rev.* **2016**, *116*, 10346–10413. [CrossRef] [PubMed]
3. Liu, L.; Meira, D.M.; Arenal, R.; Concepcion, P.; Puga, A.V.; Corma, A. Determination of the evolution of heterogeneous single metal atoms and nanoclusters under reaction conditions: Which are the working catalytic sites? *ACS Catal.* **2019**, *9*, 10626–10639. [CrossRef] [PubMed]
4. Niu, Z.; Li, Y. Removal and utilization of capping agents in nanocatalysis. *Chem. Mater.* **2014**, *26*, 72–83. [CrossRef]
5. Li, Z.; Ji, S.; Liu, Y.; Cao, X.; Tian, S.; Chen, Y.; Niu, Z.; Li, Y. Well-defined materials for heterogeneous catalysis: From nanoparticles to isolated single-atom sites. *Chem. Rev.* **2020**, *120*, 623–682. [CrossRef] [PubMed]
6. Liu, L.; Corma, A. Metal Catalysts for Heterogeneous Catalysis: From Single Atoms to Nanoclusters and Nanoparticles. *Chem. Rev.* **2018**, *118*, 4981–5079. [CrossRef] [PubMed]
7. Liu, L.; Corma, A. Evolution of isolated atoms and clusters in catalysis. *Trends Chem.* **2020**, *2*, 383–400. [CrossRef]
8. Fukamori, Y.; König, M.; Yoon, B.; Wang, B.; Esch, F.; Heiz, U.; Landman, U. Fundamental Insight into the Substrate-Dependent Ripening of Monodisperse Clusters. *ChemCatChem* **2013**, *5*, 3330–3341. [CrossRef]
9. Krishnan, G.; Al Qahtani, H.S.; Li, J.; Yin, Y.; Eom, N.; Golovko, V.B.; Metha, G.F.; Andersson, G.G. Investigation of Ligand-Stabilized Gold Clusters on Defect-Rich Titania. *J. Phys. Chem. C* **2017**, *121*, 28007–28016. [CrossRef]
10. Mousavi, H.; Yin, Y.; Howard-Fabretto, L.; Sharma, S.K.; Golovko, V.; Andersson, G.G.; Shearer, C.J.; Metha, G.F. Au101–rGO nanocomposite: Immobilization of phosphine-protected gold nanoclusters on reduced graphene oxide without aggregation. *Nanoscale Adv.* **2021**, *3*, 1422–1430. [CrossRef]
11. Al Qahtani, H.S.; Metha, G.F.; Walsh, R.B.; Golovko, V.B.; Andersson, G.G.; Nakayama, T. Aggregation Behavior of Ligand-Protected Au9 Clusters on Sputtered Atomic Layer Deposition TiO_2. *J. Phys. Chem. C* **2017**, *121*, 10781–10789. [CrossRef]
12. Al Qahtani, H.S.; Higuchi, R.; Sasaki, T.; Alvino, J.F.; Metha, G.F.; Golovko, V.B.; Adnan, R.; Andersson, G.G.; Nakayama, T. Grouping and aggregation of ligand protected Au9 clusters on TiO_2 nanosheets. *RSC Adv.* **2016**, *6*, 110765–110774. [CrossRef]
13. Ruzicka, J.-Y.; Abu Bakar, F.; Hoeck, C.; Adnan, R.; McNicoll, C.; Kemmitt, T.; Cowie, B.C.; Metha, G.F.; Andersson, G.G.; Golovko, V.B. Toward Control of Gold Cluster Aggregation on TiO2 via Surface Treatments. *J. Phys. Chem. C* **2015**, *119*, 24465–24474. [CrossRef]
14. Anderson, D.P.; Alvino, J.F.; Gentleman, A.; Qahtani, H.A.; Thomsen, L.; Polson, M.I.J.; Metha, G.F.; Golovko, V.B.; Andersson, G.G. Chemically-synthesised, atomically-precise gold clusters deposited and activated on titania. *Phys. Chem. Chem. Phys.* **2013**, *15*, 3917–3929. [CrossRef] [PubMed]
15. Anderson, D.P.; Adnan, R.H.; Alvino, J.F.; Shipper, O.; Donoeva, B.; Ruzicka, J.-Y.; Al Qahtani, H.; Harris, H.H.; Cowie, B.; Aitken, J.B.; et al. Chemically synthesised atomically precise gold clusters deposited and activated on titania. Part II. *Phys. Chem. Chem. Phys.* **2013**, *15*, 14806–14813. [CrossRef] [PubMed]
16. Campbell, C.T.; Parker, S.C.; Starr, D.E. The effect of size-dependent nanoparticle energetics on catalyst sintering. *Science* **2002**, *298*, 811–814. [CrossRef] [PubMed]
17. Ouyang, R.; Liu, J.-X.; Li, W.-X. Atomistic theory of Ostwald Ripening and disintegration of supported metal particles under reaction conditions. *J. Am. Chem. Soc.* **2013**, *135*, 1760–1771. [CrossRef] [PubMed]
18. Goodman, E.D.; Schwalbe, J.A.; Cargnello, M. Mechanistic understanding and the rational design of sinter-resistant heterogeneous catalysts. *ACS Catal.* **2017**, *7*, 7156–7173. [CrossRef]
19. Babucci, M.; Guntida, A.; Gates, B.C. Atomically dispersed metals on well-defined supports including zeolites and metal–organic frameworks: Structure, bonding, reactivity, and catalysis. *Chem. Rev.* **2020**, *120*, 11956–11985. [CrossRef]
20. Liu, L.; Corma, A. Confining isolated atoms and clusters in crystalline porous materials for catalysis. *Nat. Rev. Mater.* **2021**, *6*, 244–263. [CrossRef]
21. Wang, H.; Wang, L.; Xiao, F.-S. Metal@zeolite hybrid materials for catalysis. *ACS Cent. Sci.* **2020**, *6*, 1685–1697. [CrossRef]
22. Kosinov, N.; Liu, C.; Hensen, E.J.M.; Pidko, E.A. Engineering of transition metal catalysts confined in zeolites. *Chem. Mater.* **2018**, *30*, 3177–3198. [CrossRef] [PubMed]
23. Pagis, C.; Morgado Prates, A.R.; Farrusseng, D.; Bats, N.; Tuel, A. Hollow zeolite structures: An overview of synthesis methods. *Chem. Mater.* **2016**, *28*, 5205–5223. [CrossRef]

24. Ou, Z.; Li, Y.; Wu, W.; Bi, Y.; Xing, E.; Yu, T.; Chen, Q. Encapsulating subnanometric metal clusters in zeolites for catalysis ar their challenges. *Chem. Eng. J.* **2022**, *430*, 132925. [CrossRef]
25. Liu, L.; Lopez-Haro, M.; Calvino, J.J.; Corma, A. Tutorial: Structural characterization of isolated metal atoms and subnanometr metal clusters in zeolites. *Nat. Protoc.* **2021**, *16*, 1871–1906. [CrossRef] [PubMed]
26. Qi, G.; Davies, T.E.; Nasrallah, A.; Sainna, M.A.; Howe, A.G.R.; Lewis, R.J.; Quesne, M.; Catlow, C.R.A.; Willock, D.J.; He, Q.; et Au-ZSM-5 catalyses the selective oxidation of CH_4 to CH_3OH and CH_3COOH using O_2. *Nat. Catal.* **2022**, *5*, 45–54. [CrossRef]
27. De Graaf, J.; van Dillen, A.J.; de Jong, K.P.; Koningsberger, D.C. Preparation of highly dispersed Pt particles in zeolite Y with narrow particle size distribution: Characterization by hydrogen chemisorption, TEM, EXAFS spectroscopy, and particle modelir *J. Catal.* **2001**, *203*, 307–321. [CrossRef]
28. Zečević, J.; van der Eerden, A.M.J.; Friedrich, H.; de Jongh, P.E.; de Jong, K.P. Heterogeneities of the nanostructure of pla inum/zeolite Y catalysts revealed by electron tomography. *ACS Nano* **2013**, *7*, 3698–3705. [CrossRef]
29. Serna, P.; Gates, B.C. Molecular metal catalysts on supports: Organometallic chemistry meets surface science. *Acc. Chem. Re* **2014**, *47*, 2612–2620. [CrossRef]
30. Zeng, S.; Ding, S.; Li, S.; Wang, R.; Zhang, Z. Controlled growth of gold nanoparticles in zeolite L via ion-exchange reactions an thermal reduction processes. *Inorg. Chem. Commun.* **2014**, *47*, 63–66. [CrossRef]
31. Gu, J.; Zhang, Z.; Hu, P.; Ding, L.; Xue, N.; Peng, L.; Guo, X.; Lin, M.; Ding, W. Platinum nanoparticles encapsulated in MFI zeoli crystals by a two-step dry gel conversion method as a highly selective hydrogenation catalyst. *ACS Catal.* **2015**, *5*, 6893–690 [CrossRef]
32. Liu, L.; Zakharov, D.N.; Arenal, R.; Concepcion, P.; Stach, E.A.; Corma, A. Evolution and stabilization of subnanometric met species in confined space by in situ TEM. *Nat. Commun.* **2018**, *9*, 574. [CrossRef] [PubMed]
33. Goel, S.; Wu, Z.; Zones, S.I.; Iglesia, E. Synthesis and catalytic properties of metal clusters encapsulated within small-pore (SO GIS, ANA) zeolites. *J. Am. Chem. Soc.* **2012**, *134*, 17688–17695. [CrossRef] [PubMed]
34. Wu, Z.; Goel, S.; Choi, M.; Iglesia, E. Hydrothermal synthesis of LTA-encapsulated metal clusters and consequences for cataly stability, reactivity, and selectivity. *J. Catal.* **2014**, *311*, 458–468. [CrossRef]
35. Otto, T.; Zones, S.I.; Hong, Y.; Iglesia, E. Synthesis of highly dispersed cobalt oxide clusters encapsulated within LTA zeolites. *Catal.* **2017**, *356*, 173–185. [CrossRef]
36. Otto, T.; Zones, S.I.; Iglesia, E. Synthetic strategies for the encapsulation of nanoparticles of Ni, Co, and Fe oxides within crystallir microporous aluminosilicates. *Microporous Mesoporous Mater.* **2018**, *270*, 10–23. [CrossRef]
37. Wang, N.; Sun, Q.; Bai, R.; Li, X.; Guo, G.; Yu, J. In situ confinement of ultrasmall Pd clusters within nanosized Silicalite-1 zeoli for highly efficient catalysis of hydrogen generation. *J. Am. Chem. Soc.* **2016**, *138*, 7484–7487. [CrossRef] [PubMed]
38. Sun, Q.; Wang, N.; Bing, Q.; Si, R.; Liu, J.; Bai, R.; Zhang, P.; Jia, M.; Yu, J. Subnanometric hybrid Pd-M$(OH)_2$, M = Ni, Co cluste in zeolites as highly efficient nanocatalysts for hydrogen generation. *Chem* **2017**, *3*, 477–493. [CrossRef]
39. Chen, Q.; Wang, M.; Zhang, C.; Ren, K.; Xin, Y.; Zhao, M.; Xing, E. Selectivity control on hydrogenation of substituted nitroarene through end-on adsorption of reactants in zeolite-encapsulated platinum nanoparticles. *Chem. Asian J.* **2018**, *13*, 2077–208 [CrossRef]
40. Shan, Y.; Sui, Z.; Zhu, Y.; Zhou, J.; Zhou, X.; Chen, D. Boosting size-selective hydrogen combustion in the presence of proper using controllable metal clusters encapsulated in zeolite. *Angew. Chem. Int. Ed.* **2018**, *57*, 9770–9774. [CrossRef]
41. Liu, L.; Lopez-Haro, M.; Lopes, C.W.; Li, C.; Concepcion, P.; Simonelli, L.; Calvino, J.J.; Corma, A. Regioselective generation an reactivity control of subnanometric platinum clusters in zeolites for high-temperature catalysis. *Nat. Mater.* **2019**, *18*, 866–87 [CrossRef]
42. Liu, Y.; Li, Z.; Yu, Q.; Chen, Y.; Chai, Z.; Zhao, G.; Liu, S.; Cheong, W.-C.; Pan, Y.; Zhang, Q.; et al. A general strategy for fabricatin isolated single metal atomic site catalysts in Y zeolite. *J. Am. Chem. Soc.* **2019**, *141*, 9305–9311. [CrossRef] [PubMed]
43. Sun, Q.; Wang, N.; Zhang, T.; Bai, R.; Mayoral, A.; Zhang, P.; Zhang, Q.; Terasaki, O.; Yu, J. Zeolite-encaged single-atom rhodiur catalysts: Highly-efficient hydrogen generation and shape-selective tandem hydrogenation of nitroarenes. *Angew. Chem. Int. E* **2019**, *58*, 18570–18576. [CrossRef] [PubMed]
44. Choi, M.; Wu, Z.; Iglesia, E. Mercaptosilane-assisted synthesis of metal clusters within zeolites and catalytic consequences c encapsulation. *J. Am. Chem. Soc.* **2010**, *132*, 9129–9137. [CrossRef] [PubMed]
45. Otto, T.; Zones, S.I.; Iglesia, E. Challenges and strategies in the encapsulation and stabilization of monodisperse Au cluste within zeolites. *J. Catal.* **2016**, *339*, 195–208. [CrossRef]
46. Otto, T.; Ramallo-López, J.M.; Giovanetti, L.J.; Requejo, F.G.; Zones, S.I.; Iglesia, E. Synthesis of stable monodisperse AuPd, AuP and PdPt bimetallic clusters encapsulated within LTA-zeolites. *J. Catal.* **2016**, *342*, 125–137. [CrossRef]
47. Lee, S.; Lee, K.; Im, J.; Kim, H.; Choi, M. Revisiting hydrogen spillover in Pt/LTA: Effects of physical diluents having differer acid site distributions. *J. Catal.* **2015**, *325*, 26–34. [CrossRef]
48. Moliner, M.; Gabay, J.E.; Kliewer, C.E.; Carr, R.T.; Guzman, J.; Casty, G.L.; Serna, P.; Corma, A. Reversible transformation c Pt nanoparticles into single atoms inside high-silica Chabazite zeolite. *J. Am. Chem. Soc.* **2016**, *138*, 15743–15750. [CrossRe [PubMed]
49. Wang, Q.; Han, W.; Lyu, J.; Zhang, Q.; Guo, L.; Li, X. In situ encapsulation of platinum clusters within H-ZSM-5 zeolite for highl stable benzene methylation catalysis. *Catal. Sci. Technol.* **2017**, *7*, 6140–6150. [CrossRef]

50. Li, S.; Tuel, A.; Laprune, D.; Meunier, F.; Farrusseng, D. Transition-metal nanoparticles in hollow zeolite single crystals as bifunctional and size-selective hydrogenation catalysts. *Chem. Mater.* **2015**, *27*, 276–282. [CrossRef]
51. Ingham, B.; Lim, T.H.; Dotzler, C.J.; Henning, A.; Toney, M.F.; Tilley, R.D. How nanoparticles coalesce: An in situ study of Au nanoparticle aggregation and grain growth. *Chem. Mater.* **2011**, *23*, 3312–3317. [CrossRef]
52. Woehrle, G.H.; Hutchison, J.E. Thiol-functionalized undecagold clusters by ligand exchange: Synthesis, mechanism, and properties. *Inorg. Chem.* **2005**, *44*, 6149–6158. [CrossRef] [PubMed]
53. Thompson, R.W.; Franklin, K.C. Chapter 55—LTA Linde Type A Si(50), Al(50). In *Verified Syntheses of Zeolitic Materials*; Robson, H., Lillerud, K.P., Eds.; Elsevier: Amsterdam, The Netherlands, 2001; pp. 179–181.
54. Liu, L.; Díaz, U.; Arenal, R.; Agostini, G.; Concepción, P.; Corma, A. Generation of subnanometric platinum with high stability during transformation of a 2D zeolite into 3D. *Nat. Mater.* **2017**, *16*, 132–138. [CrossRef] [PubMed]
55. Holder, C.F.; Schaak, R.E. Tutorial on powder X-ray diffraction for characterizing nanoscale materials. *ACS Nano* **2019**, *13*, 7359–7365. [CrossRef] [PubMed]
56. Motl, N.E.; Smith, A.F.; DeSantis, C.J.; Skrabalak, S.E. Engineering plasmonic metal colloids through composition and structural design. *Chem. Soc. Rev.* **2014**, *43*, 3823–3834. [CrossRef] [PubMed]
57. Eustis, S.; El-Sayed, M.A. Why gold nanoparticles are more precious than pretty gold: Noble metal surface plasmon resonance and its enhancement of the radiative and nonradiative properties of nanocrystals of different shapes. *Chem. Soc. Rev.* **2006**, *35*, 209–217. [CrossRef] [PubMed]
58. Aikens, C.M. Chapter 9—Optical properties and chirality. In *Frontiers of Nanoscience*; Tsukuda, T., Häkkinen, H., Eds.; Elsevier: Amsterdam, The Netherlands, 2015; Volume 9, pp. 223–261.
59. Häkkinen, H. The gold–sulfur interface at the nanoscale. *Nat. Chem.* **2012**, *4*, 443–455. [CrossRef]
60. Tai, Y.; Yamaguchi, W.; Okada, M.; Ohashi, F.; Shimizu, K.-i.; Satsuma, A.; Tajiri, K.; Kageyama, H. Depletion of CO oxidation activity of supported Au catalysts prepared from thiol-capped Au nanoparticles by sulfates formed at Au–titania boundaries: Effects of heat treatment conditions on catalytic activity. *J. Catal.* **2010**, *270*, 234–241. [CrossRef]
61. Ma, G.; Binder, A.; Chi, M.; Liu, C.; Jin, R.; Jiang, D.-e.; Fan, J.; Dai, S. Stabilizing gold clusters by heterostructured transition-metal oxide–mesoporous silica supports for enhanced catalytic activities for CO oxidation. *Chem. Commun.* **2012**, *48*, 11413–11415. [CrossRef]
62. Gaur, S.; Wu, H.; Stanley, G.G.; More, K.; Kumar, C.S.S.R.; Spivey, J.J. CO oxidation studies over cluster-derived Au/TiO_2 and AUROlite™ Au/TiO_2 catalysts using DRIFTS. *Catal. Today* **2013**, *208*, 72–81. [CrossRef]
63. Nie, X.; Qian, H.; Ge, Q.; Xu, H.; Jin, R. CO oxidation catalyzed by oxide-supported $Au_{25}(SR)_{18}$ nanoclusters and identification of perimeter sites as active centers. *ACS Nano* **2012**, *6*, 6014–6022. [CrossRef]
64. Nie, X.; Zeng, C.; Ma, X.; Qian, H.; Ge, Q.; Xu, H.; Jin, R. CeO_2-supported $Au_{38}(SR)_{24}$ nanocluster catalysts for CO oxidation: A comparison of ligand-on and -off catalysts. *Nanoscale* **2013**, *5*, 5912–5918. [CrossRef] [PubMed]
65. Nuzzo, R.G.; Fusco, F.A.; Allara, D.L. Spontaneously organized molecular assemblies—Preparation and properties of solution adsorbed monolayers of organic disulfides on gold surfaces. *J. Am. Chem. Soc.* **1987**, *109*, 2358–2368. [CrossRef]
66. Toulhoat, H.; Raybaud, P.; Kasztelan, S.; Kresse, G.; Hafner, J. Transition metals to sulfur binding energies relationship to catalytic activities in HDS: Back to Sabatier with first principle calculations. *Catal. Today* **1999**, *50*, 629–636. [CrossRef]
67. Carabineiro, S.A.C.; Nieuwenhuys, B.E. Adsorption of small molecules on gold single crystal surfaces. *Gold Bull.* **2009**, *42*, 288–301. [CrossRef]
68. Angell, C.L.; Schaffer, P.C. Infrared spectroscopic investigations of zeolites and adsorbed molecules. II. Adsorbed carbon monoxide. *J. Phys. Chem.* **1966**, *70*, 1413–1418. [CrossRef]
69. Lopez, N.; Nørskov, J.K. Catalytic CO oxidation by a gold nanoparticle: A density functional study. *J. Am. Chem. Soc.* **2002**, *124*, 11262–11263. [CrossRef] [PubMed]
70. Lopez, N.; Janssens, T.V.W.; Clausen, B.S.; Xu, Y.; Mavrikakis, M.; Bligaard, T.; Nørskov, J.K. On the origin of the catalytic activity of gold nanoparticles for low-temperature CO oxidation. *J. Catal.* **2004**, *223*, 232–235. [CrossRef]
71. Min, B.K.; Friend, C.M. Heterogeneous gold-based catalysis for green chemistry: Low-temperature CO oxidation and propene oxidation. *Chem. Rev.* **2007**, *107*, 2709–2724. [CrossRef]
72. Janssens, T.V.W.; Clausen, B.S.; Hvolbæk, B.; Falsig, H.; Christensen, C.H.; Bligaard, T.; Nørskov, J.K. Insights into the reactivity of supported Au nanoparticles: Combining theory and experiments. *Top. Catal.* **2007**, *44*, 15–26. [CrossRef]
73. Liu, Y.; Jia, C.-J.; Yamasaki, J.; Terasaki, O.; Schüth, F. Highly active iron oxide supported gold catalysts for CO oxidation: How small must the gold nanoparticles be? *Angew. Chem. Int. Ed.* **2010**, *49*, 5771–5775. [CrossRef]
74. Qian, K.; Luo, L.; Bao, H.; Hua, Q.; Jiang, Z.; Huang, W. Catalytically active structures of SiO_2-supported Au nanoparticles in low-temperature CO oxidation. *Catal. Sci. Technol.* **2013**, *3*, 679–687. [CrossRef]
75. Qiao, B.; Liang, J.-X.; Wang, A.; Xu, C.-Q.; Li, J.; Zhang, T.; Liu, J.J. Ultrastable single-atom gold catalysts with strong covalent metal-support interaction (CMSI). *Nano Res.* **2015**, *8*, 2913–2924. [CrossRef]
76. Menard, L.D.; Xu, F.; Nuzzo, R.G.; Yang, J.C. Preparation of TiO_2-supported Au nanoparticle catalysts from a Au_{13} cluster precursor: Ligand removal using ozone exposure versus a rapid thermal treatment. *J. Catal.* **2006**, *243*, 64–73. [CrossRef]
77. Cargnello, M.; Chen, C.; Diroll, B.T.; Doan-Nguyen, V.V.T.; Gorte, R.J.; Murray, C.B. Efficient removal of organic ligands from supported nanocrystals by fast thermal annealing enables catalytic studies on well-defined active phases. *J. Am. Chem. Soc.* **2015**, *137*, 6906–6911. [CrossRef] [PubMed]

78. Haruta, M.; Tsubota, S.; Kobayashi, T.; Kageyama, H.; Genet, M.J.; Delmon, B. Low-temperature oxidation of CO over gol supported on TiO_2, α-Fe_2O_3, and Co_3O_4. *J. Catal.* **1993**, *144*, 175–192. [CrossRef]
79. Sankar, M.; He, Q.; Engel, R.V.; Sainna, M.A.; Logsdail, A.J.; Roldan, A.; Willock, D.J.; Agarwal, N.; Kiely, C.J.; Hutchings, G Role of the support in gold-containing nanoparticles as heterogeneous catalysts. *Chem. Rev.* **2020**, *120*, 3890–3938. [CrossRe [PubMed]
80. Meyer, R.; Lemire, C.; Shaikhutdinov, S.K.; Freund, H.J. Surface chemistry of catalysis by gold. *Gold Bull.* **2004**, *37*, 72–12 [CrossRef]
81. Ginter, D. Chapter 46—FAU Linde Type Y Si(71), Al(29). In *Verified Syntheses of Zeolitic Materials;* Robson, H., Lillerud, K.P., Eds Elsevier: Amsterdam, The Netherlands, 2001; pp. 156–158.
82. Haiss, W.; Thanh, N.T.K.; Aveyard, J.; Fernig, D.G. Determination of size and concentration of gold nanoparticles from UV−v spectra. *Anal. Chem.* **2007**, *79*, 4215–4221. [CrossRef]
83. Higaki, T.; Zhou, M.; Lambright, K.J.; Kirschbaum, K.; Sfeir, M.Y.; Jin, R. Sharp transition from nonmetallic Au_{246} to metall Au_{279} with nascent surface plasmon resonance. *J. Am. Chem. Soc.* **2018**, *140*, 5691–5695. [CrossRef]
84. Lee, S.; Fan, C.; Wu, T.; Anderson, S.L. CO oxidation on Au_n/TiO_2 catalysts produced by size-selected cluster deposition. *J. Am Chem. Soc.* **2004**, *126*, 5682–5683. [CrossRef]
85. Li, L.; Gao, Y.; Li, H.; Zhao, Y.; Pei, Y.; Chen, Z.; Zeng, X.C. CO oxidation on TiO_2 (110) supported subnanometer gold cluster Size and shape effects. *J. Am. Chem. Soc.* **2013**, *135*, 19336–19346. [CrossRef]
86. Lopez-Acevedo, O.; Kacprzak, K.A.; Akola, J.; Häkkinen, H. Quantum size effects in ambient CO oxidation catalysed b ligand-protected gold clusters. *Nat. Chem.* **2010**, *2*, 329–334. [CrossRef]
87. Haruta, M. Spiers Memorial Lecture Role of perimeter interfaces in catalysis by gold nanoparticles. *Faraday Discuss.* **2011**, *15* 11–32. [CrossRef]
88. Glemser, O.; Sauer, H. Copper, silver, gold. In *Handbook of Preparative Inorganic Chemistry*, 2nd ed.; Brauer, G., Ed.; Academic Pres inc.: London, UK, 1963; Volume 1.

MDPI AG
Grosspeteranlage 5
4052 Basel
Switzerland
Tel.: +41 61 683 77 34

Nanomaterials Editorial Office
E-mail: nanomaterials@mdpi.com
www.mdpi.com/journal/nanomaterials

www.ingramcontent.com/pod-product-compliance
Lightning Source LLC
Chambersburg PA
CBHW041403210926
51850CB00013B/3164